Curvature Scale Space Representation: Theory, Applications, and MPEG-7 Standardization

Computational Imaging and Vision

Volume 25

Curvature Scale Space Representation: Theory, Applications, and MPEG-7 Standardization

by

Farzin Mokhtarian

University of Surrey,
Guildford, U.K.

and

Miroslaw Bober

Mitsubishi Electric Visual Information Lab,
Guildford, U.K.

Springer-Science+Business Media, B.V.

A C.I.P. Catalogue record for this book is available from the Library of Congress.

ISBN 978-90-481-6270-3   ISBN 978-94-017-0343-7 (eBook)
DOI 10.1007/978-94-017-0343-7

Printed on acid-free paper

Contents

Preface

Shape analysis is a central and challenging area of research in computer vision and image understanding [207, 200, 201]. Very often it is the exploitation of some knowledge about the shapes or shape properties of objects present in the image that allows a vision system to reach some interesting conclusions or derive some non-trivial facts about the scene from the input image data. This is because object shape conveys the most useful information about objects present in analyzed images. Many important tasks in computer vision and Image Understanding include some form of shape analysis. One example is shape representation and matching where the goal is to make explicit some information about object shape which can be exploited later to recognize that object or distinguish it from other objects. Another example is image database retrieval where the objective is to search a large image database and to discover other images which contain objects similar to a query object. Other examples include:

- Image corner detection or point feature detection

- Snake or active contour localization

- Motion detection and moving region tracking

- Stereo matching and 3-D reconstruction

- Optimal view selection for multi-view 3-D object recognition

The Curvature Scale Space (CSS) technique is a powerful and general shape analysis tool which has been developed comprehensively during the last 20 years by the first author. A CSS image is a multi-scale organization of the invariant local features of a free-form 2-D contour. The features utilized in a CSS image consist of the curvature zero-crossing points recovered from the contour at multiple scales of resolution. Decreasing scales of resolution are obtained through convolutions of a parametric representation of the contour data with Gaussian filters of increasing width. The strengths of the CSS technology derive from several important properties:

Multi-scale organization results in a natural representation of shape information at multiple levels of detail, where noise and insignificant object features are filtered out at smaller scales and only the prominent shape features surviving to larger scales. Consequently, the significance of the shape features can be easily taken into consideration when matching is performed and the representation is robust to noise.

Invariant local features yield a representation which is quite robust to similarity and affine transforms as well as local shape deformations. Since no specific shape model is assumed, the representation is entirely influenced by the shape it represents and can therefore accommodate any free-form input shape very effectively.

Correspondence between features and shape parts supports hierarchical contour decomposition into concave and convex sections, and captures well the concept of shape similarity employed by humans as demonstrated by MPEG-7 experiments.

The Curvature Scale Space representation was further developed, optimised, and adapted for description of 2D contour shapes in multimedia applications by the second author, during his participation in standardiation activities of ISO - MPEG-7. MPEG conducted thorough and comprehensive testing of several shape-description techniques in a number of tests based on large image databases. The tests clearly showed the superior performance of the CSS-based descriptor, which was consequently selected for the Visual part of the Standard, for the contour-shape category. MPEG-7 became International Standard in December 2001.

MPEG testing results give an interesting insight into the properties and behaviour of various shape description techniques. Besides the CSS-based technique, the competition included several well-known shape descriptors, including:

- Wavelet-based technique

- Multi-layer eigenvector method

- Zernike Moments technique

- Polygon Approximation method

- Fourier Descriptors technique

- Angular Radial Transform (ART) method

The criteria considered during the evaluation process were: *performance* under different conditions, *matching speed* and *descriptor size*. Performance was considered the most important quality of each descriptor under consideration but they were also expected to be efficient and compact. The evaluation process consisted of four main experiments:

Robustness to scaling A database of 70 shapes was used for this experiment. Each image in the database was scaled up as well as down several times. The largest image obtained was twice the size of the original image and the smallest was one-tenth the size of the original image. This resulted in a database of 420 images. The aim of this experiment was to determine how many scaled versions of each query image could be retrieved from the database.

Robustness to rotation The same database of 70 shapes was used in this experiment. Each image in the database was rotated several times to obtain several rotated versions. This resulted in a database of 420 images. The objective of this experiment was to determine how many rotated versions of each query image could be retrieved from the database.

Similarity-based retrieval In this experiment, 1400 images were divided into 70 categories. The goal was to determine how many images from the same category as each query image could be retrieved from the database. Some of the categories used in this experiment were quite challenging for the descriptors.

Robustness to non-rigid deformations In this experiment, 200 consecutive frames of a sequence corresponding to a fish swimming in a tank were augmented by 1000 images from the SQUID marine database. The aim of this experiment was to determine the number of frames of the fish sequence retrieved using frame 0 as the query image.

CSS techniques are popular, and we are aware of the following publicly available demos which exist on the Web:

- To run an interactive demo of shape-based retrieval from the SQUID database of marine creatures, see the following web site:

`www.ee.surrey.ac.uk/Research/VSSP/imagedb/demo.html`

- A demo of the performance of the MPEG-7 contour shape descriptor on the MPEG similarity database is available at:

`www.vil.ite.mee.com/demos/`

This book is based on key publications on the Curvature Scale Space technique as well as its multiple applications and generalizations. The goal was to ensure that the reader will have access to the most fundamental results concerning the CSS method in one volume. These results have been categorized into a number of chapters to reflect their focus as well as content. The book also includes a chapter on the development of the CSS-technique within MPEG standardisation, including details of the MPEG-7 testing and evaluation processes which led to the selection of the CSS shape descriptor for the Standard.

Chapter 1 contains material on the computation of the CSS image as well as its alternative forms. It also presents results on the theoretical properties of CSS images. Chapter 2 describes two robust free-form object recognition systems based on the CSS representation. Chapter 3 investigates image database retrieval based on shape content through the CSS method. Chapter 4 discusses a number of extensions of the CSS image and studies their behaviour under affine transforms and non-rigid deformations of input shapes. Chapter 5 describes approaches to free-form 3-D object retrieval from arbitrary viewpoints. Chapter 6 covers the MPEG-7 standardization of the CSS shape descriptor. It contains:

- A description of the precise Curvature Scale Space enhancement which was selected for standardization.

- An explanation of how the image databases used for testing were generated as well as the exact procedures followed for evaluation of test results.

- A review of other shape descriptors evaluated during the competition.

Chapter 7 discusses robust image corner detection through the CSS method. Chapter 8 is concerned with efficient active contour convergence based on CSS filtering. Chapter 9 is on efficient multi-scale contour data compression and reconstruction using the CSS method. Chapter 10 explains the theory of torsion Scale Space images as multi-scale representations for free-form space curves. Finally chapter 11 presents a theory of multi-scale representations for free-form 3-D surfaces.

The material covered in this book does not exhaust all possible applications of the CSS technique. Indeed, as CSS is a powerful and general shape analysis tool, we expect that many other applications exist in computer vision and image processing. For example, the CSS method can be effectively utilized for early contour feature detection in on-line and off-line handwriting recognition [260, 301, 327, 29, 58, 281, 287, 313, 81, 88, 92, 102, 109, 131, 148, 158, 160, 173, 234, 248, 266, 274, 277], followed by the application of an HMM (Hidden Markov Model) engine. It can also be useful for matching of silhouette contours in video-based motion capture for virtual reality applications [13].

This book is expected to be of use to computer vision and image processing researchers as well as to graduate and upper undergraduate students in those areas. We expect that it will contribute to the expansion of knowledge in the area of shape analysis.

We are grateful to the following publishers and organizations for granting us permission to reproduce material from previously published papers in this book: Academic Press, BMVA, CSCSI, Danish Pattern Recognition Society, Elsevier Science, Eurasip, Eurographics Publications Board, IEE, IEEE, IRIT, Mitsubishi-Electric ITE, Springer-Verlag, and Wiley.

Farzin Mokhtarian
Miroslaw Bober
December 2002

To Roxana, Loving
companion and friend
To my Loving parents
To Neema, the sweet,
little one
- FM

To Amaia, Janek and
Mikel
- MB

Chapter 1

MULTI-SCALE REPRESENTATIONS FOR FREE-FORM PLANAR CURVES: THE CURVATURE SCALE SPACE IMAGE AND ITS PROPERTIES

This chapter presents a multiscale, curvature-based shape representation technique for planar curves that satisfies several criteria considered necessary for general-purpose shape representation methods. As a result, the representation is suitable for tasks that call for recognition of a noisy curve of arbitrary shape at an arbitrary scale or orientation.

The method rests on the concept of describing a curve at varying levels of detail using features that are invariant with respect to transformations that do not change the shape of the curve. Three different ways of computing the representation are described in this chapter. These three methods result in a family of three representation methods: the curvature scale space image, the renormalized curvature scale space image, and the resampled curvature scale space image.

The process of describing a curve at increasing levels of abstraction is referred to as the *evolution* or *arc length evolution* of that curve. Several evolution and arc length evolution properties of planar curves are described in this chapter. Some of these results show that evolution and arc length evolution do not change the physical interpretation of planar curves as object boundaries, and some characterize possible behaviors of planar curves during evolution and arc length evolution. Others impose constraints on the image location of a planar curve as it evolves. Together, these results provide a sound theoretical foundation for the representation methods introduced in this chapter.

1.   Introduction

This chapter introduces a novel theory of multiscale, curvature-based shape representation for planar curves. It should be pointed out that only the problem of representing the shape of a planar curve that has been extracted from an image or input by a user has been addressed in this chapter. We believe the problem of extracting such a curve from an image (the *segmentation problem*) is, in general, a separate problem and should not necessarily be considered to be part of the shape representation problem [273]. We also believe that the segmentation problem can be addressed effectively by making use of knowledge of the image and scene under consideration. For example, we made use of *a priori* knowledge of band 7 Landsat images of land/water scenes to arrive at a good segmentation of such an image [220]. The boundary curves thus obtained were then matched to curves from a map of the same area using their curvature scale space representations. As a result, the correct registration between the Landsat image and the map was computed.

A useful general-purpose shape representation method in computational vision should make accurate and reliable recognition of an object possible. Therefore, such a representation should necessarily satisfy a number of criteria. The following is a list of such criteria. Note that we define two planar curves as having the *same shape*, when there exists a transformation consisting of uniform scaling, rotation, and translation, which will cause one of those curves to completely overlap the other. As a result, every point of the first curve will have the same location as a point of the second curve and *vice versa*.

Invariance: If two curves have the same shape, they should also have the same representation.

Uniqueness: If two curves do not have the same shape, they should have different representations.

Stability: If two curves have a small shape difference, their representations should also have a small difference, and if two representations have a small difference, the curves they represent should also have a small shape difference.

The importance of the invariance criterion is that it guarantees that all curves with the same shape will have the same representation. It will therefore be possible to conclude that two curves have different shapes by observing that they have different representations. Without the invariance criterion, two curves with the same shape could have different representations.

The uniqueness criterion is important since it guarantees that two curves with different shapes will have different representations. It will therefore be possible to conclude that two curves have the same shape by observing that they have the same representation. Without the uniqueness criterion, two curves with different shapes may have the same representation.

The significance of the stability criterion is that it guarantees that a small change in the shape of a curve will not cause a large change in its representation, and a small difference between two representations does not indicate a large shape difference between the curves they represent. As a result, when two representations are close, the curves they represent are close in shape, and when two representations are not close, the curves they represent are not close in shape. When this criterion is satisfied, the representation can be considered to be stable with respect to noise. One way to measure the shape difference between two planar curves is the *Hausdorf distance* [110]. The computation of the Hausdorf distance between two curves C_1 and C_2 is based on finding, for each point of C_1, the closest point on C_2 and *vice versa.*

It is useful for a shape representation to satisfy a number of additional properties in order to become suitable for practical shape recognition tasks in computer vision and image processing. The following is a list of such criteria. Note that similar criteria have been proposed in [244, 220].

Local support: Very often, it is necessary to be able to recognize that the shape of a segment of a curve is the same as the shape of another curve-segment. Only a representation computed using *local* information can provide such an ability.

Efficiency: The representation should be efficient to compute and store. This is important since it may be necessary for an object recognition system to perform real-time recognition, and also because computational resources are usually scarce. By *efficient*, we mean that the computational complexity should be a low-order polynomial in time and space (and in the number of processors if a parallel computing architecture is used) with a small constant as a function of the size of the input curve.

Shape properties: It may be useful to be able to determine properties of the shape of a curve using its representation. For example, if a curve has a symmetric shape, it may be desirable to be able to determine that fact from its representation (the *symmetry* criterion). Furthermore, if the shape of a whole curve or part of a curve is the same as the shape of part of another curve, it may be useful to be able to determine that relationship using their representations (the *part/whole* criterion).

Implementation: If two or more competing representations exist which are very close in other aspects, it is advantageous to choose one of those representations such that the implementation of the computer program that computes that representation requires the least time spent on programming and debugging.

Shape representation methods for planar curves previously proposed in computational vision and image processing fail to satisfy one or more of the criteria outlined above. Note, however, that each may be quite suitable for special-purpose shape representation and recognition tasks. The *Hough transform* has been used to detect parametric structures such as lines [117], circles [70], and arbitrary shapes [25]. Edge elements in the image vote for the parameters of the objects of which they are parts. The votes are accumulated in a parameter space. The peaks of the parameter space then indicate the parameters of the objects searched for. *Chain fitting* [89, 178] techniques approximate a curve using line segments lying on a grid. *Polygonal approximations* [249] of a curve are computed by using various criteria to determine *breakpoints* that yield the *best* polygon. The *medial axis transform* [38, 175] computes the skeleton of a 2-D object by a thinning algorithm that preserves region connectivity. *Shape factors and quantitative measurements* [64] use one or more global quantitative measurements of the object such as area, perimeter, and compactness as a description of its shape. *Strip trees* [26, 66] are a set of approximating polygons ordered such that each polygon approximates the curve with less approximation error than the previous polygon. *splines* [27] represent a curve using a set of analytic and smooth curves. The *smoothing splines* [283] method parametrizes the curve to obtain two coordinate functions (see next section). Cross-validated regularization is then used to arrive at an *optimal* smoothing of each coordinate function. The smoothed functions together define a new smooth curve. *Fourier Descriptors* [252] represent a curve by the coefficients of the Fourier expansion of a parametric representation of the curve. The *curvature primal sketch* [23] technique approximates the curve using a library of analytic curves. Then, the curvature function of the approximating curve is computed and convolved with a Gaussian of varying standard deviation. The *extended circular image* [116] is the 2-D equivalent of the *extended Gaussian image*. In the extended circular image, the radius of curvature is given as a function of normal direction. *volumetric diffusion* [146] defines a geometrical object by way of its *characteristic function* $\chi(\mathbf{r})$, which equals unity when the point $\mathbf{r}$ belongs to the object and zero otherwise. The object is then blurred by requiring that its characteristic function satisfy the diffusion equation. The boundary of each blurred object is defined by the equation $\chi(\mathbf{r}) = 0.5$ or by applying the Laplacian operator to the blurred function. Richards et al. [264] located curvature extrema on a 2-D contour at multiple scales. Those extrema were then used to represent the contour shape.

A multiscale representation for 1-D functions was first proposed by Stansfield [295] and later developed by Witkin [322]. The function $f(x)$ is convolved with a Gaussian filter with variance σ^2 varying from a small to a large value. The zero crossings of the second derivative of each convolved function are extracted and marked in the x–σ plane. The result is the *scale space image* of the function.

The *curvature scale space image* was introduced by Mokhtarian and Mackworth [220] as a new shape representation for planar curves. The representation is computed by convolving a path-based parametric representation of the curve with a Gaussian function, as the standard deviation of the Gaussian varies from a small to a large value, and extracting the curvature zero-crossing points of the resulting curves. The representation is essentially invariant under rotation, uniform scaling, and translation of the curve. This and a number of other properties makes it suitable for a number of applications including recognition of a noisy curve of arbitrary shape at any scale or orientation. The process of describing a curve at increasing levels of abstraction is referred to as the *evolution* of that curve. The evolution of a planar curve and the curvature scale space image are described in section 2.

Mackworth and Mokhtarian [170] also introduced a modification of the curvature scale space image referred to as the *renormalized curvature scale space image*. This representation is computed in a similar fashion, but the curve is reparametrized by arc length after convolution. As was demonstrated in [170], the renormalized curvature scale space image is more suitable for recognizing a curve with non-uniform noise added (see figure 1.10). Section 3 contains a brief description of the renormalized curvature scale space image.

The *resampled curvature scale space image* is a substantial refinement of the curvature scale space based on the concept of *arc length evolution*. It is shown that the resampled curvature scale space image is more suitable than the renormalized curvature scale space image for recognition of curves with added nonuniform noise or when local shape differences exist [222]. The arc length evolution of a planar curve and the resampled curvature scale space image are described in detail in section 4.

Section 5 contains descriptions of the evolution and arc length evolution properties of planar curves and discusses the significance of each of those properties. Almost all these properties are shown to be true of both evolution and arc length evolution. Together, these properties provide a theoretical foundation for the representation methods proposed in this chapter. The proofs of the theorems of section 5 are given in appendix A.

Section 6 presents an additional experiment carried out to demonstrate the stability of the curvature scale space image with respect to noise. It also presents a table comparing the representations introduced here and an evaluation of those representations according to the criteria of section 1.

Section 7 presents the conclusions of this chapter.

2.   The Curvature Scale Space Image

A planar curve is a set of points whose position vectors are the values of a continuous, vector-valued function. It can be represented by the parametric vector equation

$$\mathbf{r}(u) = (x(u), y(u)).$$

The function $\mathbf{r}(u)$ is a parametric representation of the curve. A planar curve has an infinite number of distinct parametric representations. A parametric representation in which the parameter is the arc length s is called a *natural* parametrization of the curve. A natural parametrization can be computed from an arbitrary parametrization using the following equation:

$$s(u) = \int_0^u |\dot{\mathbf{r}}(v)|\, dv,$$

where $\dot{\ }$ represents the derivative, i.e., $\dot{\mathbf{r}} = d\mathbf{r}/dv$. For any parametrization

$$\dot{\mathbf{r}}(u) = (\dot{x}(u), \dot{y}(u)),$$

$$|\dot{\mathbf{r}}(u)| = (\dot{x}^2 + \dot{y}^2)^{1/2},$$

$$\mathbf{t}(u) = \frac{\dot{\mathbf{r}}}{|\dot{\mathbf{r}}|} = \left(\frac{\dot{x}}{(\dot{x}^2 + \dot{y}^2)^{1/2}}, \frac{\dot{y}}{(\dot{x}^2 + \dot{y}^2)^{1/2}} \right),$$

$$\mathbf{n}(u) = \left(\frac{-\dot{y}}{(\dot{x}^2 + \dot{y}^2)^{1/2}}, \frac{\dot{x}}{(\dot{x}^2 + \dot{y}^2)^{1/2}} \right),$$

where $\mathbf{t}(u)$ and $\mathbf{n}(u)$ are the tangent and normal vectors at u, respectively. For any planar curve, the vectors $\mathbf{t}(u)$ and $\mathbf{n}(u)$ must satisfy the simplified Serret–Frenet vector equations [95]:

$$\dot{\mathbf{t}}(s) = \kappa(s)\mathbf{n}(s)$$

$$\dot{\mathbf{n}}(s) = -\kappa(s)\mathbf{t}(s),$$

where $\kappa(s)$ is the curvature of the curve at s and is defined as

$$\kappa(s) = \lim_{h \to 0} \frac{\phi}{h},$$

ϕ is the angle between $\mathbf{t}(s)$ and $\mathbf{t}(s+h)$. Now, observe that

$$\dot{\mathbf{t}}(s) = \frac{d\mathbf{t}}{ds} = \frac{d\mathbf{t}}{du}\frac{du}{ds}.$$

Therefore

$$\frac{d\mathbf{t}}{du} = \frac{ds}{du}\kappa\mathbf{n} = |\dot{\mathbf{r}}|\kappa\mathbf{n}.$$

Hence

$$\kappa(u) = \frac{\dot{\mathbf{t}}\mathbf{n}}{|\dot{\mathbf{r}}|}.$$

Differentiating the expression for $\mathbf{t}(u)$, we obtain

$$\dot{\mathbf{t}}(u) = \left(\frac{-\dot{y}(\dot{x}\ddot{y} - \ddot{x}\dot{y})}{(\dot{x}^2 + \dot{y}^2)^{3/2}}, \frac{\dot{x}(\dot{x}\ddot{y} - \ddot{x}\dot{y})}{(\dot{x}^2 + \dot{y}^2)^{3/2}} \right).$$

It now follows that

$$\kappa(u) = \frac{\dot{x}(u)\ddot{y}(u) - \dot{y}(u)\ddot{x}(u)}{(\dot{x}(u)^2 + \dot{y}(u)^2)^{3/2}}.$$

Therefore, it is possible to compute the curvature of a planar curve from its parametric representation. Special cases of the parametrization yield simplifications of these formulas. If w is the normalized arc length parameter, then

$$k(w) = \dot{x}(w)\ddot{y}(w) - \ddot{x}(w)\dot{y}(w).$$

Given a planar curve

$$\Gamma = \{(x(w), y(w)) \mid w \in [0, 1]\},$$

where w is the normalized arc length parameter, an *evolved* version of that curve is defined by

$$\Gamma_\sigma = \{(\mathcal{X}(u, \sigma), \mathcal{Y}(u, \sigma)) \mid u \in [0, 1]\},$$

where

$$\mathcal{X}(u, \sigma) = x(u) \otimes g(u, \sigma)$$
$$\mathcal{Y}(u, \sigma) = y(u) \otimes g(u, \sigma).$$

$g(u, \sigma)$ denotes a Gaussian of width σ [176] defined by

$$g(u, \sigma) = \frac{1}{\sigma\sqrt{2\pi}} e^{-u^2/2\sigma^2}$$

and $\otimes$ denotes convolution. Functions $\mathcal{X}(u,\sigma)$ and $\mathcal{Y}(u,\sigma)$ are given explicitly by

$$\mathcal{X}(u,\sigma) = \int_{-\infty}^{\infty} x(v)\frac{1}{\sigma\sqrt{2\pi}}\, e^{-(u-v)^2/2\sigma^2}\, dv,$$

$$\mathcal{Y}(u,\sigma) = \int_{-\infty}^{\infty} y(v)\frac{1}{\sigma\sqrt{2\pi}}\, e^{-(u-v)^2/2\sigma^2}\, dv.$$

The curvature of Γ_σ is given by

$$\kappa(u,\sigma) = \frac{\mathcal{X}_u(u,\sigma)\mathcal{Y}_{uu}(u,\sigma) - \mathcal{X}_{uu}(u,\sigma)\mathcal{Y}_u(u,\sigma)}{(\mathcal{X}_u(u,\sigma)^2 + \mathcal{Y}_u(u,\sigma)^2)^{3/2}},$$

where

$$\mathcal{X}_u(u,\sigma) = \frac{\partial}{\partial u}(x(u) \otimes g(u,\sigma)) = x(u) \otimes g(u,\sigma),$$

$$\mathcal{X}_{uu}(u,\sigma) = \frac{\partial^2}{\partial u^2}(x(u) \otimes (u,\sigma)) = x(u) \otimes g_{uu}(u,\sigma),$$

$$\mathcal{Y}_u(u,\sigma) = y(u) \otimes g_u(u,\sigma),$$

$$\mathcal{Y}_{uu}(u,\sigma) = y(u) \otimes g_{uu}(u,\sigma).$$

The process of generating the ordered sequence of curves $\{\Gamma_\sigma \mid \sigma \geq 0\}$ is referred to as the *evolution* of P.

Figure 1.1 shows a planar curve depicting the shoreline of Africa. Figure 1.2 shows several evolved versions of that curve for increasing values of σ. It can be observed that the evolution process results in a continuous fine-to-coarse description of the shape of a planar contour. Increasing values of σ remove noise as well as small features from the original contour leaving only the major shape components. Eventually all shape structures disappear at a sufficiently large value of σ and the evolved contour becomes convex.

Note that when a planar curve evolves according to the process defined above, its total arc length shrinks. The amount of shrinkage is directly proportional to the value of σ. In certain applications, this may be an undesirable feature. For example, the evolution process defined above may be used to smooth edges extracted from an image by an edge detector. However, it may be advantageous to have the smoothed edges as close as possible to the physical location of the original edges. This can be accomplished by estimating the amount of movement at each point on the smoothed edges and adding a vector to the location vector

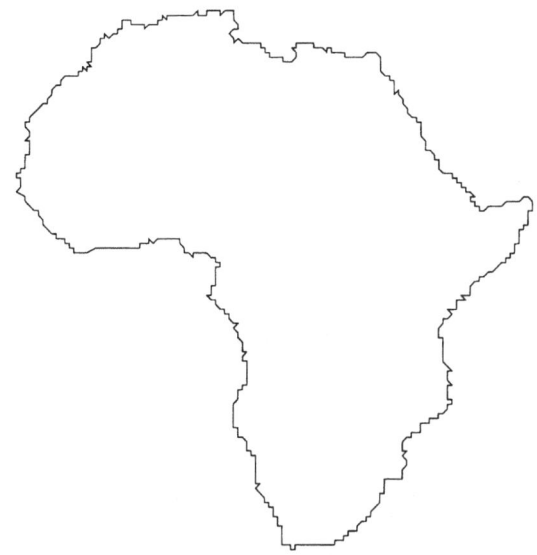

Figure 1.1. Shoreline of Africa

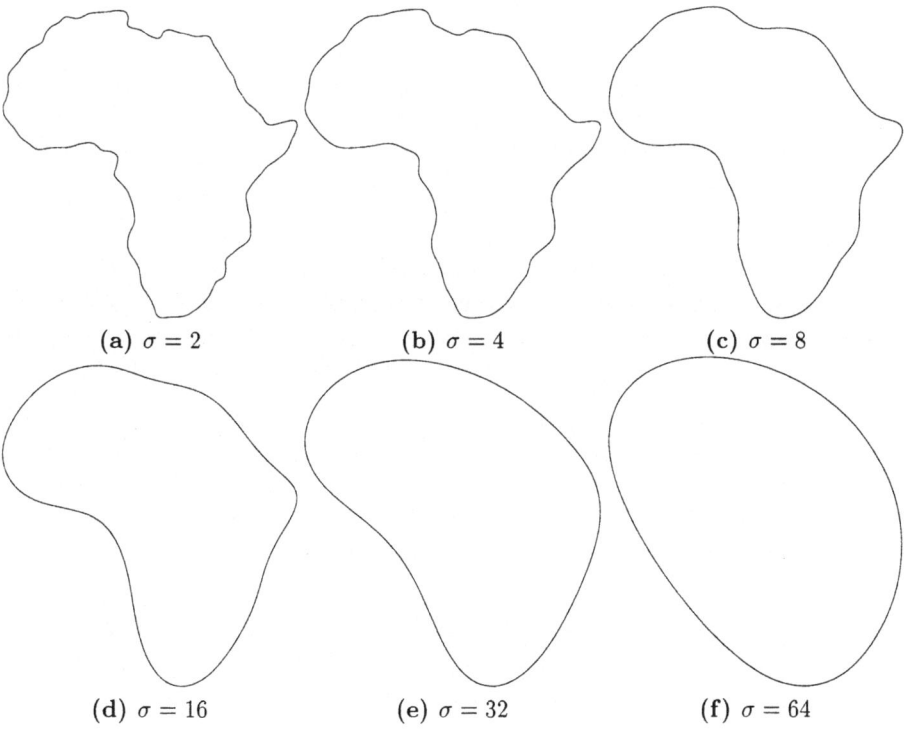

(a) $\sigma = 2$ **(b)** $\sigma = 4$ **(c)** $\sigma = 8$

(d) $\sigma = 16$ **(e)** $\sigma = 32$ **(f)** $\sigma = 64$

Figure 1.2. Africa during evolution

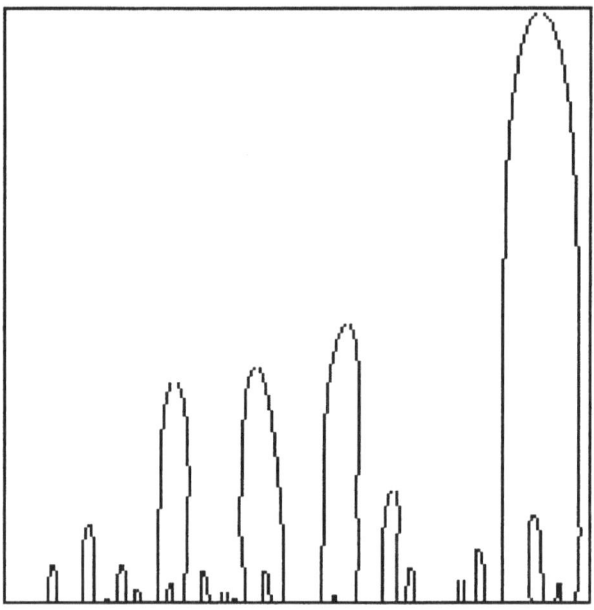

Figure 1.3.   Curvature Scale Space image of Africa

of that point to compensate for that movement [166]. The result is a smoothed curve that is physically closer to the original curve.

The solution of the equation

$$\kappa(u, \sigma) = 0$$

forms the *curvature scale space* image of Γ. Figure 1.3 shows the curvature scale space image of the curve of Figure 1.1. Note that each of the arch-shaped contours in the CSS image corresponds to a feature on the original contour with the size of the arch being proportional to the size of the corresponding feature. Note also that a small gap can often be observed at the top of each arch. These gaps exist due to the discrete sampling of the CSS image. Figure 1.4 shows the shoreline of Hokaido, and figure 1.5 shows several of its evolved versions. Figure 1.6 shows the curvature scale space image of Hokaido. Figure 1.7 shows a design from a Persian carpet, and Figure 1.8 shows several of its evolved versions. Figure 1.9 shows the curvature scale space image of the carpet design.

We carried out two experiments to test the stability of the curvature scale space image under conditions of noise. Figure 1.10(c) shows the coastline of Africa with a significant amount of uniform, random noise added to it. Figure 1.10(d) shows the curvature scale space image of Africa with uniform noise. As expected, the CSS images shown in Fig-

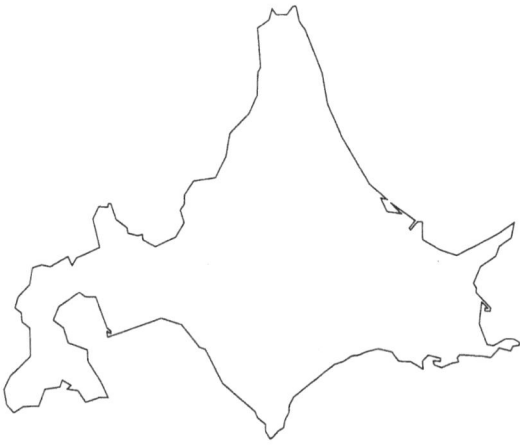

Figure 1.4. Shoreline of Hokaido

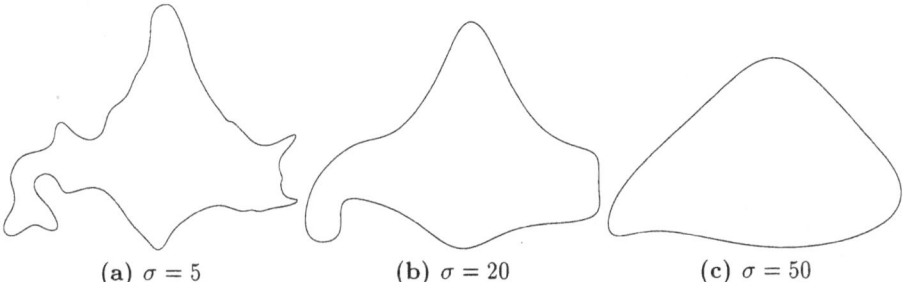

(a) $\sigma = 5$ (b) $\sigma = 20$ (c) $\sigma = 50$

Figure 1.5. Hokaido during evolution

ures 1.10(b) and (d) show differences in detail. However, a remarkable similarity in the basic structures of the two CSS images can be observed. This experiment shows that the curvature scale space image is very robust and stable even when a significant amount of uniform noise corrupts the shape of the input curve. The next experiment demonstrates the behavior of the curvature scale space image under severe noise conditions. Figure 1.10(e) shows the coastline of Africa with severe, uniform noise added to it. Figure 1.10(f) shows the curvature scale space image of Africa with severe noise. Even with the presence of severe noise, a very close similarity can be observed between the two CSS images shown in figures 1.10(b) and (f).

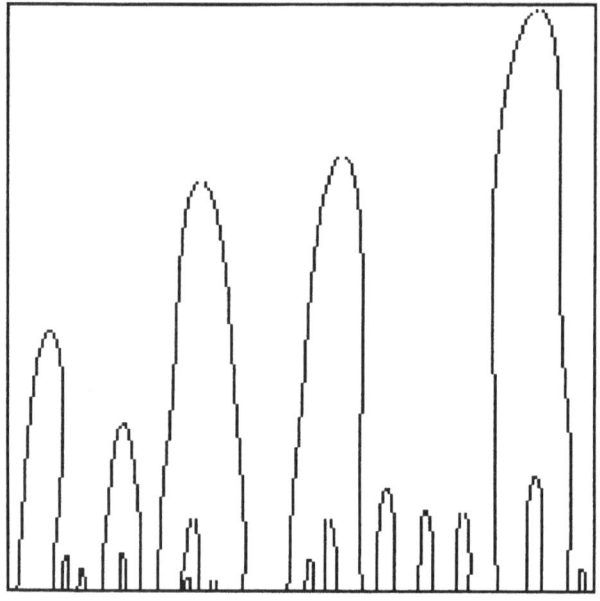

Figure 1.6.   Curvature Scale Space image of Hokaido

Figure 1.7.   Design from a Persian carpet

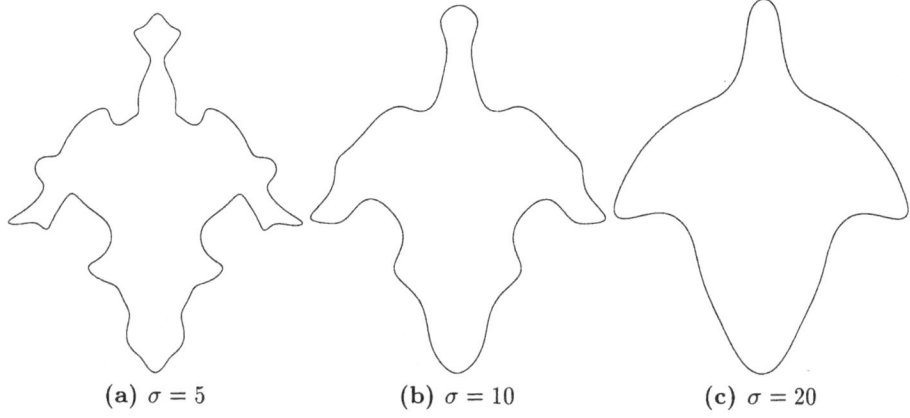

(a) $\sigma = 5$ (b) $\sigma = 10$ (c) $\sigma = 20$

Figure 1.8.   Carpet design during evolution

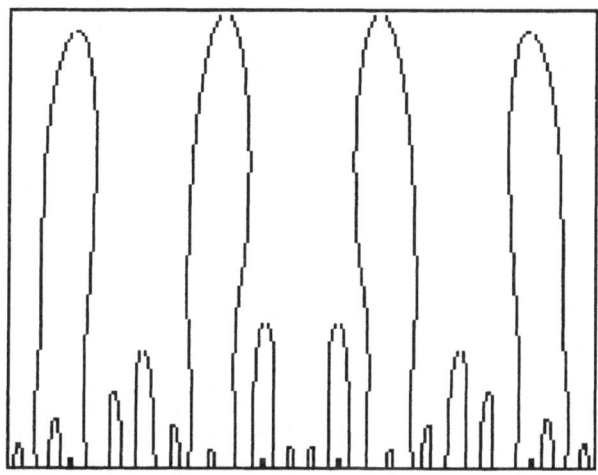

Figure 1.9.   Curvature Scale Space image of carpet design

3.   The Renormalized Curvature Scale Space Image

Mackworth and Mokhtarian [170] observed that although w is the normalized arc length parameter on the original curve Γ, the parameter u is *not*, in general, the normalized arc length parameter on the evolved curve Γ_σ. Figure 1.10(g) shows the shoreline of Africa with noise added to its lower half. Figure 1.10(h) shows the curvature scale space of that curve. A comparison of Figures 1.10(b) and (h) shows that a good match of one curvature scale space image to the other does not exist.

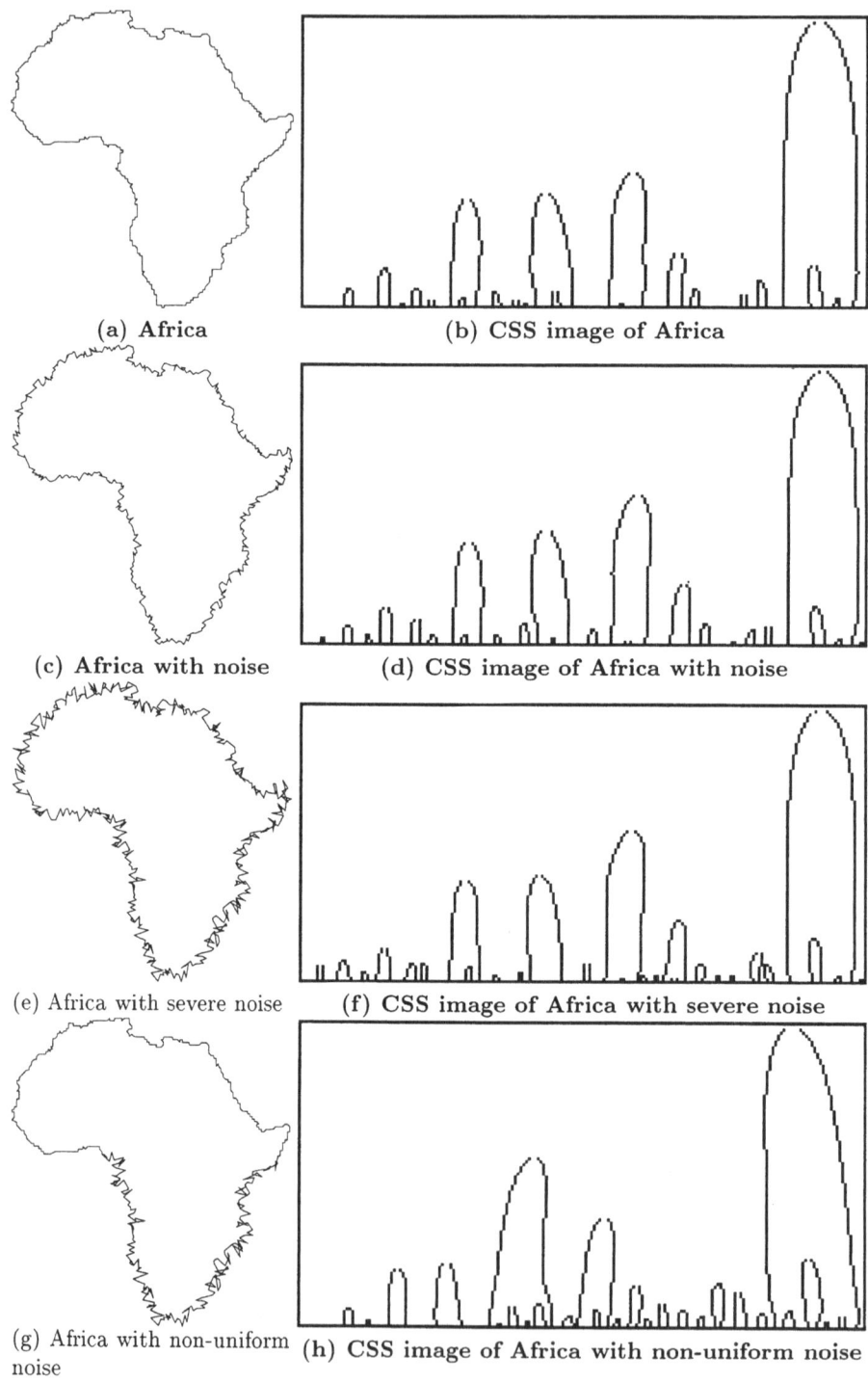

(a) Africa   (b) CSS image of Africa

(c) Africa with noise   (d) CSS image of Africa with noise

(e) Africa with severe noise   (f) CSS image of Africa with severe noise

(g) Africa with non-uniform noise   (h) CSS image of Africa with non-uniform noise

Figure 1.10.   Experiments on Africa and its CSS image under noise conditions

To overcome this problem, Mackworth and Mokhtarian [170] proposed the *renormalized* curvature scale space image. Let

$$\mathbf{R}(u, \sigma) = (\mathcal{X}(u, \sigma), \mathcal{Y}(u, \sigma))$$

and

$$w = \Phi_\sigma(u) = \frac{\int_0^u |\mathbf{R}_v(v, \sigma)| \, dv}{\int_0^1 |\mathbf{R}_v(v, \sigma)| \, dv}.$$

Now, define

$$\hat{\mathcal{X}}(w, \sigma) = \mathcal{X}(\Phi_\sigma^{-1}(w), \sigma)$$

$$\hat{\mathcal{Y}}(w, \sigma) = \mathcal{Y}(\Phi_\sigma^{-1}(w), \sigma)$$

that is, each evolved curve Γ_σ is reparametrized by its normalized arc length parameter w. Notice that

$$\Phi_\sigma(0) = 0$$

$$\Phi_\sigma(1) = 1$$

and

$$\frac{d\Phi_\sigma(u)}{du} = \frac{|\mathbf{R}_u(u, \sigma)|}{\int_0^1 |\mathbf{R}_v(v, \sigma)| \, dv} > 0 \qquad \text{at nonsingular points.}$$

In addition,

$$\Phi_0(u) = u.$$

$\Phi_\sigma(u)$ deviates from the identity function

$$\Phi_0(u) = u$$

only to the extent that the arc length parameter on the original curve is distorted by the smoothing process.

Once we have changed parameters, the curvature of the curve with the normalized path length parameter is given by

$$\kappa(w, \sigma) = \hat{\mathcal{X}}_w(w, \sigma)\hat{\mathcal{Y}}_{ww}(w, \sigma) - \hat{\mathcal{X}}_{ww}(w, \sigma)\hat{\mathcal{Y}}_w(w, \sigma).$$

The solution of the equation

$$\kappa(w, \sigma) = 0$$

forms the *renormalized* curvature scale space image of Γ. For examples of the renormalized CSS image, see [222].

4.   The Resampled Curvature Scale Space Image

Note that as a planar curve evolves according to the process defined in section 2, the parametrization of its coordinate functions $x(u)$ and $y(u)$ does not change. In other words, the function mapping values of the parameter u of the original coordinate functions $x(u)$ and $y(u)$ to the values of the parameter u of the smoothed coordinate functions $X(u, \sigma)$ and $Y(u, \sigma)$ is the identity function.

For both theoretical and practical reasons, it is interesting to generalize the definition of evolution so that the mapping function can be different from the identity function. Again, let Γ be defined by

$$\Gamma = \{(x(w), y(w)) \mid w \in [0, 1]\}.$$

The generalized evolution that maps Γ to Γ_σ is now defined by

$$\Gamma \rightarrow \Gamma_\sigma = \{(\mathcal{X}(\mathcal{W}, \sigma), \mathcal{Y}(\mathcal{W}, \sigma)) \mid \mathcal{W} \in [0, 1]\},$$

where

$$\mathcal{X}(\mathcal{W}, \sigma) = x(\mathcal{W}) \otimes g(\mathcal{W}, \sigma)$$
$$\mathcal{Y}(\mathcal{W}, \sigma) = y(\mathcal{W}) \otimes g(\mathcal{W}, \sigma).$$

Note that

$$\mathcal{W} = \mathcal{W}(w, \sigma).$$

Furthermore, $\mathcal{W}(w, \sigma_0)$, where σ_0 is any value of σ, is a continuous and monotonic function of w. This condition is necessary to ensure physical plausibility since $\mathcal{W}$ is the parameter of the evolved curve Γ.

An especially interesting case is when $\mathcal{W}$ always remains the arc length parameter as the curve evolves. When this criterion is satisfied, the evolution of Γ is referred to as *arc length evolution*. An implicit equation for $\mathcal{W}$ can be derived [91]. Let

$$\mathbf{R}(\mathcal{W}, \sigma) = (\mathcal{X}(\mathcal{W}, \sigma), \mathcal{Y}(\mathcal{W}, \sigma)).$$

The Frenet equations for a planar curve are given by

$$\frac{\partial \mathbf{t}}{\partial u} = \left| \frac{\partial \mathbf{R}}{\partial u} \right| \kappa \mathbf{n},$$

$$\frac{\partial \mathbf{n}}{\partial u} = \left| \frac{\partial \mathbf{R}}{\partial u} \right| \kappa \mathbf{t},$$

Let $t = \sigma^2 / 2$. Observe that

$$\frac{\partial}{\partial t} \left(\left| \frac{\partial \mathbf{R}}{\partial u} \right| \right)^2 = \frac{\partial}{\partial t} \left(\frac{\partial \mathbf{R}}{\partial u} \cdot \frac{\partial \mathbf{R}}{\partial u} \right) = 2 \left(\frac{\partial \mathbf{R}}{\partial u} \cdot \frac{\partial^2 \mathbf{R}}{\partial u \partial t} \right).$$

Note that

$$\frac{\partial \mathbf{R}}{\partial u} = \left| \frac{\partial \mathbf{R}}{\partial u} \right| \mathbf{t}$$

$$\frac{\partial \mathbf{R}}{\partial t} = \kappa \mathbf{n}$$

since the Gaussian function satisfies the heat equation. It follows that

$$\begin{aligned}
\frac{\partial}{\partial t} \left(\left| \frac{\partial \mathbf{R}}{\partial u} \right|^2 \right) &= 2 \left(\left| \frac{\partial \mathbf{R}}{\partial u} \right| \mathbf{t} \cdot \frac{\partial}{\partial u} (\kappa \mathbf{n}) \right) \\
&= 2 \left(\left| \frac{\partial \mathbf{R}}{\partial u} \right| \mathbf{t} \cdot \left(\frac{\partial \kappa}{\partial u} \mathbf{n} - \left| \frac{\partial \mathbf{R}}{\partial u} \right| \kappa^2 \mathbf{t} \right) \right) \\
&= -2 \left| \frac{\partial \mathbf{R}}{\partial u} \right|^2 \kappa^2.
\end{aligned}$$

Therefore

$$2 \left| \frac{\partial \mathbf{R}}{\partial u} \right| \cdot \frac{\partial}{\partial t} \left| \frac{\partial \mathbf{R}}{\partial u} \right| = -2 \left| \frac{\partial \mathbf{R}}{\partial u} \right|^2 \kappa^2$$

or

$$\frac{\partial}{\partial t} \left| \frac{\partial \mathbf{R}}{\partial u} \right| = - \left| \frac{\partial \mathbf{R}}{\partial u} \right| \kappa^2.$$

Let L denote the length of the curve. Now, observe that

$$\frac{\partial L}{\partial t} = \int_0^L \frac{\partial}{\partial t} \left| \frac{\partial \mathbf{R}}{\partial u} \right| du = - \int_0^L \left| \frac{\partial \mathbf{R}}{\partial u} \right| \kappa^2 \, du = - \int_0^1 \kappa^2 \, dw.$$

Since the value w_0 of the normalized arc length parameter w at a point P measures the length of the curve from the starting point to point P, it follows that

$$\frac{\partial \mathcal{W}}{\partial t} = - \int_0^{\mathcal{W}} \kappa^2 (\mathcal{U}, t) \, d\mathcal{U}$$

and therefore

$$\mathcal{W}(w, t) = - \int_0^t \int_0^{\mathcal{W}} \kappa^2 (\mathcal{U}, \mathcal{T}) \, d\mathcal{U} \, d\mathcal{T} + w. \tag{1.1}$$

Note that

$$\mathcal{W}(w, 0) = w.$$

Note further that for any given value t_0 of t, $\mathcal{W}(w, t_0)$ is a monotonically increasing function of w. To see this, observe that after changing the order of integration in equation 1.1 and applying the chain rule for derivatives, we obtain

$$\frac{\partial \mathcal{W}(w,t)}{\partial w} = \left(-\int_0^t \kappa^2(\mathcal{W}, \mathcal{T}) \, d\mathcal{T}\right) \frac{\partial \mathcal{W}(w,t)}{\partial w} + 1$$

or

$$\frac{\partial \mathcal{W}}{\partial w} \left(1 + \int_0^t \kappa^2(\mathcal{W}, \mathcal{T}) \, d\mathcal{T}\right) = 1$$

or

$$\frac{\partial \mathcal{W}}{\partial w} = \frac{1}{1 + \int_0^t \kappa^2(\mathcal{W}, \mathcal{T}) \, d\mathcal{T}}$$

which is always positive for any t. Therefore, $\mathcal{W}(w,t)$ is a monotonically increasing function of w for any t.

The solution of the equation

$$\kappa(\mathcal{W}, \sigma) = 0$$

forms the *resampled* curvature scale space of Γ. Since the function $\kappa(\mathcal{W}, t)$ in equation 1.1 is unknown, $\mathcal{W}(w,t)$ can not be computed directly from equation 1.1. However, the resampled curvature scale space can be computed in a simple way. A Gaussian filter based on a small value of the standard deviation is computed. The curve Γ is parametrized by the normalized arc length parameter and convolved with the filter. The resulting curve is reparametrized by the normalized arc length parameter and convolved again with the same filter. This process is repeated until the curve is convex and no longer has any curvature zero-crossing points. The curvature zero crossings of each curve are marked in the resampled curvature scale space image. Note that the standard deviation of the Gaussian chosen above should be small enough so that the deviation from arc length parametrization after each iteration is negligible. Then, the entire process can be considered to model arc length evolution.

We shall next show that this approximation process converges to the solution of equation 1.1. Let ϵ be the maximum error in the location of any point of Γ when the arc length evolution of Γ is approximated through the process described above using a Gaussian with standard deviation $\Delta\sigma$. Observe that at a point P of Γ

$$\epsilon \approx |(\mathbf{r} + \kappa\mathbf{n}) - (\mathbf{r} + \Delta\mathbf{r}_g)| = |\kappa\mathbf{n} - \Delta\mathbf{r}_g|,$$

where κ and $\mathbf{n}$ are the curvature and the normal vector at P, $\mathbf{r}$ is the position vector of point P, and $\Delta\mathbf{r}_g$ is the amount of change in the position vector of P after Gaussian approximation. Let

$$\Delta\mathbf{r}_g = \kappa_g\mathbf{n}_g,$$

where $\mathbf{n}_g$ is a unit vector with the same direction as that of $\Delta\mathbf{r}_g$, and κ_g is equal to the length of $\Delta\mathbf{r}_g$. $\mathbf{n}_g$ and κ_g can be thought of as the normal vector at P and the curvature of an arc of a circle going through P. As $\Delta\sigma \to 0$, the curve segment covered by the Gaussian filter can be approximated by a circular segment with constant curvature. It is easily seen that on an arc of a circle, regular evolution causes the same shrinkage rate at every point and is therefore equivalent to arc length evolution. It follows that for a small $\Delta\sigma$, $\kappa_g \to \kappa$, and $\mathbf{n}_g \to \mathbf{n}$. Therefore

$$\epsilon \approx |\kappa\mathbf{n} - \kappa_g\mathbf{n}_g| \to 0.$$

After j iterations of smoothing, total error is given by $j\epsilon$ which is small.

Therefore, the approximation process described above converges to the solution of equation (1.1). Figure 1.11(b) shows the resampled curvature scale space of Africa, and Figure 1.11(d) shows the resampled curvature scale space of Africa with nonuniform noise. Note that a very close match can be observed when matching Figures 1.11(b) and (d).

5. Evolution and Arc Length Evolution Properties of Planar Curves

This section presents a number of important results in the form of theorems on evolution and arc length evolution of planar curves as defined in sections 2 and 4. It also discusses the practical significance of each of those results. The proofs of these theorems are given in appendix A.

The first five theorems express a number of fundamental properties of evolution and arc length evolution.

THEOREM 1.1 *The order of application of evolution or arc length evolution and a shape preserving transformation (consisting of rotation, uniform scaling and translation) to a planar curve does not change the final result.*

It follows from theorem 1.1 that the regular, renormalized, and resampled curvature scale space images of a planar curve satisfy the *invariance* property. The invariance property is essential since it makes it possible to match a planar curve to another of similar shape that has undergone

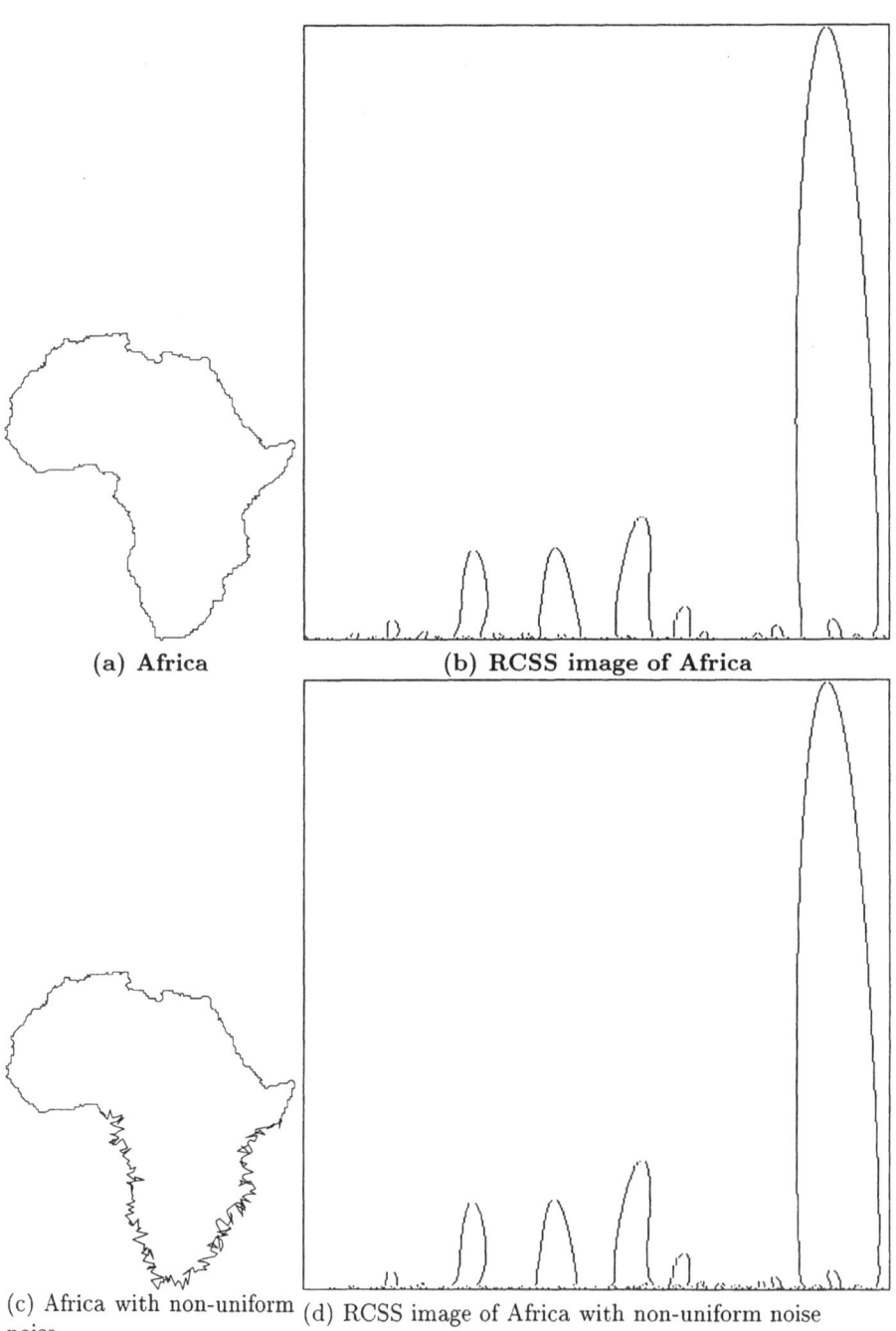

(a) Africa

(b) RCSS image of Africa

(c) Africa with non-uniform noise

(d) RCSS image of Africa with non-uniform noise

Figure 1.11. Experiments on the resampled Curvature Scale Space image

a transformation consisting of arbitrary amounts of rotation, uniform scaling, and translation.

THEOREM 1.2 *A closed planar curve remains closed during its evolution and arc length evolution.*

THEOREM 1.3 *A connected planar curve remains connected during its evolution and arc length evolution.*

Theorems 1.2 and 1.3 show that connectedness and closedness of a planar curve are preserved during evolution and arc length evolution. These theorems show that evolution and arc length evolution do not change the physical interpretation of a planar curve as the boundary of a 2-D object. Consider a closed, connected planar curve that represents the boundary of a 2-D object. If such a curve is not closed and connected after evolution or arc length evolution, then it can no longer be viewed as the boundary of a 2-D object.

THEOREM 1.4 *The center of mass of a planar curve does not move during evolution and arc length evolution of that curve.*

THEOREM 1.5 *Let Γ be a closed planar curve and let $\mathcal{G}$ be its convex hull. Γ remains inside $\mathcal{G}$ during evolution and arc length evolution.*

Theorem 1.4 shows that the center of mass of a planar curve remains stationary as the curse evolves, and theorem 1.5 shows that a planar curve remains inside its convex hull during evolution and arc length evolution. Together, theorems 1.4 and 1.5 impose constraints on the physical location of a planar curve as it evolves. These constraints become useful whenever the physical location of curves in an image or their locations with respect to each other is important. A possible application area is stereo matching, in which it is advantageous to carry out matching at coarser levels of detail first and then match at fine detail levels to increase accuracy.

Theorem 1.6 shows that the mapping from a planar curve to its curvature scale space image is an invertible one.

THEOREM 1.6 *Let Γ be a planar curve in C_2. The high-order derivatives at a single point on one curvature zero-crossing contour in the regular, renormalized, or resampled curvature scale space image of Γ determine Γ uniquely up to uniform scaling, rotation, and translation (except on a set of measure zero).*

Theorem 1.6 shows that the curvature scale space images of a planar curve in fact satisfy the *uniqueness* property. This property ensures that curves of different shapes do not have the same representation. Note that the proof of this theorem is not meant to be used as a practical reconstruction scheme.

Theorem 1.7 states that under certain conditions, new curvature zero-crossing points are not created during evolution and arc length evolution of planar curves. In other words, only the curvature zero-crossings which exist on the original contour can exist at higher scales.

THEOREM 1.7 *Let Γ be a planar curve in C_2. If all evolved and arc length evolved curves Γ_σ are in C_2, then all extrema of contours in the regular, renormalized, and resampled curvature scale space images of Γ are maxima.*

Theorem 1.8 locally characterizes the behavior of planar curves during evolution and arc length evolution just before the creation of a cusp point.

THEOREM 1.8 *Let $\Gamma = (x(u), y(u))$ be a planar curve in C_1, and let $x(u)$ and $y(u)$ be polynomial functions of u. Let Γ_σ be an evolved or arc length evolved version of Γ with a cusp point at u_0. There is a $\delta > 0$ such that $\Gamma_{\sigma-\delta}$ intersects itself in a neighborhood of point u_0.*

The following theorem holds only for arc length evolution

THEOREM 1.9 *Simple (not self-intersecting) curves remain simple during arc length evolution.*

Theorem 1.10 locally characterizes the behavior of a planar curve during evolution and arc length evolution just after the creation of a cusp point.

THEOREM 1.10 *Let $\Gamma = (x(u), y(u))$ be a planar curve in C_1, and let $x(u)$ and $y(u)$ be polynomial functions of u. Let Γ_σ be an evolved version of Γ with a cusp point at u_0. There is a $\delta > 0$ such that $\Gamma_{\sigma+\delta}$ has two new curvature zero crossings in a neighborhood of u_0.*

Theorems 1.8 and 1.10 together locally characterize the behavior of a planar curve just before and just after the formation of a cusp point during evolution and arc length evolution. This behavior can be used to detect any cusp points that form during evolution or arc length evolution of a planar curve. Such cusp points can then be used effectively to facilitate matching since they provide us with a set of distinctive and easily recognizable features. These theorems also show that self-intersecting curves are described in a natural way by our representation technique.

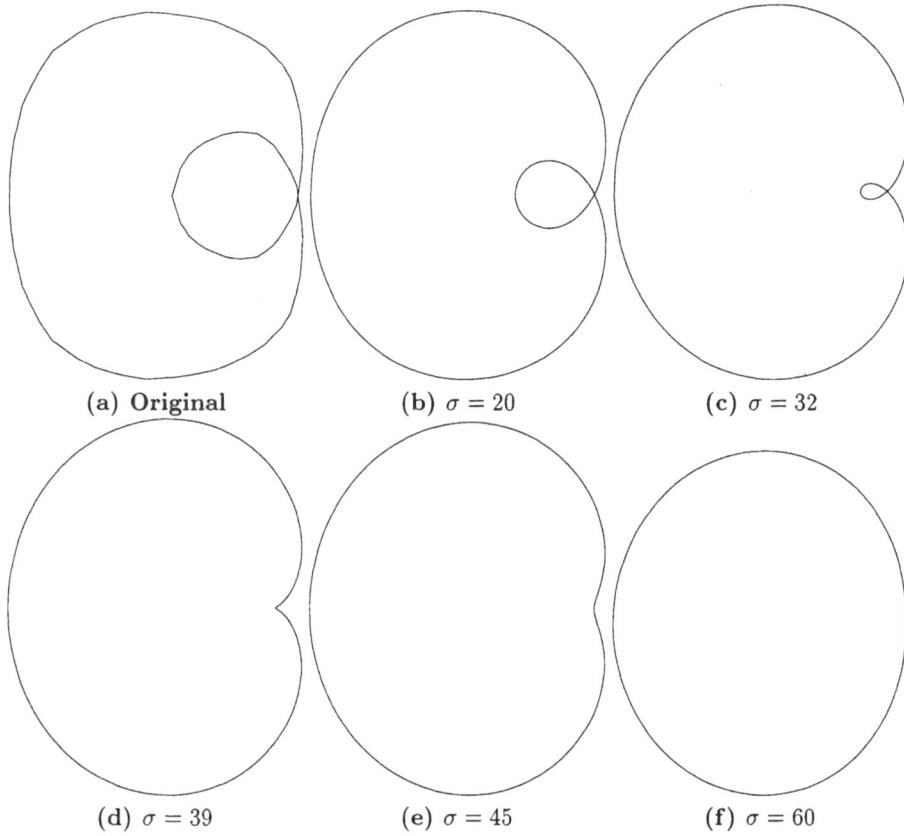

(a) Original   (b) $\sigma = 20$ (c) $\sigma = 32$

(d) $\sigma = 39$ (e) $\sigma = 45$ (f) $\sigma = 60$

Figure 1.12. Convex, self-crossing curve during evolution

The self-intersection loop gradually grows smaller until it turns into a cusp point and vanishes. In contrast, Asada and Brady's method [23] enlarges the smaller loop until it becomes as large as the larger loop. Figures 1.12 and 1.13 show two self-intersecting curves during evolution. The self-intersection is resolved through the formation of a cusp point, after which the curve becomes simple. Figure 1.14 shows another self-crossing curve which does not form any cusp points during evolution.

Theorem 1.7 showed that if a planar curve remains smooth during evolution or arc length evolution, then no new curvature zero crossings will be observed in its curvature scale space images. Theorem 1.8 showed that every planar curve intersects itself during evolution or arc length evolution just before the formation of a cusp point, and theorem 1.9 showed that simple curves remain simple during arc length evolution. Combining theorems 1.7, 1.8 and 1.9, we conclude that no new curvature zero-crossing points are created during arc length evolution of simple

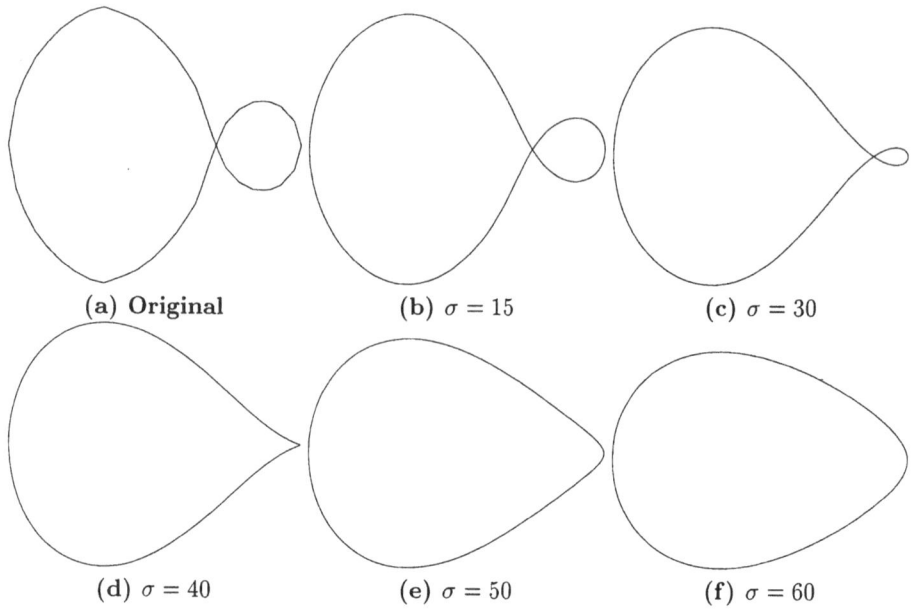

Figure 1.13.   Concave, self-crossing curve during evolution

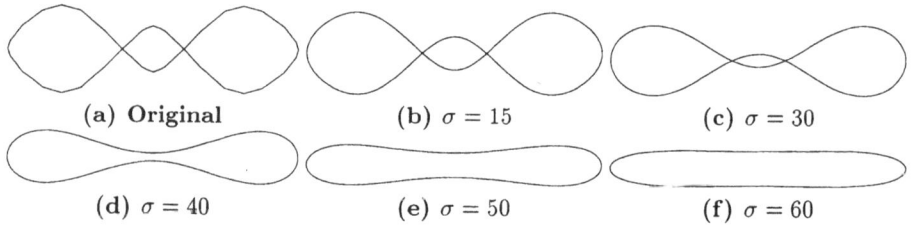

Figure 1.14.   Self-crossing curve during evolution without cusp formation

curves. This is an important result since it indicates that new *structure* is not created in the curvature scale space representations of simple curves [177]. Note that a subclass of self-crossing curves also shares this property.

The result stated by theorem 1.9 is also significant. Simple planar curves usually represent the boundaries of 2-D objects. Arc length evolved versions of those curves can only have physical plausibility if they are also simple. Theorem 1.9 shows that this is in fact the case. Figure 1.15 shows a simple curve and its evolved versions. It can be seen that the curve intersects itself during evolution. Figure 1.16 shows the same curve and its arc length evolved versions. As expected, the curve remains simple during arc length evolution.

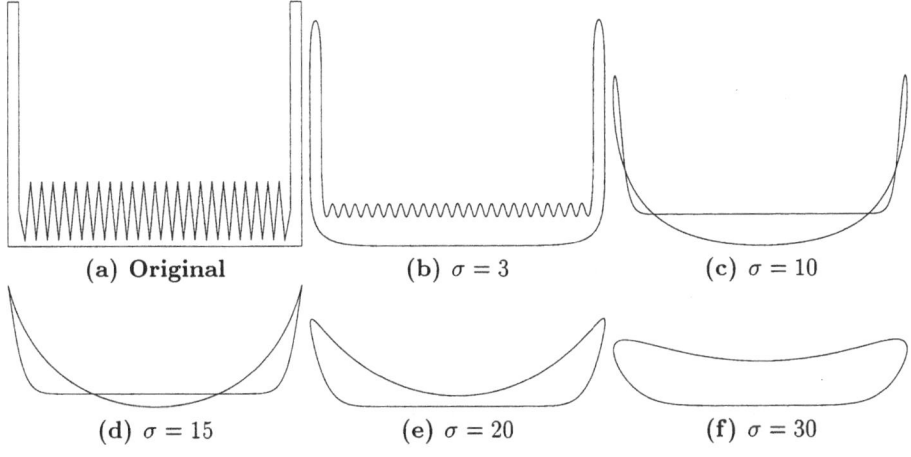

Figure 1.15.   Simple curve during regular evolution

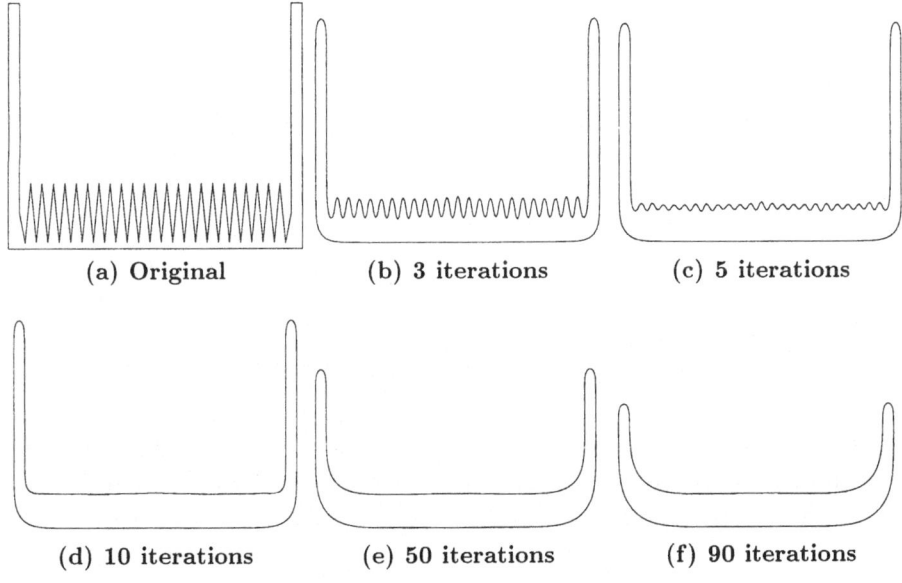

Figure 1.16.   Simple curve during arc-length evolution

The final theorem is on the convergence properties of curvature scale space representations.

THEOREM 1.11 *Let* Γ *be a closed planar curve.* Γ *becomes simple and convex during evolution or arc length evolution and remains simple and convex.*

Theorem 1.11 describes a very important property of curvature scale space representations since it shows that the computation of such a representation always has a clearly defined termination point.

6.   Experiments, Discussion, and Evaluation

This section contains additional experiments to demonstrate the stability of the curvature scale space representation. It also contains a discussion of the regular, renormalized, and resampled curvature scale space representations and an evaluation of those representations according to the criteria proposed in section 1.

Since the representation methods proposed in this chapter involve identification of curvature zero-crossing points on planar curves, it may be conjectured that they are not suitable for application to curves with straight segments on them. However, it should be noted that while the presence of straight line segments on a curve might introduce instabilities at the finest scale levels, after a small number of iterations, the originally straight segments will have nonzero curvature, and the computations will stabilize. Figure 1.17(a) shows a planar curve made up of straight line segments, and Figure 1.17(c) shows the curve of Figure 1.17(a) with added random noise. Figure 1.17(b) shows the curvature scale space image of the curve of Figure 1.17(a), and Figure 1.17(d) shows the curvature scale space image of the curve of Figure 1.17(c). It can be seen that while there is disagreement between the two representations at the finest scale levels, a very close match exists at the higher levels of the CSS images.

Three different multiscale representation techniques for planar curves were described in this chapter. These three are the regular curvature scale space image, the renormalized curvature scale space image, and the resampled curvature scale space image. Each representation technique is suitable for specific applications. When uniform noise exists on the curve, the regular curvature scale space image can be used. However, when there is nonuniform noise on the curve or when there are local shape differences between the model curves and the image curves, either the renormalized or the resampled curvature scale space images should be used. The renormalized curvature scale space image is the most computationally intensive. Observations indicate that when there are local shape differences, the resampled curvature scale space images show the best overall match, whereas the renormalized curvature scale space images match well at high scales but are more influenced by the shape differences at lower scales. Therefore, the choice of the representation technique should depend on the scale level of the curve features that

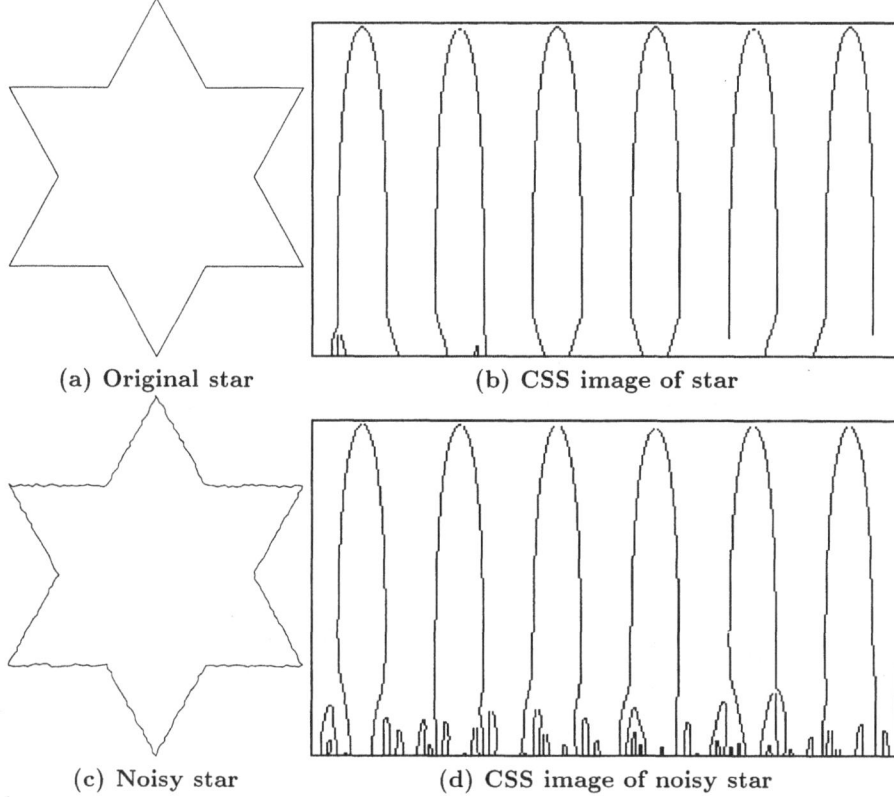

(a) Original star   (b) CSS image of star

(c) Noisy star   (d) CSS image of noisy star

Figure 1.17. Star and noisy star with CSS images

Representation technique	Advantages	Disadvantages
The Regular Curvature Scale Space Image	Suitable for transformations consisting of uniform scaling, rotation and translation. Suitable when uniform, noise corrupts the curve.	Nonuniform noise or local difference in shape can cause problems.
The Renormalized Curvature Scale Space Image	More suitable when there is nonuniform noise on the curve or local shape differences exist.	Most computationally intensive.
The Resampled Curvature Scale Space Image	Most suitable when there is high-intensity, nonuniform noise or local shape differences exist.	De-emphasizes shape differences at fine scales.

Table 1.1. Comparison of regular, renormalized, and resampled CSS images.

one wishes to emphasize. Table 1.1 summarizes the advantages and disadvantages of each representation technique.

We now present an evaluation of the curvature scale space representations according to the criteria proposed in section 1.

Criterion: *Invariance*

Recall that by *invariance*, we meant that the representation for the shape of a curve should not change when shape-preserving transformations (rotation, uniform scaling, and translation) are applied to that curve.

Translation of the curve causes no change in the curvature scale space representations proposed here. Uniform scaling causes the curvature scale space representations to undergo uniform scaling as well. If the represented curves are closed, then their curvature scale space representations can be normalized, and invariance with respect to uniform scaling will also be achieved. If the represented curves are open, changes due to uniform scaling can be handled by a matching algorithm such as the one used in [220].

Rotation causes only a horizontal shift in the curvature scale space representations. However, due to the multiscale nature of those representations, a matching algorithm can determine the shift difference between two matching curvature scale space representations.

Criterion: *Uniqueness*

The *uniqueness* criterion required that two curves with different shapes be mapped to different representations. This is necessary in order to be able to recognize two or more curves with the same shape by observing that their representations are the same.

As argued earlier, theorem 1.6 shows that a planar curve can be reconstructed from any of its curvature scale space representations, and therefore, the curvature scale space representations satisfy the uniqueness criterion.

The only arbitrary choice to be made when computing curvature scale space representations is the starting point for parametrization on a closed curve. This only causes a horizontal shift in the curvature scale space representations, but it causes no structural change.

Criterion: *Stability*

The *stability* criterion requires that a small change in the shape of a curve leads to a small change in its representation and vice versa. Theorem 1.3 shows that planar curves remain connected during evolution and arc length evolution, and therefore, their curvature scale space representations can always be constructed. Furthermore, our observations show that while a planar curve evolves, a small change in the standard deviation of the Gaussian filter always results in a small change in the locations of the curvature zero crossings on that curve. The experiments of this chapter also show that curvature scale space representations are stable with respect to significant uniform and nonuniform noise on the curves they represent and therefore satisfy the stability criterion.

Criterion: *Local support*

CSS computations use finite Gaussian filters and finite neighborhoods. Therefore a CSS representation can also be computed for an open contour and, except near the endpoints of the curve, will resemble the corresponding representation for a closed contour of which it is a part. It is therefore believed that the representations also satisfy the local support criterion.

Criterion: *Efficiency*

The computation of the representations proposed here calls for evaluation of a large number of convolutions. This process can be rendered efficient using one or more of the following techniques:

- Fast Fourier transforms,

- Parallelism,

- Expression of convolutions involving Gaussians of large widths in terms of convolutions involving Gaussians of small widths only,

- Tracking curvature zero-crossing points across multiple scales; when it is known that new curvature zero crossings will not be created at higher scales, convolutions can be carried out only in a small neighborhood of the existing zero crossings in order to find the zero crossings at the next higher level.

Furthermore, curvature scale space representations can be stored very efficiently as encoded binary images. An alternative is to store only a selected subset of points from those scale space representations that will be used for matching. In general, all algorithms proposed are efficient in that their complexities are low-order polynomials in the size of the input.

Criterion: *Shape Properties*

The curvature scale space representations of symmetric curves are also symmetric since a symmetric curve also has symmetric curvature zero crossings across scales. Therefore, the symmetry criterion is satisfied. Furthermore, curvature scale space computations are carried out using finite Gaussian filters and making use of finite-sized neighborhoods. Therefore, a curvature scale space representation can also be computed for an open curve and, except near the endpoints of the curve, will resemble the corresponding representation for a closed curve of which it is a part. It is therefore believed that the representations also satisfy the part/whole criterion.

Criterion: *Implementation*

The procedures needed to compute curvature scale space images are not difficult to implement. Convolutions with Gaussian filters are at the heart of the computations. These are standard and well-understood procedures in the computational vision and image processing areas. It follows that this criterion is also satisfied.

Finally, note that volumetric diffusion [146] and reactive and diffusive deformations of shape [141] are shape representation techniques that are the most similar to ours since they also compute deformations of shapes of planar curves. It may therefore be suggested that an alternative way to compute the curvature scale space representations is to use one of the above techniques to compute deformations of the curves to be represented and then locate curvature zero-crossing points on each deformed curve and map them to the appropriate representations. However, the application of each of the techniques mentioned above might result in disconnected curves. In such cases, it will no longer be possible to construct the curvature scale space representations. Furthermore, our technique combines the curve deformation and the computation of curvature into one step and is therefore more numerically accurate than the aforementioned techniques that separate the processes of curve deformation and computation of curvature.

It follows that the curvature scale space representations satisfy the criteria considered necessary for a general-purpose shape representation method.

7.   Concluding Remarks

This chapter introduced a novel shape representation technique far planar curves and proposed a number of criteria considered necessary for any general-purpose shape representation scheme. Those criteria are *invariance, uniqueness, stability, local support, efficiency, shape properties*, and *implementation*.

Three different ways of computing the representation were described. Each method relies on extracting features of the curve that are invariant under shape-preserving transformations at varying scales. These methods result in the curvature scale space image, the renormalized curvature scale space image, and the resampled curvature scale space image. It was shown that each of those representations is suitable for a specific application.

A number of theoretical properties of those representations were also investigated. These properties together provide a sound foundation for the representations proposed in this chapter. Finally, it was shown that the proposed representations satisfy nearly all the criteria introduced earlier.

Chapter 2

ROBUST FREE-FORM OBJECT RECOGNITION THROUGH CURVATURE SCALE SPACE

Two complete and practical systems for free-form object recognition have been developed which are very robust with respect to noise and local deformations of shape (due to weak perspective distortion, segmentation errors and non-rigid material) as well as scale, position and orientation changes of the objects. The systems have been tested on a wide variety of free-form 3-D objects. Industrial applications are envisaged where a fixed camera and a light-box are utilized to obtain images. Within the constraints of the systems, every rigid 3-D object can be modeled by a limited number of classes of 2-D contours corresponding to the object's resting positions on the light-box. The contours in each class are related to each other by a 2-D similarity transformation.

1. Introduction

Object representation and recognition is one of the central problems in computer vision and image processing. This chapter describes two complete, working vision system which recognize isolated and occluded objects in images reliably using their curvature scale space representations [226, 190, 189, 220, 221, 222]. The CSS representation is based on the *scale space image* concept introduced in [295] and popularized by [322]. The recognition systems developed here may be used for recognition of occluded 3-D objects. In particular, it is assumed that a number of 3-D objects are resting on a plain flat surface S with a fixed camera looking straight down at them. The distance from the camera to S is assumed to be much larger than the height of any object resting on S. Note that S may be a light-box or another surface which can provide good contrast. A preliminary version of this work appeared in [192]. It is believed that this task is interesting for the following reasons:

- No constraints have been placed on object shapes or types. Even the simpler ones have been treated as free-form objects. Furthermore, environment constraints are not difficult to satisfy in many object recognition tasks (such as in industrial settings).

- Within the constraints defined above, every rigid 3-D object can be represented by a limited number of classes of 2-D contours such that the contours within each class are related to each other by a 2-D similarity transformation.

- Recognition can become challenging due to arbitrary shapes of objects, noise, and local deformations of shape which can be caused by weak perspective distortions, segmentation errors and the non-rigid material used in some objects.

The existing literature on shape representation and recognition is quite large. A survey of some recent work can be found in [296]. Shape polynomials were used to represent and compare shapes in [303] but only a limited class of objects with simple shapes can be recognized using polynomials. Polygonal approximations were used for recognition of partially occluded 2-D objects in [24] but polygons fitted to free-form objects tend to be not reliable for object recognition since the location of vertices tend to be arbitrary. Alignment was used for recognizing 3-D objects in 2-D images in [15]. An object recognition system was developed for bin-picking tasks in [123] which organized faces in an aspect graph. A parallel implementation of a 2-D model-based recognition system was presented in [47]. Registration of 3-D shapes was addressed in [36]. Sparse 3D position and orientation measurements were utilized for object recognition in [99]. The 3-D profile method was utilized for object recognition in [105]. The generalized Hough transform was used to detect arbitrary shapes in [25] but this approach can suffer from an explosion in the parameter space. Probabilistic 3-D object recognition was presented in [285] but the system was demonstrated on only one image of simple polyhedral objects. Interest features were used to create a hash-table in [154] which was utilized in conjunction with a generalized Hough transform technique. Parametrized 3-D models were fitted to images in [167] but the system was demonstrated on only one such model. Overlapping parts were localized in [98] by searching the interpretation tree. An adaptive 3-D object recognition system was developed in [280] using multiple views. Rapid object indexing and recognition using enhanced geometric hashing was utilized in [155] but the system was demonstrated on simple polyhedral objects. Recognition of geons by parametric deformable contour models was developed in [255] but the geons appear to be applicable only to 3-D objects with quite simple shapes. Object recognition by alignment was proposed in [121] and [122]. In [121], the authors describe a procedure for scale-space segmentation of 2-D contours but they select three specific scales without making it clear how those scales were chosen. Clearly, the specific scales utilized for the segmentation process can affect the outcome. A polynomial-time object recognition system in presence of occlusion and uncertainty was developed in [48] but results show only one object model and one input image. The curvature primal sketch proposed in [23] makes use of curvature information to represent contours but each contour has to be divided into a set of "primitive parametrized curvature discontinuities" such as *corners*, *ends*, *cranks*, *smooth joins*, and *bumps*. This makes the representation unsuitable for free-form contours since a natural division of it into such primitives will not be possible.

As this review shows, other object recognition systems utilizing 2-D views for recognition have been developed. The major advantage of the system described here is that free-form 3-D objects are represented using invariant, multi-scale features. Another advantage is that the CSS image is completely data-driven, and does not assume that the data fits any specific shape models. This makes it unnecessary to use inappropriate models such as polygons, arcs of circles, splines and other primitives which do not suit objects with arbitrary shapes. As a result, the system has added robustness and reliability. Keeping in mind the constraints outlined above, the system presented here is believed to be more robust than other comparable systems.

The organization of the remainder of this chapter is as following. Section 2 contains a description of a silhouette-based isolated object recognition system. Section 3 presents an explanation of a silhouette-based occluded object recognition system, and section 4 contains the concluding remarks.

2.   Silhouette-Based Isolated Object Recognition

A complete, efficient and practical isolated object recognition system has been developed which is very robust with respect to scale, position and orientation changes of the objects as well as noise and local deformations of shape (due to perspective projection, segmentation errors and non-rigid material used in some objects). The system has been tested on a wide variety of 3-D objects with different shapes and material and surface properties. A light-box setup is used to obtain silhouette images which are segmented to obtain the physical boundaries of the objects which are classified as either convex or concave. Convex curves are recognized using their four high-scale curvature extrema points. Curvature

Scale Space Representations are computed for concave curves. The CSS representation is a multi-scale organization of the natural, invariant features of a curve (curvature zero-crossings or extrema) and useful for very reliable recognition of the correct model since it places no constraints on the shape of objects. A three-stage, coarse-to-fine matching algorithm prunes the search space in stage one by applying the CSS aspect ratio test. The maxima of contours in CSS representations of the surviving models are used for fast CSS matching in stage two. Finally, stage 3 verifies the best match and resolves any ambiguities by determining the distance between the image and model curves. Transformation parameter optimization is then used to find the best fit of the input object to the correct model.

2.1.   Curvature Scale Space Matching

The basic idea behind the CSS matching algorithm is to obtain a coarse-level match using the structural features of the input curves. Such a match can be found quickly and reliably since at the high scales of CSS images, there are relatively few features to be matched. The actual features used for matching are the maxima of the curvature zero-crossing contours. The reason for using the maxima as features is that they are the most significant points of zero-crossing contours: the CSS coordinates of a maximum convey information on both the location and the scale of the corresponding contour whereas the *body* of the contour is, in general, similar in shape to those of other contours. Furthermore, the maxima are isolated point features and therefore solving the feature correspondence problem is relatively simple. This is specially true at the high scales of the CSS image where the maxima are sparse.

It follows that the task of the matching algorithm is to find the correct correspondence between two sets of maxima: one from each CSS image. The allowed transformation from one set to the other set is mere horizontal translation. The translation parameter is computed when the first image curve CSS maximum is mapped to the first model curve CSS maximum and then used to map each of the remaining image curve CSS maxima to the model curve CSS. The corresponding model curve CSS maximum for each mapped image curve CSS maximum should then be the *closest* model curve CSS maximum (and the associated cost is the Euclidean distance between them). Many candidates may have to be considered since the correspondence between the first pair of maxima can be made in possibly many ways. This matching problem can be solved using a *best-first* matching strategy [Winston 1979] which will gradually expand a number of candidate matches in parallel (always selecting the best partial match) until the lowest-cost complete match is found.

The CSS matching algorithm is therefore as follows:

1 For each of the input CSS images, carry out the following: Extract the maxima of each CSS image. Record the coordinates of each maximum in a feature list as it is encountered. When the algorithm ends, this list will be sorted by the scale coordinate of the maxima. Normalize those coordinates so that the horizontal coordinate u varies in the range [0,1].

2 Create a number of nodes corresponding to the possible match of the highest-scale maximum of the image curve CSS and each maximum of the model CSS which has a σ-coordinate *close* (within 90%) to that of the highest model maximum. Initialize the *cost* of each node to zero.

3 For each node created in the previous step, compute a CSS shift parameter α using the following formula:

$$u_m = u_i + \alpha$$

where u_i is the horizontal coordinate of the image curve CSS maximum and u_m is the horizontal coordinate of the model curve CSS maximum.

4 Create two lists for each node created in step 2. The first list will contain the image curve CSS maxima matched within that node at any point during program execution and the second list will contain the corresponding model curve CSS maxima. Initialize the first list of each node to contain the highest-scale image curve CSS maximum. Initialize the second list of each node to contain the corresponding model curve CSS maximum determined in step 2.

5 Expand each node created in step 2 one step using the procedure described in the next step.

6 To expand a node, select the highest-scale, image curve CSS maximum (which is *not* in its first list) and apply that node's CSS shift parameter computed in step 3 to map that maximum to the model CSS image. Locate the nearest model curve CSS maximum (which is *not* in the node's second list). The cost of match is defined as the straight line distance in the model CSS image between the two maxima. If there are no more image curve CSS maxima left, define cost of match as the height of the highest model curve CSS maximum *not*

in the node's second list. Likewise, if there are no more model curve CSS maxima left, define cost of match as the height (after mapping) of the selected image curve CSS maximum. Add the match cost to the node cost. Update the two lists associated with the node.

7 Select the lowest-cost node. If there are no more model or image curve CSS maxima that remain unmatched within that node, then return that node as the lowest-cost node. Otherwise, go to step 6 and expand the lowest-cost node.

2.2.   A Silhouette-based Object Recognition System

It is now possible to give an overall description of the system designed and implemented for silhouette-based object recognition through the curvature scale space representation. The system was designed so that both *convex* and *concave* curves can be recognized. Two hundred (200) sample points are used on all model and image contours.

The system can be divided into an *off-line* and an *on-line* component. The off-line part in completed first and is itself divided into two stages: *model acquisition* and *model representations.*

In the model acquisition stage, model contours can be obtained in two ways: manually entering the coordinates of points on the contours or obtaining an image of the model object, segmenting the image and recovering the contour. The former technique can result in less noise on the contour. Both techniques were used here to obtain the model contours. One contour must be obtained for each stable position of the model object. Each model contour is normalized so that it touches a square of size one on at least three sides.

In the model representations stage, if the model contour is concave, the following procedure is followed: the CSS representation of the model contour is computed off-line. The aspect ratio of the model contour CSS image is also computed. The maxima of the CSS representation are then extracted. The final representation for the model contour is simply the CSS coordinates of each of those maxima. If the model contour is convex, that contour is smoothed until only 4 curvature extrema remain on the contour. The arc length coordinates of those 4 extrema are then recorded. The ratio of the Euclidean distance between the curvature maxima to the distance between the curvature minima is also computed for each convex contour.

The on-line part can be divided into several stages as following:

1 The input image is first segmented to obtain a contour from the image. The contour is smoothed slightly to remove noise and normalized

so that it touches a square of size one on at least three sides. Curvature is then computed at each point along the contour. If curvature is nearly constant, the system concludes that the contour is circular and stops. Otherwise, if curvature is positive everywhere or very close to zero but possibly negative (this situation can be caused by noise on a convex contour), the system concludes that the input contour is convex and follows steps 7 through 10. Otherwise, the input contour is considered to be concave and steps 2 through 6 are followed.

2 The CSS representation of the image curve is computed.

3 The maxima of the image curve CSS contours are extracted.

4 The aspect ratio of the image curve CSS is computed. Any model curve CSS image whose aspect ratio is close (within 10% but this figure can be higher) to the aspect ratio of the image curve CSS is accepted for step 5. Otherwise, it is rejected. This step can also be implemented using a hash table.

5 CSS matching is applied to the surviving models.

6 The best 20% (but this figure can be higher) of the matches in step 5 are selected for verification. For each model curve which is selected, image curve transformation parameters are computed using their best CSS match and used to map the image curve to the model curve. The image-model curve distance is then computed. The model curve with the lowest image-model curve distance is chosen as the best matching model. Transformation parameter optimization is then used to find the best fit of the image curve to the chosen model (see sections 3.5, 3.6, and 3.7). In situations where there is little difference between some model curves, parameter optimization can be used during the recognition process to find the best matching model.

7 The input convex contour is smoothed until only 4 curvature extrema remain on the contour.

8 The ratio of the Euclidean distance between the curvature maxima on the contour to the distance between the curvature minima on the contour is computed. Any convex model curve whose corresponding ratio is close (within 10% but this figure can be higher) to this ratio is accepted for step 9. Otherwise, it is rejected. This step can also be implemented using a hash table.

9 The first curvature maximum on the image curve can map to either of the two curvature maxima on each of the surviving model curves.

A set of image curve transformation parameters is computed for each case and applied to the image curve. The image-model curve distance is computed in each case (see sections 3.5 and 3.6).

10 The model curve with the lowest image-model curve distance is chosen as the best matching model. Transformation parameter optimization is then used to find the best fit of the image curve to the chosen model. In situations where there is little difference between some model curves, parameter optimization can be used during the recognition process to find the best matching model (see section 3.7).

2.3.   Results and Discussion

The recognition system described in the previous subsection was implemented in *C* and ran on a *Silicon Graphics IRIS Crimson* workstation. It was tested using a total of 22 model curves and 19 images. All model contours were concave except the calculator, the glue stick and the spray can lid. Figure 2.17 shows all the model contours used to evaluate system performance.

Due to the light-box setup used, the images obtained had high contrast. As a result, thresholding was successful in properly segmenting each input image after which the bounding contours were recovered. Figure 2.1 shows four input images and the contours recovered from those images. Figure 2.2 shows all the remaining input images.

Each one of the 19 input objects was recognized correctly by the system in less than one second (this includes time for computation of image curve CSS representation). In most cases, CSS matching was sufficient to correctly recognize the input object. When ambiguities remained after CSS matching, curve distance computation successfully resolved those ambiguities.

Each input object was recognized in less than one second by the system. The matching program is fast because it has been designed specifically for the task of matching closed curves using maxima of curvature scale space contours. These maxima (excluding the ones corresponding to very small CSS contours) are small in number (from 2 to 6 for the curves used as input to our system) but have a high multi-scale discriminative power. As a result, we believe they are the ideal feature points for the matching task considered here. It should also be pointed out that the CSS matching algorithm is a best-first algorithm which finds the best match of two CSS representations. It can be used in a serial fashion to find the best match of the image curve CSS to each of the surviving model curve CSS representations. It can also be run on parallel machines to find the best match to each model curve CSS

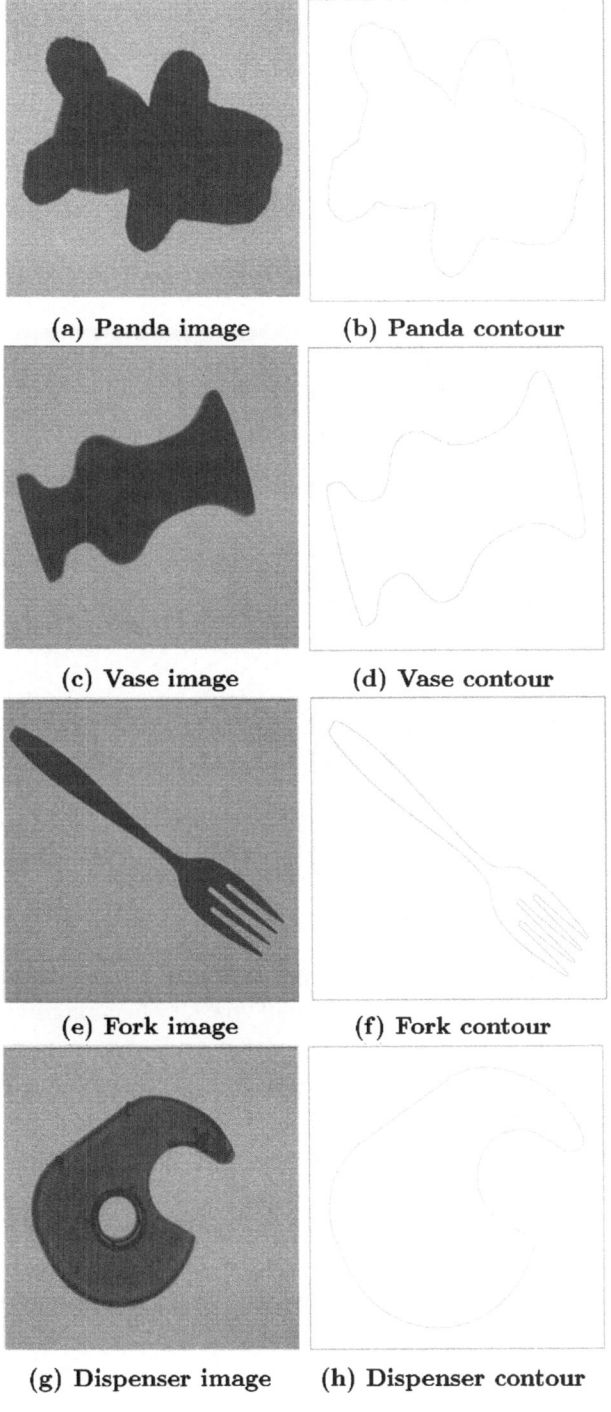

(a) Panda image (b) Panda contour

(c) Vase image (d) Vase contour

(e) Fork image (f) Fork contour

(g) Dispenser image (h) Dispenser contour

Figure 2.1. Four input images and recovered contours

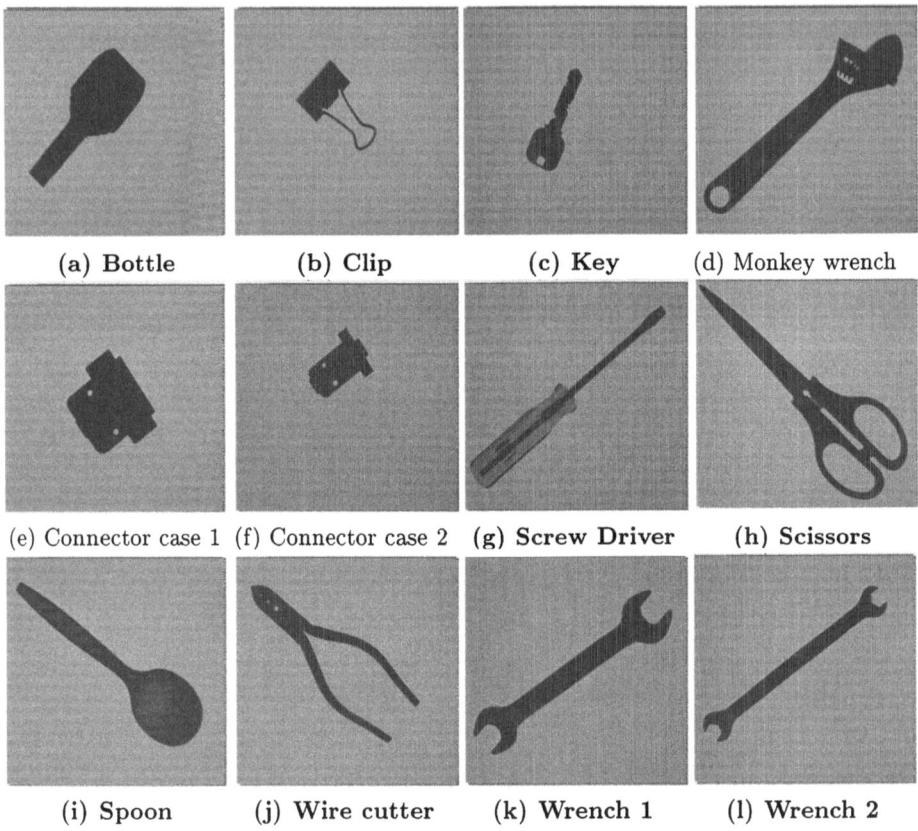

(a) **Bottle**　　(b) **Clip**　　(c) **Key**　　(d) Monkey wrench

(e) Connector case 1　(f) Connector case 2　(g) **Screw Driver**　(h) **Scissors**

(i) **Spoon**　　(j) **Wire cutter**　　(k) **Wrench 1**　　(l) **Wrench 2**

Figure 2.2.　Remaining input images

representation on a separate machine. Finally, it can be embedded in another best-first algorithm which will match the image curve CSS to each model curve CSS one step at a time and always pursue only the best match. The latter technique would be the fastest serial implementation of the algorithm. Here, the first technique was used since the algorithm is already quite fast. Furthermore, the coarse-to-fine matching technique used makes efficient use of recognition power: models pass through *recognition filters* which become increasingly fine until the best-matching model is selected.

The thresholds used in the tests which act as filters between the various stages of the system were drastically changed several times. In one test, those filters were effectively removed. It was verified that there were no effects on the output of the system; the filters exist only for efficiency reasons. The system was very robust in each case despite the presence of noise and local deformations of shape due to:

- Perspective projection of actual 3-D objects with depth (such as the vase).

- Segmentation errors near smooth physical boundaries of objects.

- Non-rigid material (such as cloth or soft plastic) used in some input objects.

It was discovered that, in general, if two contours had the same rough, coarse-level shape structure, then they would also have a close CSS match value. For example, this was true about the bottle and the clip contours. In such cases, image-model curve distance computation resolved the ambiguities. It was also discovered that a single model can be used to represent a class of similar looking objects. For example, the model screwdriver was longer than the screwdriver used as input to the system.

The following are some examples of the matches found by the system. In each case, the image curve (drawn using a thin line) has been mapped to the model curve (drawn using a thick line). Figure 2.3 shows the key matched to the model key. The CSS images of both contours are also shown. Note that a considerable amount of noise exists on the image curve. Methods that rely on extracting features at a single scale would encounter difficulties with this example. Figure 2.4 shows one of the connector cases matched to its model. The CSS images of both curves are also displayed. The effect of segmentation error can be observed on the image contour. Figure 2.5 shows the Panda matched to its model. Note that the Panda was made of cloth and therefore did not have a very rigid shape. The CSS images of the two curves are also shown. Figure 2.6 shows the vase matched to the model vase. The local mismatch that can be observed is due to the fact that the model curve corresponds to an orthogonal projection of the vase whereas the image curve corresponds to its perspective projection. The CSS images of the two curves are also shown.

The initial fit of the image curve to the model curve may not be very good due to noise or local shape deformations. Transformation parameter optimization will guarantee the best possible global fit. Figures 2.3(b), 2.4(b), 2.5(b), and 2.6(b) show the matches shown in figures 2.3(a), 2.4(a), 2.5(a), and 2.6(a) after transformation parameter optimization.

3.   Silhouette-Based Occluded Object Recognition

This section describes a complete and practical system for occluded object recognition which is quite robust to noise and local deformations of shape [193, 196].

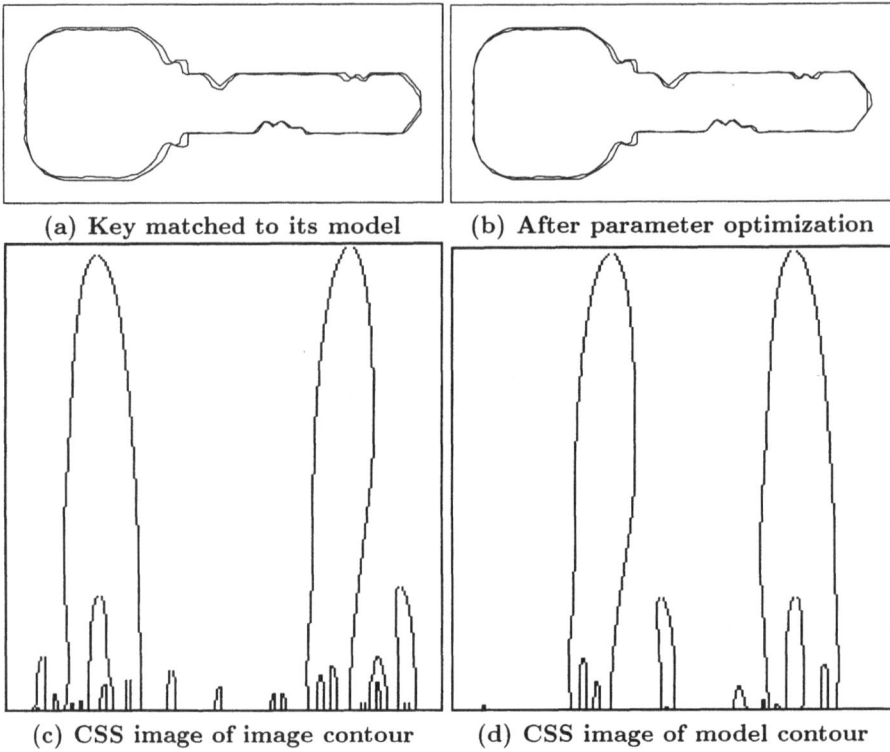

(a) Key matched to its model	(b) After parameter optimization
(c) CSS image of image contour	(d) CSS image of model contour

Figure 2.3.   Matching the key to its model contour

3.1.   Image Segmentation

The same threshold value T was used to effectively segment all input images. A salt-and-pepper noise removal procedure was applied to the resulting binary image in order to remove isolated noise. The next step was to apply a process of region growing followed by shrinking to the image in order to fill in cracks and small holes. The resulting binary image always had only one connected region of 1-pixels which corresponded to the objects. The final step was to recover the image coordinates of the boundary points of that region.

3.2.   Multi-Scale Segmentation of 2-D Contours

The purpose of the multi-scale segmentation procedure described here [195, 199] is to divide a 2-D contour into basic segments to be used by a local matching algorithm for occluded object recognition. Curvature zero-crossing points are used as inherent feature points to segment the contour since their locations are invariant to rotation, uniform scaling, and translation of the contour. (Note, however, that corners (or maxima

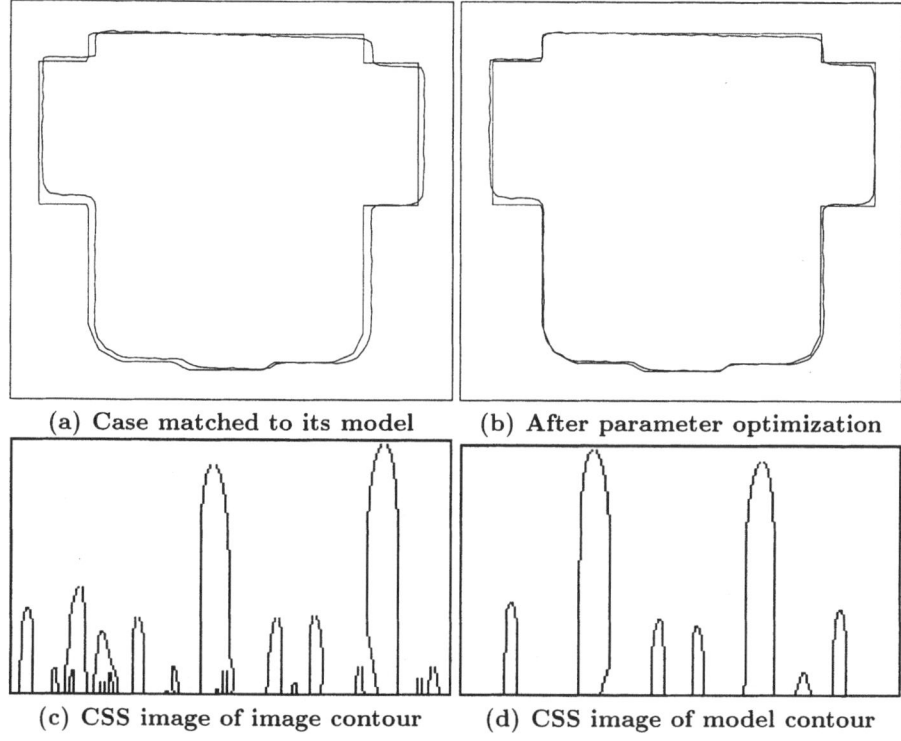

(a) Case matched to its model   (b) After parameter optimization

(c) CSS image of image contour   (d) CSS image of model contour

Figure 2.4.   Matching the connector case to its model contour

of the absolute value of curvature) are also inherent feature points which can be combined with curvature zero-crossing points, if necessary, to obtain a richer segmentation.) Since curvature zero-crossing points can be extracted from a contour at different scales, the choice of an appropriate scale to be used for feature detection is a crucial one. On free-form contours, features exist at different scales and therefore attempts to discover a *natural* scale to be used for feature extraction will be misguided. Some heuristics may be used to select a specific scale but if the scale chosen is too small, the segmentation will be affected by noise and local distortions of shape, and if it is too large, important structure on the contour may be lost.

The solution used here was motivated by the main underlying concept of the curvature scale space representation: utilize information from multiple scales rather than prefer a single scale. The adaptation of that concept to the problem considered here necessitates the extraction of curvature zero-crossing segments from different scales. This process would ensure robustness to noise and local shape distortions as well

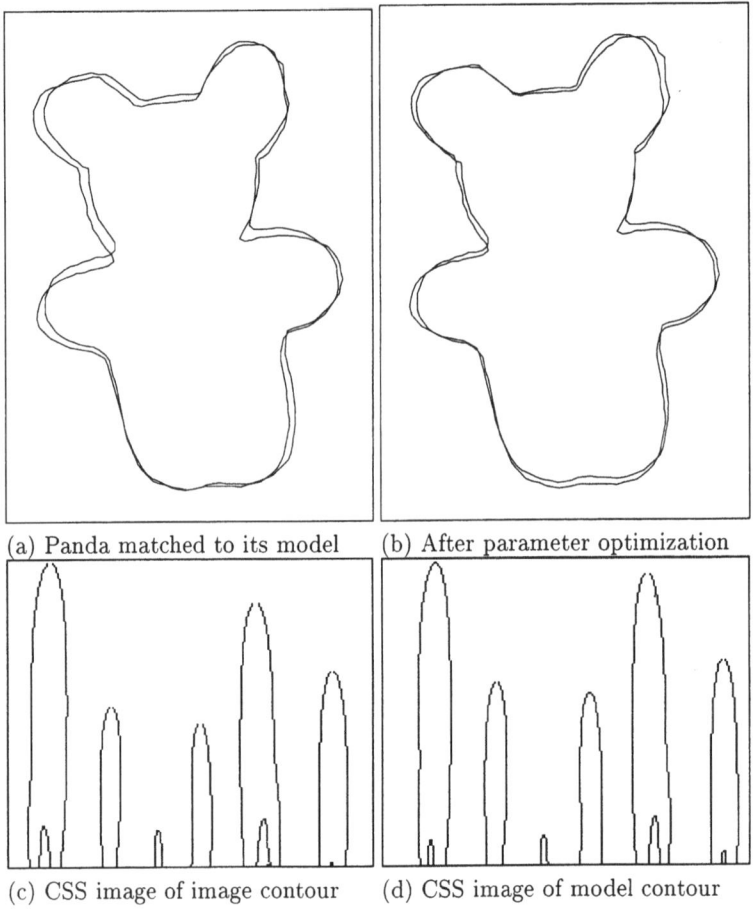

(a) Panda matched to its model   (b) After parameter optimization

(c) CSS image of image contour   (d) CSS image of model contour

Figure 2.5.   Matching the panda to its model contour

as small features that may be part of the model object but missing from a similar (but physically distinct) input object. The multi-scale segmentation procedure is as follows:

- Choose a low scale to start the segmentation process. This scale should be selected such that most or all of the camera as well as discretization noise is smoothed out. Its value was experimentally determined: input contours were sampled at 300 points and the starting scale was $\sigma = 2.0$ with the filter covering 10σ sampled points on the input contour.

- Locate all curvature zero-crossing points on the input contour at the starting scale and add all zero-crossing segments to the main segment-list L_m. Also add each such segment to the auxiliary segment-list L_a

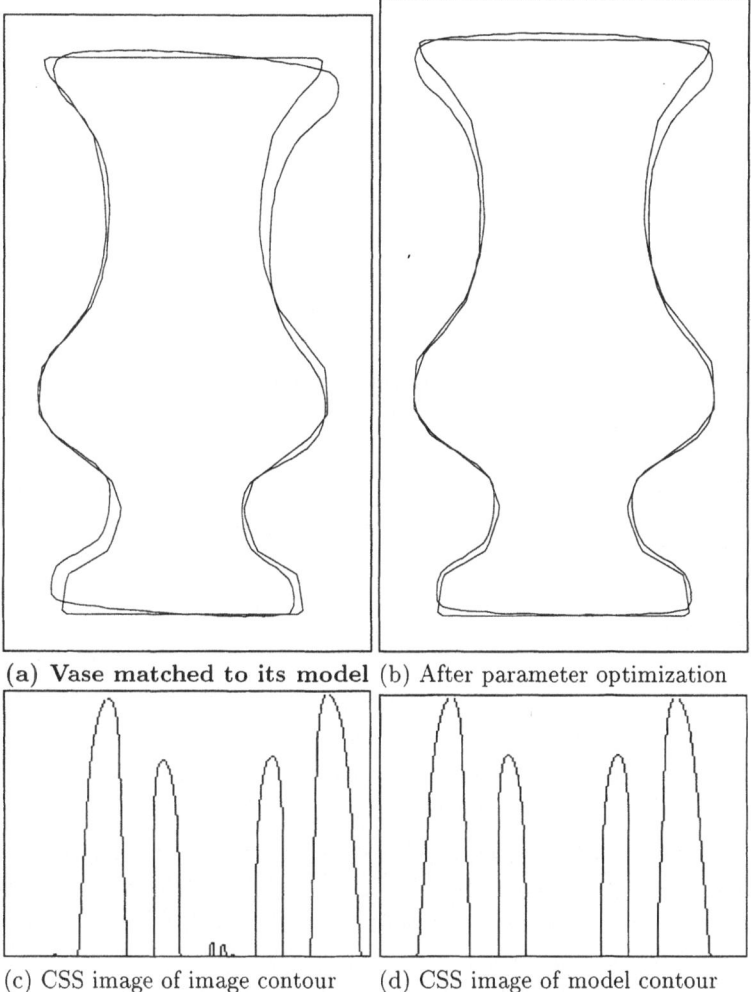

(a) **Vase matched to its model** (b) After parameter optimization

(c) CSS image of image contour   (d) CSS image of model contour

Figure 2.6. Matching the vase to its model contour

(to be explained shortly). A zero-crossing segment is defined as a segment on the input contour delimited by two consecutive curvature zero-crossing points referred to as its *endpoints*. Each endpoint is specified uniquely by the value of the arc-length parameter at that point which is relative to the starting point for arclength parametrization.

- Repeat the following steps until no more curvature zero-crossing points are detected at a high scale: the evolving contour becomes simple and convex.

– Determine the next higher scale for segment extraction. This can be achieved by adding a small increment $\Delta\sigma$ (about 0.1) to the previous value of σ.

– Detect all curvature zero-crossing points at the current scale. Note that since new curvature zero-crossings will not be created at higher scales on a simple contour, zero-crossing detection can be accomplished more efficiently by *tracking* the zero-crossing points detected at the previous scale: the corresponding filter is convolved with the data only in a small neighborhood (two points on each side) of each existing zero-crossing.

– Add each of the zero-crossing segments detected at the current scale to L_m if it does not already exist in that list. In order to establish existence, two zero-crossing segments are considered to be the same if the difference between the arc-length values of their corresponding endpoints is small. Note however, that curvature zero-crossing points can move across scales and for the existence test to work properly, that movement must be accounted for. Therefore an auxiliary segment-list L_a is also maintained. Each time a new zero-crossing segment is added to L_m, it is also added to L_a (this also happens at the starting scale). The purpose of L_a is to keep a record of the last known locations of the endpoints of each segment added to L_m as that segment is tracked across scales. So the existence test proceeds as follows:

– Determine whether each of the curvature zero-crossing segments detected at the current scale is the same as an existing segment in L_a. Let the segment under consideration be $S[e_1, e_2]$ with endpoints e_1 and e_2, and let $L_a = \{S_1, S_2, ..., S_n\}$ where S_i has endpoints S_{i1} and S_{i2}. Segment S is considered to already exist in L_a (and therefore L_m) if $|e_1 - e_{i1}| < j$ and $|e_2 - e_{i2}| < j$ for some $i \in [1, n]$. If this is the case, simply update the locations of the endpoints of S_i in L_a to be the same as those of segment S. Otherwise, segment S is a new segment: it is added to L_m as well as L_a. In both lists, its endpoints will be e_1 and e_2. Note that the value of j was chosen to be 3 which will work well even if the number of samples points varies as long as $\Delta\sigma$ is kept small (about 0.1).

When the final scale of the curvature scale space representation is reached, the multi-scale segmentation process terminates and the segments recorded in L_m are written to the output.

Figure 2.7 shows all the curvature zero-crossing segments detected at the starting scale on the outline of the panda. Figure 2.8 shows the additional curvature zero-crossing segments detected at all higher scales. These curvature zero-crossing segments have been displayed by marking their endpoints on the *original* contour. Note that most segments added at higher scales will be convex segments. Concave segments can be added at higher scales only when a curvature zero-crossing contour that is nested inside another zero-crossing contour disappears. In particular, the segment added in figure 2.8(a) is concave, in figure 2.8(b) it is concave, in figure 2.8(c) it is convex, in figure 2.8(d) it is concave, and all segments added in figures 2.8(e) to 2.8(h) are convex. This example demonstrates that the initial segmentation of the panda is not satisfactory since parts of the object are oversegmented but that useful segments are added at higher scales. Starting the segmentation at a higher scale may have removed some useful segments as well as noise.

Figure 2.9 shows all the curvature zero-crossing segments detected at the starting scale on the coastline of Africa. These curvature zero-crossing segments have again been displayed by marking their endpoints on the *original* contour. Figure 2.10 shows the additional curvature zero-crossing segments detected at all higher scales. They are again displayed by marking their endpoints on the original contour. Note that if two curvature zero-crossing segments are detected at different scales, then they are displayed on separate contours. Note that figures 2.10(a) and 2.10(c) show that two segments are added at each of the scales $\sigma = 2.3$ and $\sigma = 2.8$. As figures 2.9 and 2.10 show, most curvature zero-crossing segments are detected at the starting scale with a few added at higher scales. Again, most of the segments added at higher scales are convex, including the segments shown in figures 2.10(m) to 2.10(p). This example also demonstrates that multi-scale segmentation is immune to both noise and possible loss of useful structure in parts of the input contour due to over-smoothing.

3.3. Efficient Termination of Multi-Scale Segmentation

The algorithm described in the previous subsection recovers curvature zero-crossing segments from an input contour at multiple scales until the final scale of the corresponding CSS image is reached at which the number of curvature zero-crossings found drops to zero. So the termination of the multi-scale segmentation process is well-defined. Note however that some of the final segments recovered can be very long and can in fact cover most of the input contour. Examples are segments shown in figures 2.8(h) and 2.10(p). Very long segments do not contribute to object recognition systems since the presence of occluding objects makes them redundant. Moreover, they reduce the efficiency of the recogni-

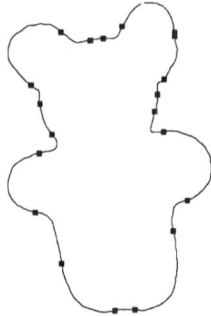

Figure 2.7.   Curvature zero-crossing segments of panda at the initial scale

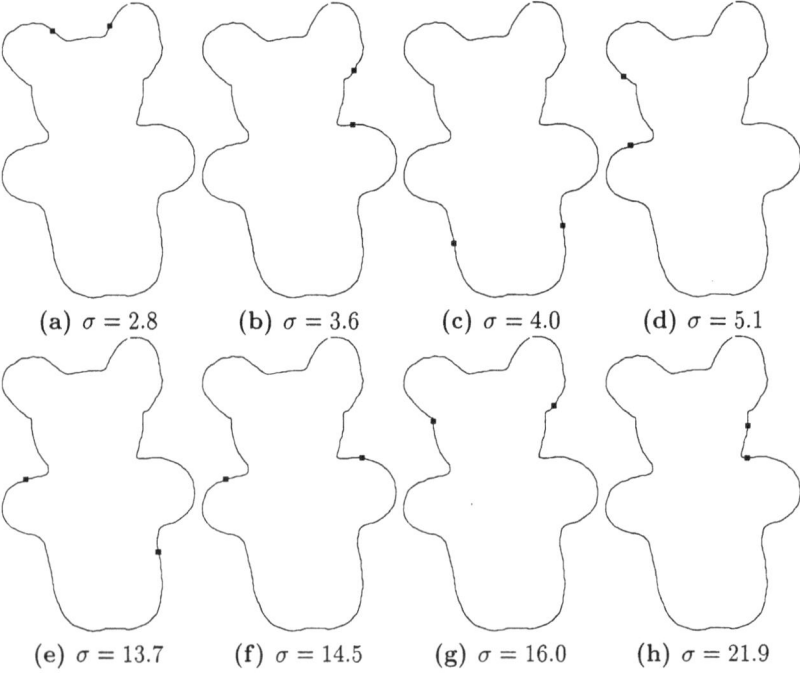

(a) $\sigma = 2.8$ (b) $\sigma = 3.6$ (c) $\sigma = 4.0$ (d) $\sigma = 5.1$

(e) $\sigma = 13.7$ (f) $\sigma = 14.5$ (g) $\sigma = 16.0$ (h) $\sigma = 21.9$

Figure 2.8.   Curvature zero-crossing segments of panda at higher scales

tion system since it will have to process more segments. This section proposes three algorithms for efficient termination of the multi-scale segmentation process so that only the small to medium scales are covered. The termination criteria are determined automatically, eliminating the need to use arbitrary thresholds. The following is a description of each algorithm:

Figure 2.9. Curvature zero-crossing segments of Africa at the initial scale

- The underlying idea of the first algorithm is to determine the longest scale interval in which the number of curvature zero-crossing points remains constant. The starting scale of that interval is then defined as the final scale used for the multi-scale segmentation process. An initial pass through the CSS image is required in order to build a histogram of the number of curvature zero-crossing points found at each scale. The longest interval with a constant number of curvature zero-crossing points is then recovered from that histogram.

- The second algorithm incorporates the idea that only curvature zero-crossing segments whose lengths are below a reasonable ratio (when divided by the length of the input contour) should be accepted. The multi-scale segmentation process described in the previous section is first applied to recover segments at all scales of the CSS image. For each of those segments, a length-ratio is defined by dividing its length by the full length of the input contour. The next step is to build a histogram of the number of curvature zero-crossing segments as a function of length-ratio. The longest interval with a constant number of segments is then recovered from that histogram. The starting value of length-ratio corresponding to that interval is then utilized as the cutoff ratio.

- The third approach is to construct a histogram of the number of curvature zero-crossing points as a function of scale (as in the first approach) or a histogram of the number of curvature zero-crossing segments as a function of length-ratio (as in the second approach). However, rather than finding the longest constant interval in each of those histograms, each histogram is smoothed to remove noise. smoothing is accomplished through convolution with a Gaussian filter

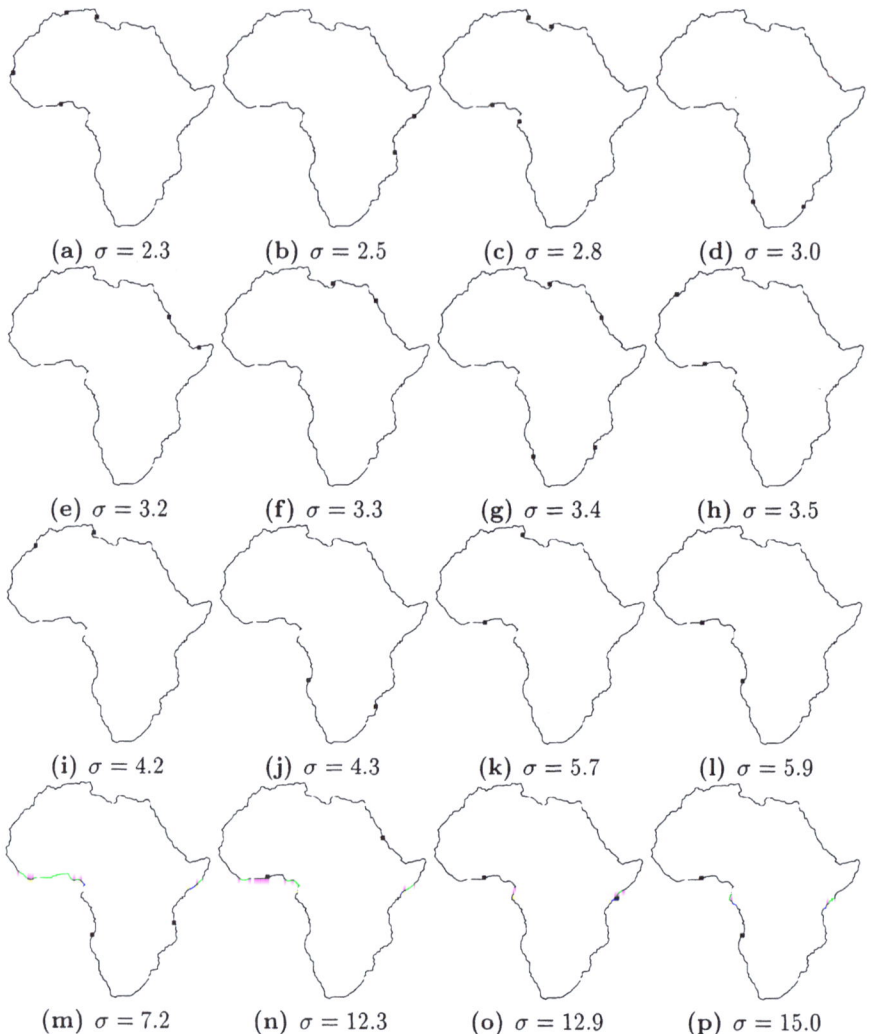

Figure 2.10.   Curvature zero-crossing segments of Africa at higher scales

and continues until only one maximum of the absolute value of the second derivative of the smoothed histogram remains. The location of that maximum is then defined as the final scale or as the largest length-ratio for multi-scale segmentation. Let $h(u)$ be the histogram under consideration, and let

$$\mathcal{S} = \{maxima\ of\ |g_{uu}(u, \sigma) \otimes h(u)|\}$$

For some $\sigma = \sigma_f$, $|\mathcal{S}| = 1$. The remaining maximum will exist at $u = u_m$. The point u_m can be considered the *bending point* of the

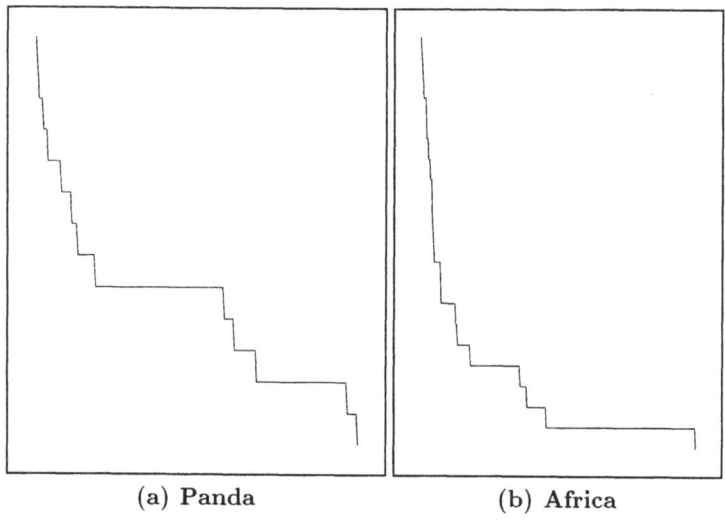

(a) **Panda**   (b) **Africa**

Figure 2.11.   Number of curvature zero-crossings vs. scale

histogram: it is the point at which the histogram switches from *mostly vertical* to *mostly horizontal.*

The histogram of the number of curvature zero-crossings on the panda contour as a function of scale is shown in figure 2.11(a). The longest interval with a constant number of curvature zero-crossings (10) starts at $\sigma = 5.1$, and ends at $\sigma = 13.7$. As a result, $\sigma = 5.1$ is selected as the final scale for multi-scale segmentation. Therefore, the segments shown in figures 2.8(e) to 2.8(h) are removed from the final list of segments. The histogram of the number of curvature zero-crossings on Africa contour as a function of scale is shown in figure fig:histo1(b). The longest interval with the same number of curvature zero-crossings (2) starts at $\sigma = 15.0$, and ends at $\sigma = 30.2$. As a result, $\sigma = 15.0$ is chosen as the final scale for multi-scale segmentation. However, no segments are removed from the final segment list in this case.

The histogram of the number of curvature zero-crossing segments on the panda contour as a function of the ratio r of their lengths to the length of the whole contour is shown in figure 2.12(a). The longest interval with the same number of curvature zero-crossing segments (25) starts at $r = 0.55$ and ends at $r = 0.95$. As a result, $r = 0.55$ is selected as the cutoff length ratio. It can be said that, in this case, this procedure has eliminated curvature zero-crossing segments that are longer than about half the length of the input contour. This results in the removal of the segment shown in figure 2.8(h) from the final list of segments. The histogram of the number of curvature zero-crossing

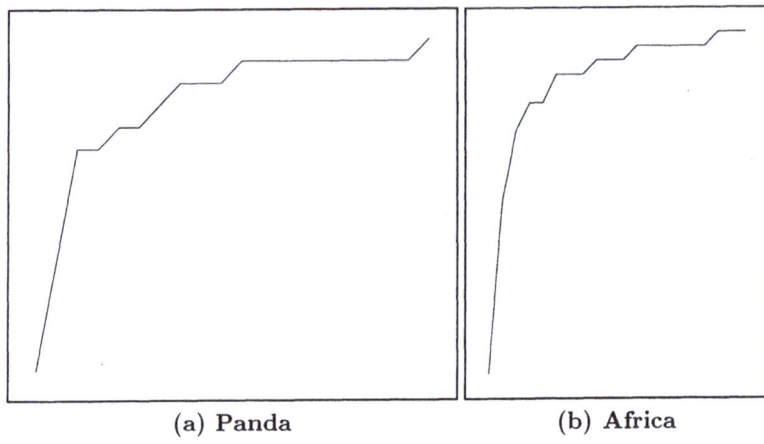

(a) **Panda**   (b) **Africa**

Figure 2.12.   Number of curvature zero-crossing segments vs. length ratio

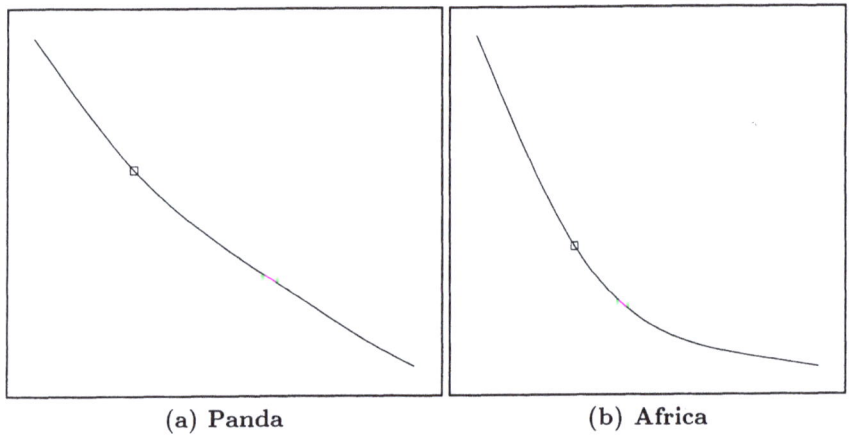

(a) **Panda**   (b) **Africa**

Figure 2.13.   Smoothed histograms of figure 2.11

segments on Africa contour as a function of the ratio r of their lengths to the length of the whole contour is shown in figure 2.12(b). The longest interval with a constant number of curvature zero-crossing segments (58) starts at $r = 0.6$ and ends at $r = 0.85$. Hence, $r = 0.6$ is chosen as the cutoff length ratio. The result is the removal of the segments shown in figure 2.10(o) and 2.10(p) from the final segment list.

Based on these two examples, it appears that the second algorithm performs better than the first one in terms of removing very long segments but preserving the ones detected at small to medium scales.

Figure 2.13 shows the histograms of figure 2.11 smoothed until only one maximum of the absolute value of the second derivative of the

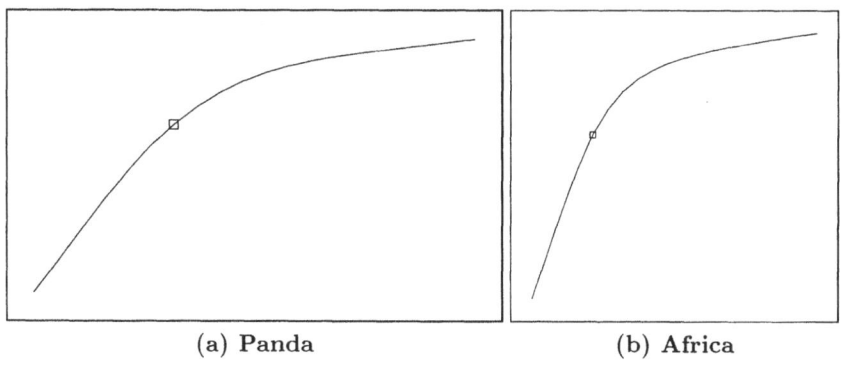

|(a) Panda|(b) Africa|

Figure 2.14.   Smoothed histograms of figure 2.12

smoothed histogram remains. Figure 2.14 shows the histograms of figure 2.12 smoothed in a similar fashion. The location of each surviving maximum is shown on each smoothed histogram.

In figure 2.13(a), the maximum occurs at $\sigma = 6.8$. As a result, the segments shown in figures 4(e) to 4(h) are removed from the final list of segments. In figure 2.13(b), the maximum occurs at $\sigma = 10.2$. As a result, the segments shown in figures 5(m), 5(n) and 5(o) are removed from the final list of segments.

In figure 2.14(a), the maximum occurs at $r = 0.35$. Therefore, the segments shown in figures 2.8(e), 2.8(f) and 2.8(h) are removed from the final segment-list. In figure 2.14(b), the maximum occurs at $r = 0.25$. As a result, the segments shown in figures 2.10(j) and 2.10(l) to 2.10(p) are removed from the final segment-list.

In conclusion, it can be observed that the third algorithm is the most effective one in terms of eliminating long segments that do not contribute to matching. This algorithm is also able to preserve the useful segments detected at small to medium scales.

3.4.   Local Matching through CSS

Due to occlusion, the matching algorithm employed is a local one and consists of several stages. The subsections of this section describe various stages of the matching algorithm. The following is a step-by-step description of the algorithm:

1 Acquire model contours and apply rescaling to them off-line.

2 Apply CSS-based multi-scale segmentation (as described in section to the model contours offline: curvature zero-crossing segments are extracted at multiple scales.

3 Process input image to recover image contour and apply rescaling to it.

4 Apply CSS-based multi-scale segmentation to the image contour.

5 Generate initial candidates while applying the object indexing scheme. Filter those candidates using an admissible. Each candidate generated consists of a model segment, a matched image segment, the average distance of those segments, and the transformation parameters computed using Least-Squares and then optimized.

6 Merge compatible candidates. Average distance and transformation parameters are recomputed for each new candidate. Merging continues as long as compatible candidates can be found.

7 Extend the new candidates as much as possible to determine accurately the object intersection points. Average distance and transformation parameters are recomputed for each extended candidate.

8 Group the remaining candidates into compatible but disjoint candidates. This accounts for parts of an object that are disconnected in the input image.

9 Compute a cost for each candidate and select the lowest-cost candidates until all the data is accounted for. Display the selected models registered to the data.

3.4.1   Rescaling

Model contours are rescaled so that they reflect the relative sizes of model objects when viewed at the same distance with the largest one just fitting in a unit square. Their aspect ratios are not changed in the process. Model contour rescaling is carried out off-line. The image contour is also rescaled so that it just fits in a unit square. The image contour touches the square at 3 points or more. This is because the image contour is made as large as possible while fitting inside the unit square. By a combination of scaling and translation, it will always touch at least 3 points of the square. The aspect ratio of the contour is not changed in the process. As a result of image and model contour rescaling, the possible scale changes from model contours to the image contour become predictable which helps to define an *admissible space* for the scale-factors. The consequence of this is to make the subsequent search more efficient. The multi-scale segmentation procedure is then used to segment the model contours and the image contour.

3.4.2   Candidate Generation and Filtering

Due to occlusion, all possible local matches must be considered (note however, that very small segments on either the image contour or the model contours are discarded). In order to avoid an exhaustive search of all model contour segments, *object indexing* [154, 296, 263] is employed to render the initial search more efficient. After segmentation, each model contour segment is rescaled so that each has the same length $\mathcal{L}$ (subject to constraints imposed by the admissible space defined in the previous subsection: $\mathcal{L}$ is chosen such that the scaling applied to each model contour segment is compatible with the admissible space). Average curvature is then computed for each of those segments and used to create an index-table for all the model contour segments. Computation of curvature is carried out after smoothing the contour to remove noise which makes the process robust. All the computation is carried out off-line.

Once the segmentation of the image contour is completed, each image contour segment is also rescaled so that each has the same length $\mathcal{L}$ (again subject to constraints imposed by the admissible space). Average curvature is also computed for each of those segments. The average curvatures now serve as indices into the model contour segment index-table to recover a more likely (and smaller) set of potentially matching model contour segments. A candidate is generated corresponding to the possible match of each image contour segment and the model contour segments recovered from the index-table for that image contour segment. The admissible space defined earlier is used to further prune the candidates: the scaling factor required to map the model contour segment to the image contour segment must be inside that admissible space. Note that a candidate is stored as a data structure consisting of an image segment, a matched model segment, the corresponding transformation parameters, and the average distance between the two segments.

The initial values for the transformation parameters are computed through least-squares estimation. Transformation parameter optimization is then applied to the generated candidates in order to refine the initial estimate of those parameters (see sections 3.5, 3.6, and 3.7). *This step is crucial since the accuracy of segment distance calculation depends greatly on the accuracy of the transformation parameters.* For each candidate, *segment-dist* is defined as the average point distance between the image-model contour segments and used as a measure of the goodness of fit between the two segments. Candidates with relatively low *segment-dist* values are then selected for further processing.

3.4.3   Candidate Merging

Initial candidates correspond to simple segments delimited by neighboring curvature zero-crossing points. Nevertheless, it is possible for the visible boundary of an object in the input image to be divided into several neighboring or even overlapping segments. It is therefore necessary to merge those initial candidates which satisfy several criteria intended to measure candidate compatibility. It follows that two candidates c_1 and c_2 will be merged if they satisfy the following criteria:

- c_1 and c_2 must be valid (not previously merged) and different candidates.

- c_1 and c_2 must correspond to the same model.

- The transformation parameters of c_1 and c_2 should be *roughly* the same. It must be emphasized that the test used here was **not** strict since local matches can result in significantly different parameters even for compatible candidates. This is specially true when the corresponding segments are relatively short.

- The corresponding segments of c_1 and c_2 must be neighboring or overlapping.

- The scale factor associated with the new merged candidate must be in the admissible space.

- The new candidate must have a low *segment-dist* value.

When two candidates are merged, the corresponding segments will be the union of the old segments. The old candidates are invalidated. New transformation parameters (again computed using Least-Squares followed by optimization) and a new value of *segment-dist* are computed for the new candidate. The algorithm described above will continue merging candidates until no two candidates can be found which satisfy the merging criteria.

3.4.4   Candidate Extension

The image contour segment that is fitted to its coresponding model should be as long as possible to obtain the best possible fit. The quality of fit itself is used to rank various candidates and choose among them. Competing candidates can be very close and the ones which are supported by (or can account for) the largest number of data points should be selected. As a result, it is important to find the exact location of object boundary intersection points.

In general, the intersection point of two object boundaries in the input image does not coincide with an endpoint of a curvature zero-crossing segment. Therefore in order to find the exact location of such intersection points, it is necessary to gradually extend the image and model contour segments associated with the merged candidates as long as a good fit between the image and model segments can be observed. Extension is first carried out at the right endpoint until mismatch error is too large and then carried out at the left endpoint. In general, object intersection points are a subset of the curvature maxima on the image contour. First, all curvature extrema are located on a slightly smoothed version of the input image contour (The degree of smoothing is experimentally determined and is proportional to the level of noise present in the data. A perfect output is not required since the candidate extension phase can locate the most appropriate curvature maxima). Then, the following procedure is applied at each endpoint of each candidate:

- Extend the image contour segment to the next curvature maximum.

- The corresponding model contour segment is extended accordingly.

- Determine new transformation parameters and the new value of *segment-dist* for the candidate being extended.

- Determine the number of points k (on the model contour segment) in a small neighborhood of the endpoint which are far from the image contour.

Extension stops if at least one of the following conditions comes true:

- New candidate no longer has a low *segment-dist* value.

- New value of *segment-dist* rises sharply compared to previous value.

- k rises above an acceptable limit.

When extension stops, tests are carried out to detect a borderline case (k is just above the acceptable limit or value of *segment-dist* is just above the cut-off threshold). If so, the current endpoint becomes the final endpoint. Otherwise, the previous endpoint becomes the final endpoint. New transformation parameters and a new value of *segment-dist* are computed for the new candidate.

In order to discover the correct scale, location and pose of each model matching the data, the system must consider all possible local matches (as indicated by the index-table) between all models and the data occurring at any scale in the admissible space. In doing so, quite frequently the

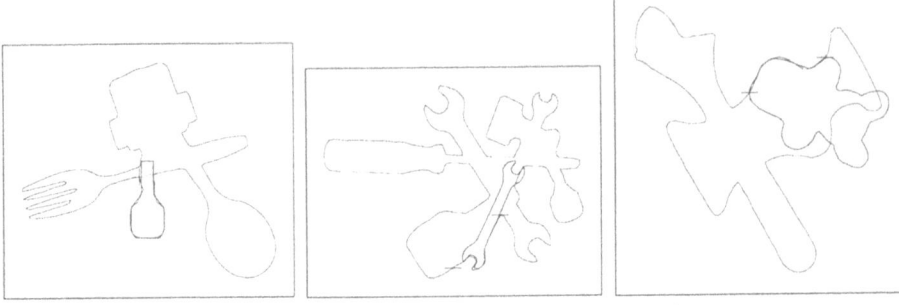

(a) Incorrect match: bottle   (b) Incorrect match: wrench (c) Incorrect match: panda

Figure 2.15.   Locally plausible but incorrect matches

system discovers locally plausible matches which are globally incorrect. Such situations are, in general, unavoidable and make the recognition task more challenging. Figure 2.15 illustrates this point by showing three intermediate candidates generated by the system. Figure 2.15(a) shows the bottle matched at an incorrect scale to an image contour. The correct model is the key. Figure 2.15(b) shows a wrench again incorrectly matched to another image contour. Figure 2.15(c) shows the panda incorrectly matched to another image contour. This example shows that even two objects (such as panda and vase) which appear to have very different shapes, can match well locally at the right scale and orientation. *Note however that these were not the lowest-cost candidates chosen by the system. The lowest-cost candidates were in fact the correct ones, but the three matches illustrated in figure 2.15 ranked quite close to the lowest-cost candidates.*

3.4.5   Candidate Grouping

The next step in matching is to group compatible but disjoint candidates. The tests applied to determine compatibility are the same as the tests used for candidate merging except that the fourth test is not applied. It is certainly possible that, due to occlusion, an object in the scene may appear as two or more disjoint components in the input image. The goal of this step is to identify such situations to aid in the process of recognition.

3.4.6   Candidate Selection

What remains is to select the *best* candidates using an appropriate criterion. As stated earlier the value of *segment-dist* for each candidate is the average point distance between the contour segments associated with that candidate. This is a suitable measure of how well the shapes of those contour segments match. Another measure of the significance of a candidate is its *support*. Candidate support is defined as the length of the image contour segment associated with the candidate (note that if two disjoint candidates are found to be compatible, the support of each candidate is increased by the length of the image contour segment associated with the other candidate). Hence the *cost* of each candidate is defined as:

$$candidate - cost = \frac{segment - dist}{candidate - support}.$$

Note that a candidate with a lower cost is a *better* candidate. The following procedure is then used to select the best candidates:

- Determine the cost of each candidate.

- Select the *valid* candidate with the lowest cost. Note that a valid candidate is one which has not yet been disqualified.

- Disqualify all candidates whose corresponding image contour segment overlaps with the image contour segment of the chosen candidate or the image contour segment of any candidate compatible with the chosen candidate.

- Find any image contour segments delimited by negative curvature minima (which are a subset of maxima of absolute value of curvature) which do not overlap with the image contour segments associated with any chosen candidates or candidates compatible with them, and which fit well with the model associated with the chosen candidate. Examples are straight line segments which do not occur in valid candidates.

- Disqualify all candidates whose corresponding image contour segment overlaps with any of the image contour segments discovered in the previous step.

- Determine the final fit of the model associated with the chosen candidates using all relevant image contour segments and map it to the image space. The relevant image contour segments are those corresponding to the chosen candidate and any candidates that were grouped with it.

- Disqualify the chosen candidate and all candidates compatible with it.

- If any valid candidates remain, go to the second step above, otherwise STOP.

Note that this procedure stops automatically and is independent of the number of objects in the input image.

3.5.   Solving for the Transformation Parameters

When mapping a model curve segment to an image curve segment, it is possible to obtain many pairs of points on those segments in order to compute an initial approximation for the transformation parameters. This is because the correspondence between arc length values on the curve segments is known (in turn because the arc length values at each of the corresponding pairs of endpoints of the two segments are known). The transformation to be solved for consists of scaling, rotation and translation in x and y. Let

$$\mathcal{X} = (x_j, y_j)$$

be a set of η points on the image curve and let

$$\Xi = (\xi_j, \psi_j)$$

be the set of corresponding points on the model curve. The parameters of the following system:

$$x_j = a\xi_j + b\psi_j + c \qquad\qquad y_j = -b\xi_j + a\psi_j + d \qquad\qquad (2.1)$$

must be solved for. A *least-squares estimation* method is used to estimate values of a, b, c and d. Let the *dissimilarity measure* Ω which measures the difference between the model curve segment and the image curve segment be defined by:

$$\Omega = \sum_{j=1}^{\eta} (x_j^t - x_j^c)^2 + (y_j^t - y_j^c)^2$$

where (x_j^c, y_j^c) is the closest point on the image curve to transformed model curve point (x_j^t, y_j^t). Using equation (2.1) to eliminate x_j^t and y_j^t yields:

$$\Omega = \sum_{j=1}^{\eta} (a\xi_j + b\psi_j + c - x_j^c)^2 + (-b\xi_j + a\psi_j + d - y_j^c)^2.$$

Let

$$\mathcal{P} = (a, b, c, d)$$

be the vector defined by the transformation parameters. The solution of

$$\frac{\partial \Omega}{\partial \mathcal{P}} = 0$$

is the least-squares estimate of those parameters. To compute that estimate, determine the partial derivatives of Ω with respect to each of a, b, c and d and set those partial derivatives to zero. The result is a linear system of four equations in four unknowns which is solved to obtain estimates for a, b, c and d:

$$a = \frac{\sum \xi_j x_j^c + \sum \psi_j y_j^c - \frac{1}{\eta} \sum x_j^c \sum \xi_j - \frac{1}{\eta} \sum y_j^c \sum \psi_j}{\sum \xi_j^2 + \sum \psi_j^2 - \frac{1}{\eta} \sum \xi_j \sum \xi_j - \frac{1}{\eta} \sum \psi_j \sum \psi_j}$$

$$b = \frac{\sum \psi_j x_j^c - \sum \xi_j y_j^c + \frac{1}{\eta} \sum y_j^c \sum \xi_j - \frac{1}{\eta} \sum x_j^c \sum \psi_j}{\sum \xi_j^2 + \sum \psi_j^2 - \frac{1}{\eta} \sum \xi_j \sum \xi_j - \frac{1}{\eta} \sum \psi_j \sum \psi_j}.$$

$$c = \frac{\sum x_j^c - a \sum \xi_j - b \sum \psi_j}{\eta}$$

$$d = \frac{\sum y_j^c + b \sum \xi_j - a \sum \psi_j}{\eta}.$$

3.6. Measuring Image-Model Curve Distances

Once an estimate of the transformation parameters is available, it is possible to map the model curve to the space of the image curve. It is then useful to measure the image-model curve segment distance for two reasons:

- As described earlier, different model curves are mapped to the image curve in order to determine which model curve is locally closest to the image curve. This is accomplished by measuring image-model curve segment distances.

- The computation of the image-model curve segment distance is essential to transformation parameter optimization.

The following procedure is used to determine the image-model curve segment distance:

1 Let $k = 1$. Let $\eta = $ number of vertices on model curve segment. Let $\delta = 0.0$.

2 Determine the closest point on the image curve segment (not necessarily a vertex) to vertex k of the model curve using the procedure of step 3.

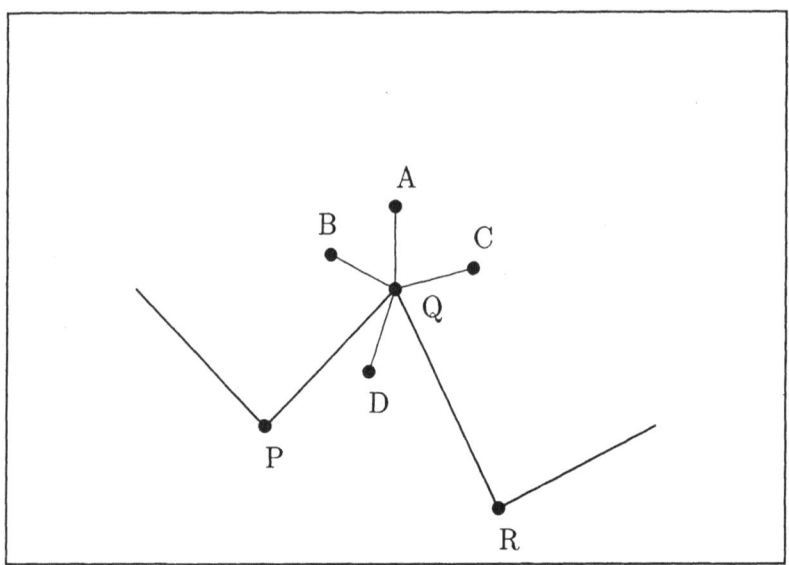

Figure 2.16.   Closest point detection

3 Locate the closest vertex of the image curve segment to vertex k of the model curve. (All vertices of the image curve can be considered but to speed up the algorithm, only a small number in a neighborhood on the image curve segment into which vertex k maps, are considered.) Let that be point Q (figure 2.16). Let points P and R be the image curve vertices which occur before and after Q on the image curve segment respectively. Let point X be vertex k on the model curve segment. X can be any one of the points A, B, C or D in figure 2.16. Cosines of the angles $\theta_1 = \angle XQP$ and $\theta_2 = \angle XQR$ can be computed using the definitions for the dot product of two vectors. Depending on the signs of the cosines, four cases are possible:

- $\cos(\theta_1)$ and $\cos(\theta_2)$ are both negative. Point X is the same as point A. In this case, the closest image point is point Q.

- $\cos(\theta_1)$ is positive and $\cos(\theta_2)$ is negative. Point X is the same as point B. In this case, the closest image point lies on line segment PQ. Since $\cos(\theta_1)$ is known, the location of that point as well as its distance to point B can be computed.

- $\cos(\theta_1)$ is negative and $\cos(\theta_2)$ is positive. Point X is the same as point C. In this case, the closest image point lies on line segment QR. Since $\cos(\theta_2)$ is known, the location of that point and its distance to point C can be computed.

- $\cos(\theta_1)$ and $\cos(\theta_2)$ are both positive. Point X is the same as point D. In this case, the closest image point lies on either PQ or QR. Compute the distances to both segments and choose the shorter distance and the corresponding point.

4 Let $\delta = \delta +$ distance computed in the previous step. Let $k = k + 1$. If $k > \eta$ then return δ/η as the average point distance between the image and model curve segments and STOP. Otherwise, go to step 2.

3.7. Optimizing the Transformation Parameters

The least-squares estimate of the transformation parameters is, in general, *not* the optimal estimate. This is because the image-model point correspondences are not precise due to noise and local shape distortions. Nevertheless, it is possible to optimize those parameters using the following procedure:

- Let $D_p = \infty$.

- Compute the least-squares estimate of the parameters and use it to map the model curve to the image curve.

- Determine a new set of corresponding points on the image curve and compute the new image-model curve distance D_n.

- If $D_p - D_n < \varepsilon$, then STOP.

- Let $D_p = D_n$ and go to the second step above.

In this system, it was possible to compute the optimal parameters with less than 1% error using at most 10 iterations of the procedure described above.

3.8. Results and Discussion

A total of 15 model objects (18 model contours) and six input images were used to evaluate the object recognition system described in this chapter. All of those images and the system's corresponding output are shown here. Figure 2.17 shows the model objects used to test the system. Note that each of the wrenches (models (f), (g) and (h) in figure 2.17) was modelled using two contours (one for each side). Some model contours were acquired by manually reading and entering the coordinates of points on the contour. Others were acquired automatically by obtaining an image of the isolated model object, segmenting the image and recovering the contour. Each model contour was represented by 200

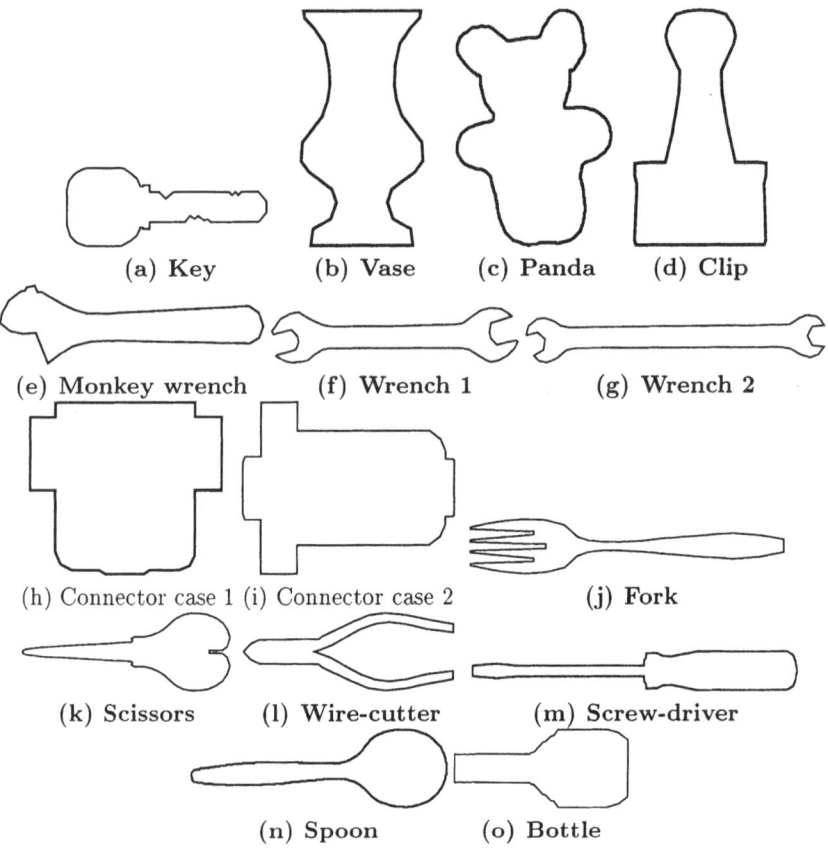

Figure 2.17. Model objects used to test the system

points. The segmentation of each model contour was also computed off-line. The exact same starting scale and final scale were used to compute the segmentation of each model contour. About 10-20 segments were extracted from each model contour.

Due to the light-box setup used, the images obtained had high contrast. As a result, thresholding followed by preprocessing was successful in properly segmenting each of the input images after which the bounding contours were recovered. Figures 2.18 through 2.23 show the initial processing of six input images. In each of those figures, sub-figure (a) shows the original input image, and sub-figure (b) shows only the outermost contour recovered from that image after thresholding, processing and boundary detection. Note that only the outermost contours were used by the system to arrive at recognition results even though the inner contours are visually significant to human viewers (demonstrated by the fact that without the inner contours, some visible object boundaries

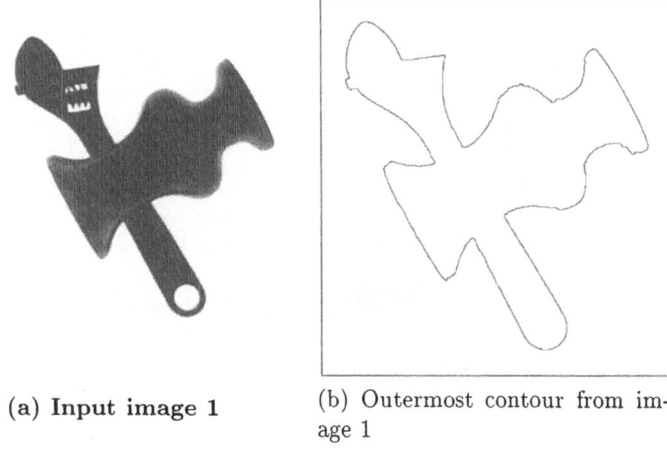

(a) Input image 1

(b) Outermost contour from image 1

Figure 2.18. Initial processing of input image 1

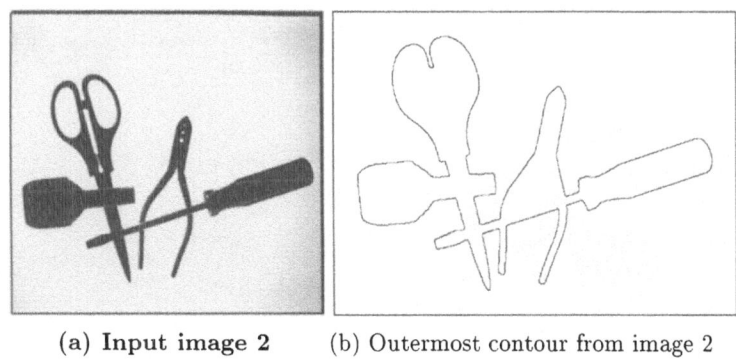

(a) Input image 2   (b) Outermost contour from image 2

Figure 2.19. Initial processing of input image 2

are lost and as a result, recognition becomes slower and more difficult). This was done to demonstrate the recognition ability of the system. If necessary, the system can be extended trivially to also make use of the inner contours. However, for the examples shown here, the results would be the same. Each image contour was represented by 300 points. The exact same starting scale and final scale were used to compute the segmentation of each image contour. About 20-40 segments were extracted from each image contour.

The input images depicted scenes of varying complexity. Figure 2.18(a) depicts a simple scene with 2 objects. Figures 2.19(a) and 2.20(a) each show 4 objects and can be considered to be of medium complexity. Figures 2.21(a), 2.22(a) and 2.23(a) which show 6, 7 and 8 objects respectively can be thought of as difficult images. The system was tested on

(a) Input image 3   (b) Outermost contour from image 3

Figure 2.20.   Initial processing of input image 3

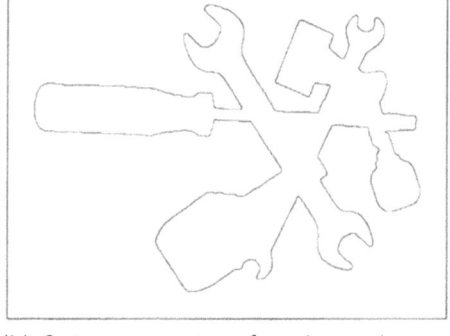

(a) Input image 4   (b) Outermost contour from image 4

Figure 2.21.   Initial processing of input image 4

each of the images 1 through 6. Each of the objects in each input image was recognized correctly by the system which also determined the correct scale, location and pose of each object. *Note that none of the internal parameters of the program were modified from one run to the next: the exact same system produced the correct result for each input image.* The system was implemented in *C* and ran on a *SiliconGraphics Crimson* workstation.

Figures 2.24(a) through 2.29(a) show the recognition results reached by the system for the six input images. Note that in each figure the model contours are shown using a thin line and the image contour is shown using a thick line. The system was very robust in each case despite the presence of noise and local deformations of shape due to segmentation

 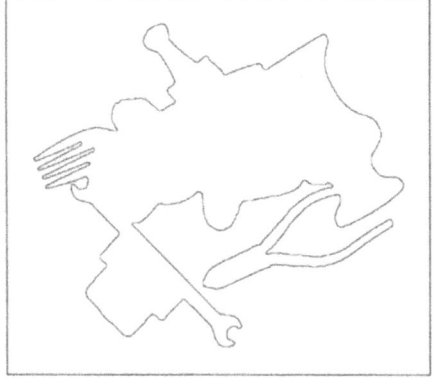

(a) **Input image 5** (b) Outermost contour from image 5

Figure 2.22. Initial processing of input image 5

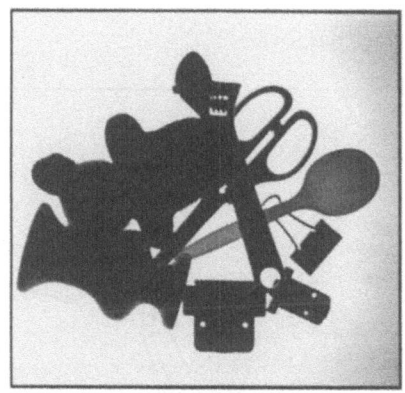 

(a) **Input image 6** (b) Outermost contour from image 6

Figure 2.23. Initial processing of input image 6

errors, non-rigid material in some objects (such as Panda and the plastic fork, spoon and connector cases), and weak perspective distortion. Figures 2.24(b) through 2.29(b) again show the same recognition results with segmentation points also displayed. In almost all cases, the system was able to determine the exact locations of segmentation points.

4. Concluding Remarks

This chapter presented two complete and practical systems for free-form object recognition which are very robust with respect to noise and local deformations of shape (due to weak perspective distortion, non-rigid material used in some objects (such as Panda, and plastic fork,

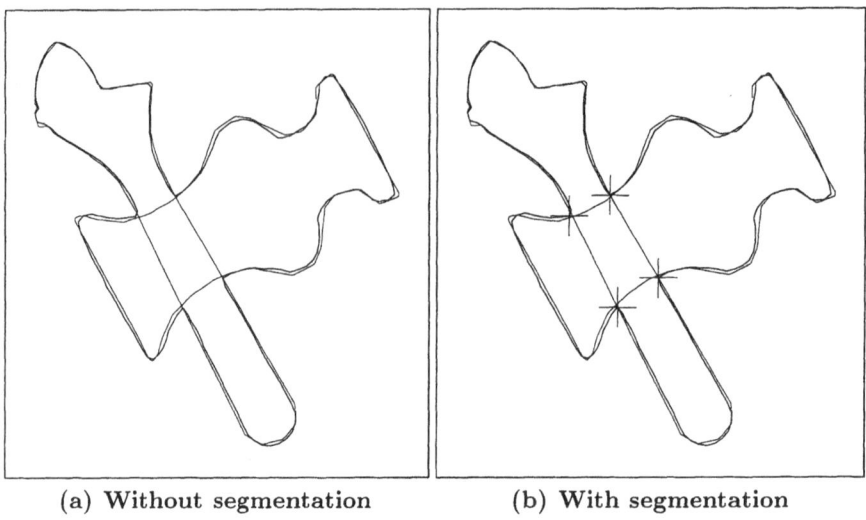

(a) **Without segmentation**   (b) **With segmentation**

Figure 2.24. Matching result for scene 1

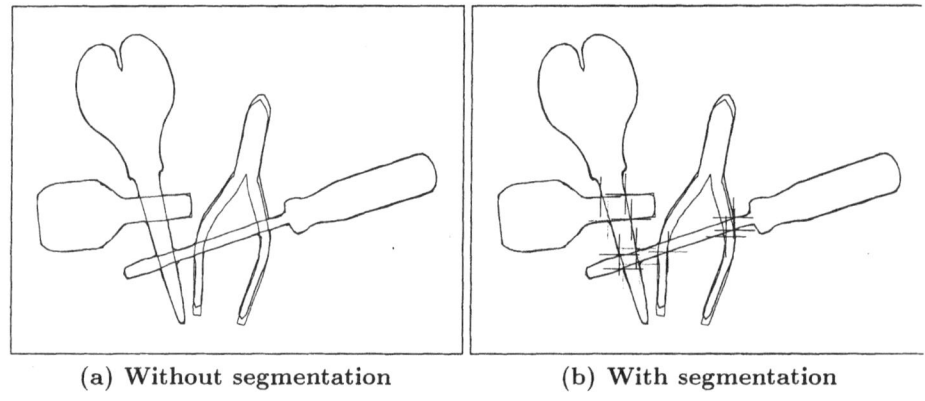

(a) **Without segmentation**   (b) **With segmentation**

Figure 2.25. Matching result for scene 2

spoon and connector cases) and segmentation errors) as well as scale, position and orientation changes of the objects. The systems were tested on a wide variety of free-form 3-D objects with different shapes and surface properties. A light-box setup was used to obtain silhouette images which are segmented to obtain object boundaries. In the second system, the *Curvature Scale Space* technique was then used to obtain a multiscale segmentation of the image contour and the model contours using curvature zero-crossing points. This method made the system robust with respect to noise and local shape differences. A local matching algorithm applied *candidate generation, selection, merging, extension* and

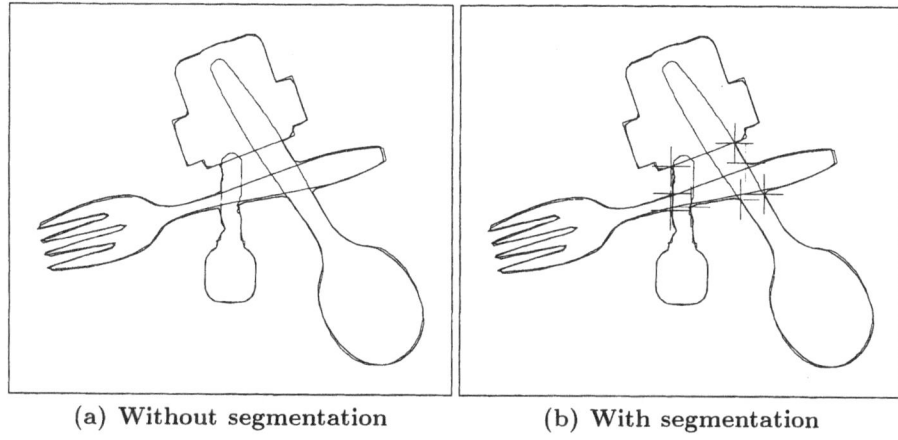

(a) **Without segmentation**   (b) **With segmentation**

Figure 2.26.   Matching result for scene 3

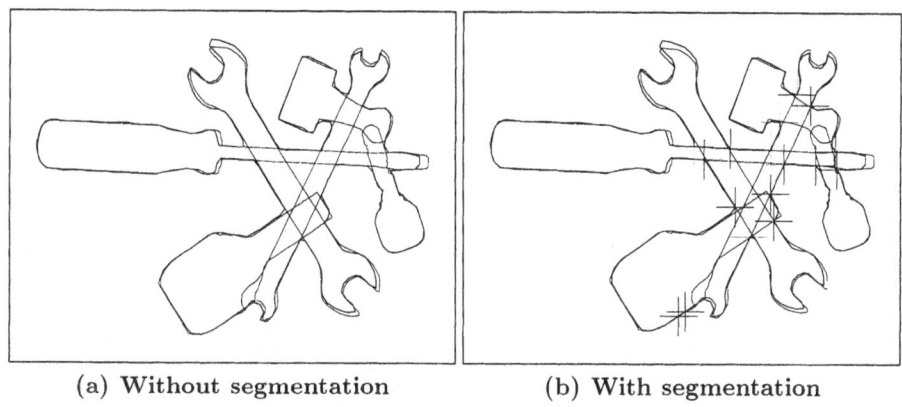

(a) **Without segmentation**   (b) **With segmentation**

Figure 2.27.   Matching result for scene 4

grouping to select the best matching models. Efficient transformation parameter optimization was used to map candidate models to the image space and directly measure the model-data quality of match. It was also used to compute the optimal pose for selected models.

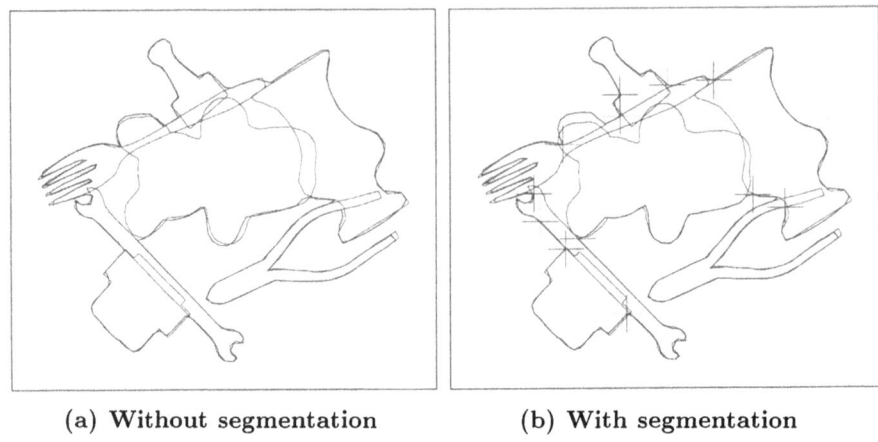

(a) **Without segmentation**   (b) **With segmentation**

Figure 2.28.   Matching result for scene 5

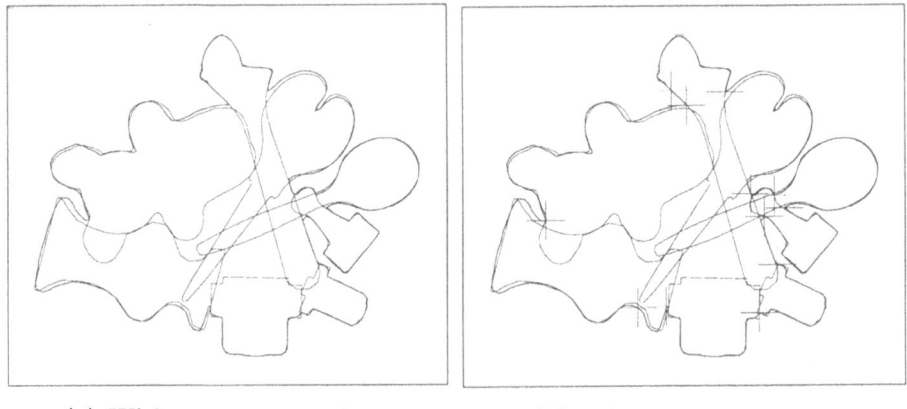

(a) **Without segmentation**   (b) **With segmentation**

Figure 2.29.   Matching result for scene 6

Chapter 3

IMAGE DATABASE RETRIEVAL BASED ON SHAPE CONTENT USING THE CURVATURE SCALE SPACE METHOD

A content based image database system employs features such as shape, color and texture to find images similar to the input image. We address the problem of shape similarity retrieval which can be considered as a part of a complete content based database system.

The Curvature Scale Space image is a multi-scale organisation of the inflection points of a closed planar curve as it is smoothed. It consists of several arch shape contours, each related to a concavity or a convexity of the curve. The maxima of these contours have been used as shape descriptors to find similar shapes in large image databases. In this chapter, we propose enhancements to the CSS image designed to boost its performance.

We tested the proposed methods on a database of 1100 images of marine creatures. To evaluate the proposed methods, we created a small classified image database. We then measured the performance of the system on this database. The quantified results of this test also provided supporting evidence for the performance superiority of the proposed methods.

This chapter also includes a section on the application of our system to Chrysanthemum leaf classification.

1. Introduction

Advances in memory technologies and processing speed have made it feasible to store a large number of images in computers. This has given rise to the problem of organising them for rapid access to their content. The users may need to search through a large number of images, namely *image databases*, to find images of interest. An *image database system* aims to help people in this regard and enable them to find their desired images as quickly as possible [50]. In many applications, the user of an image database system points to an image, and wishes to retrieve similar images from the database. Computer vision researchers have tried to capture image information in feature vectors which describe shape, texture and color properties of the image [243, 254]. These vectors are indexed or compared to one another during query processing to find images from the database.

Considerable amount of information exists in 2-D boundaries of objects. This enables us to recognise objects without using further information. As a result, shape properties play an important role in *content based image database retrieval systems* devised by computer vision and image processing researchers [67, 129, 179]. We address the problem of image retrieval based on shape properties and we use the Curvature Scale Space image to represent the boundaries of image objects.

To construct the CSS image of a digital curve, its curvature zero crossing points should be determined at different levels of smoothing. In Gaussian smoothing, there will be two curvature zero crossings on every concave or convex part of the shape and as the curve becomes smoother these points approach each other. The locations of each pair of zero crossings at different levels of scale create a contour in the CSS image. At a certain level of smoothing when the segment is filled, the two points join and represent the maximum of the relevant contour. The height of this contour then reflects the depth and size of the concavity or convexity. The deeper and larger the segment, the higher the maximum.

The ambiguities of CSS matching are mainly due to the problem of shallow concavities on the shape. It can be shown that the shallow and deep concavities may create the same large contours in the CSS image. Therefore, a shallow concavity may be matched with a deep one during CSS matching.

In object recognition applications, the ambiguities of CSS matching were resolved by a verification stage where the actual distance between the image and model curves is determined and compared. In shape similarity retrieval [209, 208, 210], two global parameters, eccentricity and circularity, were also used in conjunction with the maxima of the CSS image to represent the boundaries of objects. As a result, the speed of the system improved and CSS matching between shapes with shallow concavities and shapes with deep concavities was avoided in most cases.

In this chapter, different approaches based on modifications of the CSS image are presented [211, 212]. The fundamental idea is to enrich the CSS image with additional information about the curvature value at different levels of scale, rather than just rely on the information about curvature zero crossings. This will be carried out by partitioning the curve and determining the corresponding segment to each contour of the CSS image, extracting some information about each segment and using this information to enrich the representation. It will then be possible to distinguish between low curvature shallow segments and high curvature deep segments. The process leads to the introduction of a new CSS representation for shape similarity retrieval. Two other approaches are also introduced and the relevant results are presented and discussed.

The organisation of the remainder of this chapter is as follows. Section 2 is devoted to the matching algorithm which compares two sets of maxima and computes a matching value as the measure of similarity between the corresponding shapes. Section 3 describes the global parameters employed in our original system. Section 4 explains the method used for the performance evaluation of the system. Section 5 presents the results for the original method. Section 6 compares our system to other popular techniques. Section 7 explains the problem posed by shallow concavities to CSS images, and describes three enhancements to the CSS image (referred to as the Height-Adjusted method, the average-Curvature method, and the Mean-Distance method) designed to overcome that problem. Section 8 presents results on the evaluation of the enhanced methods. Section 9 discusses an application of the system to Chrysanthemum leaf classification. Section 10 contains the concluding remarks for this chapter.

2.   The CSS Matching Algorithm

Every object in the database is represented by the locations of the maxima of its CSS image. In this section we first explain the basic concepts of our matching algorithm which compares two sets of maxima and assigns a matching value to them. The matching value represents the similarity measure between the actual boundaries of objects. A more complete description of the CSS matching algorithm then follows.

Each contour of the CSS image corresponds to a concavity or a convexity of the relevant object as presented in Fig. 3.1. It is obvious that regions 6 and 1 of the first object must be matched with regions 7 and 8 of the second object, respectively. Observing the locations of the relevant maxima on the first and second rows of Fig. 3.1, we realize that they are in quite different positions. This is due to different starting points. If we change the starting points properly, then the locations of corresponding maxima on CSS images will be close to each other. This can be observed on the third row of Fig. 3.1.

Therefore, the first step in CSS matching is to shift one of the two sets of maxima so that the effect of a randomly selected starting point is compensated for. Since the exact value of required shift is not available,

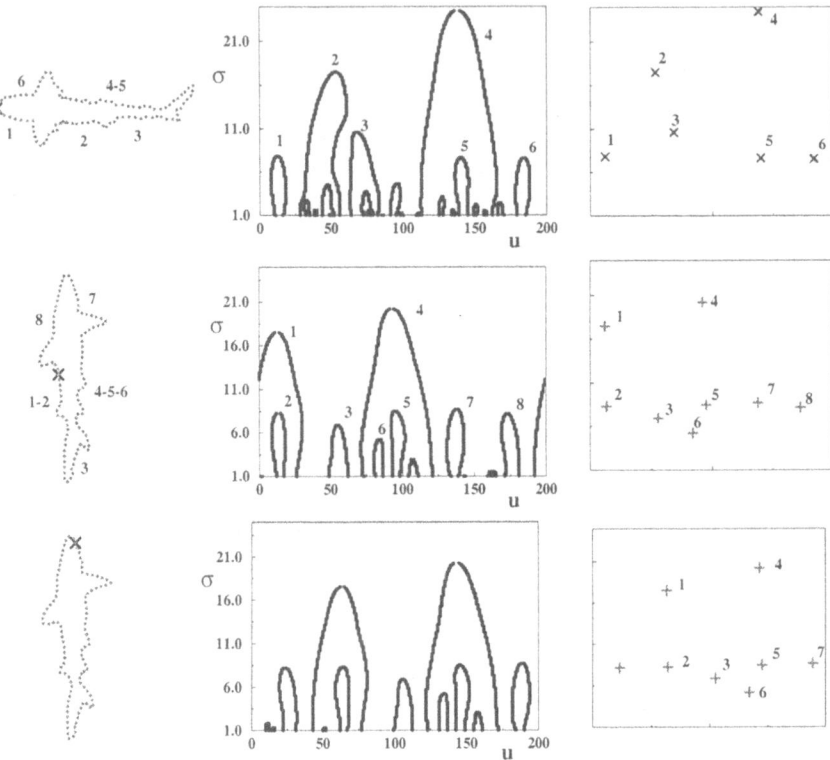

Figure 3.1. CSS image and its maxima, left: re-sampled boundary with the marked starting point, middle: CSS image, right: normalised maxima of CSS images.

we choose several values for it and then find the best match among them. The best choice is a value that shifts one CSS image so that its major maximum covers the major maximum of the other CSS image. Other possible choices are those values which accomplish the same with the second and possibly the third major maxima.

For the two sets of maxima shown in Fig. 3.1, four choices are shown in Fig. 3.2. Considering this Figure, one can quickly realize that the first one is the best. Every maximum of the first CSS image is matched with a maximum of the second one, and two maxima remain unmatched. The matching value will be the summation of the straight line distances between the matched pairs plus the vertical coordinates of the unmatched maxima.

For convenience, from now on, we refer to the input as *image* and the images in the database as *models*. The maxima of every model are

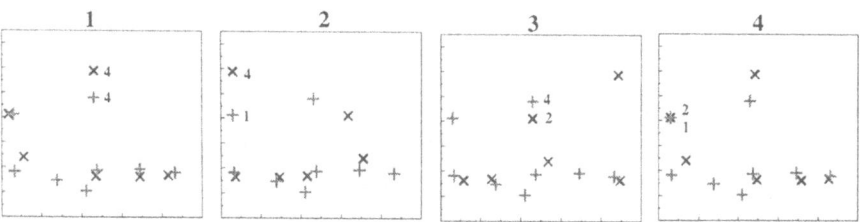

Figure 3.2. Four possible choices for matching of the two sets of maxima related to first and second rows of the previous Figure.

sorted according to their σ-coordinates during the process of maxima extraction.

It should be noted that in addition to the CSS maxima, we also use several *global parameters* to discard dissimilar shapes, prior to the matching. These parameters and the way we use them are described in section 3.

After extracting the maxima of every model, they are normalised so that the horizontal coordinate u varies in the range $[0, 1]$. The maxima of every model are sorted according to their normalised σ-coordinates during the process of maxima extraction. The complete matching algorithm which compares the two sets of maxima, one from the image and the other from the model is as follows. Note that there are some subtle differences between this algorithm and the one presented in chapter 2.

1 Create a node consisting of the largest scale maximum of the image and the largest scale maximum of the model. Initialise the *cost* of this node to the absolute difference of σ-coordinates of the image and the model. Compute a CSS shift parameter α for each node:

$$\alpha = U_m - U_i$$

where U is the horizontal coordinate of a maximum, and i and m refer to image and model respectively. This parameter is used to compensate for the effect of different start points or change in orientation.

2 If there are more than one maximum in the model which have a σ-coordinate close (within 80 percent) to the largest scale maximum of the image, create extra nodes consisting of the largest scale maximum of the image and the additional maxima of the model. Also create the same nodes for the second largest scale maximum of the image and the respective maxima of the model. Initialise the cost and compute the CSS shift parameter for each node accordingly.

3 Create two lists for each node obtained in steps 1 and 2. The first list will contain the image curve maxima and the second list will contain the model curve maxima matched within that node at any point of the matching procedure. Initialise the first and second list of each node by the corresponding maxima determined in the first two steps.

4 Expand each node created in steps 1 and 2 using the procedure described in step 5.

5 To expand a node, select the largest scale image curve CSS maximum (which is not in the first list) and apply that node's shift parameter α to map that maximum to the model CSS image. Locate the nearest model curve CSS maximum (which is not in the second list). If the two maxima are within a reasonable horizontal distance (0.2 of the maximum possible distance), define the cost of the match as the straight line distance between the two maxima. Otherwise, define the height of the image curve CSS maximum as the cost of the match. If there are no more image curve CSS maxima left, define the cost of match as the height of the highest model curve CSS maximum *not* in the node's second list. Likewise, if there are no more model curve CSS maxima left, define the cost of match as the height of the selected image curve maximum. Add the match cost to the node cost. Update the two lists associated with the node.

6 Select the lowest cost node. If there are no more model or image curve CSS maxima that remain unmatched within that node, then return that node as the lowest cost node. Otherwise, go to step 5 and expand the lowest cost node.

7 Interchange the image and the model and repeat steps 1 to 6 to find the lowest cost node in this case.

8 Consider the lowest-cost node as the final matching cost between the image and the model.

If a model in the database is similar to the mirror-image of the input, the CSS image of the model may also be similar to the mirror-image of the CSS image of the input. Since by applying only a circular shift, it is not possible to map the corresponding maxima in this case, the above mentioned algorithm will fail to discover the similarity between the input and the model. Therefore, the mirror-image of the input should also be compared to the existing models of the database. Using the input maxima, we can easily calculate a new set of maxima which corresponds to the CSS image of the mirror-image of the input. We can then either repeat steps 1 to 8 for the new set and consider the lowest matching cost between the two or construct new nodes in step 1 for the new set and expand all nodes simultaneously.

3. Global Parameters

The matching algorithm is not applied to all models of the database. We use a number of global parameters to reject shapes dissimilar to the input prior to the matching process.

To reject the dissimilar images based on the global parameters, we first calculate α_e, α_c and α_a as follows:

$$\alpha_e = \frac{|e_i - e_m|}{max(e_i, e_m)}$$

$$\alpha_c = \frac{|c_i - c_m|}{max(c_i, c_m)}$$

$$\alpha_a = \frac{|a_i - a_m|}{max(a_i, a_m)}$$

where e and c represent the *eccentricity* and *circularity* of the boundary and a represents the *aspect ratio* of the CSS image, while i and m stand for image and model respectively.

Circularity is the ratio of perimeter squared to the area. Eccentricity has been widely used as a shape feature [243, 74]. It is a region based parameter and illustrates how the region points are scattered around the centroid of the region. The central moments of a region are defined as:

$$\mu_{p,q} = \sum_x \sum_y (x - \overline{x})^p (y - \overline{y})^q$$

where x and y are coordinates of the region points and $(\overline{x}, \overline{y})$ is the *centre of gravity* or *centroid*. The *principal moments* of a region are eigenvalues of the matrix:

$$\begin{bmatrix} \mu_{2,0} & \mu_{1,1} \\ \mu_{1,1} & \mu_{0,2} \end{bmatrix}$$

and the eccentricity of the region is:

$$eccentricity = \sqrt{\frac{\lambda_{max}}{\lambda_{min}}}$$

where λ_{max} and λ_{min} are the eigenvalues of the matrix above.

According to their definition, α_e, α_c and α_a are between zero and one. We need to choose a threshold for each of these parameters so that if one of them is above the relevant threshold, the corresponding model is rejected. For our system we chose all threshold values, α_{et}, α_{ct} and α_{at} as 0.3. Lower threshold values may result in missing similar shapes, while with higher values, the number of candidates increases and using global parameters will become ineffective. However, it should be noted that the performance of the system is not sensitive to small changes of these values and this will be demonstrated later in section 5.2.

4. Performance Evaluation

Shape similarity retrieval is involved with the notion of similarity which can not be measured. As a result, the evaluation of system performance turns out to be a difficult task. A subjective evaluation involving human subjects is presented in [208]. The subjects are asked to find similar shapes to a number of queries from a small database. The results are then compared to the results of the system.

Here we present an objective evaluation which involves a classified subset of the SQUID database. There are 17 classes in this database, each consisting of about 8 objects. The whole database is presented in Fig. 3.3. These objects are selected carefully so that the within-class similarity is reasonably high. There are also particular characteristics in each group to distinguish it from other groups. We have made every effort to perform a fair classification, and have paid more attention to the whole appearance of shapes rather than taking into account the shape features which are used by our method. We believe that this classified database provides a good basis to compare the performance of different methods we have tried.

The procedure used for evaluating the performance of the system is as follows:

- Choose one of the objects in class one as the input query, and determine the first n outputs of the system. These are the most similar images of the database to the input according to the system. $n = 15$ is chosen for this test.

- Count the number of outputs which are in the same class as the input. Divide this number by the number of members of this class,

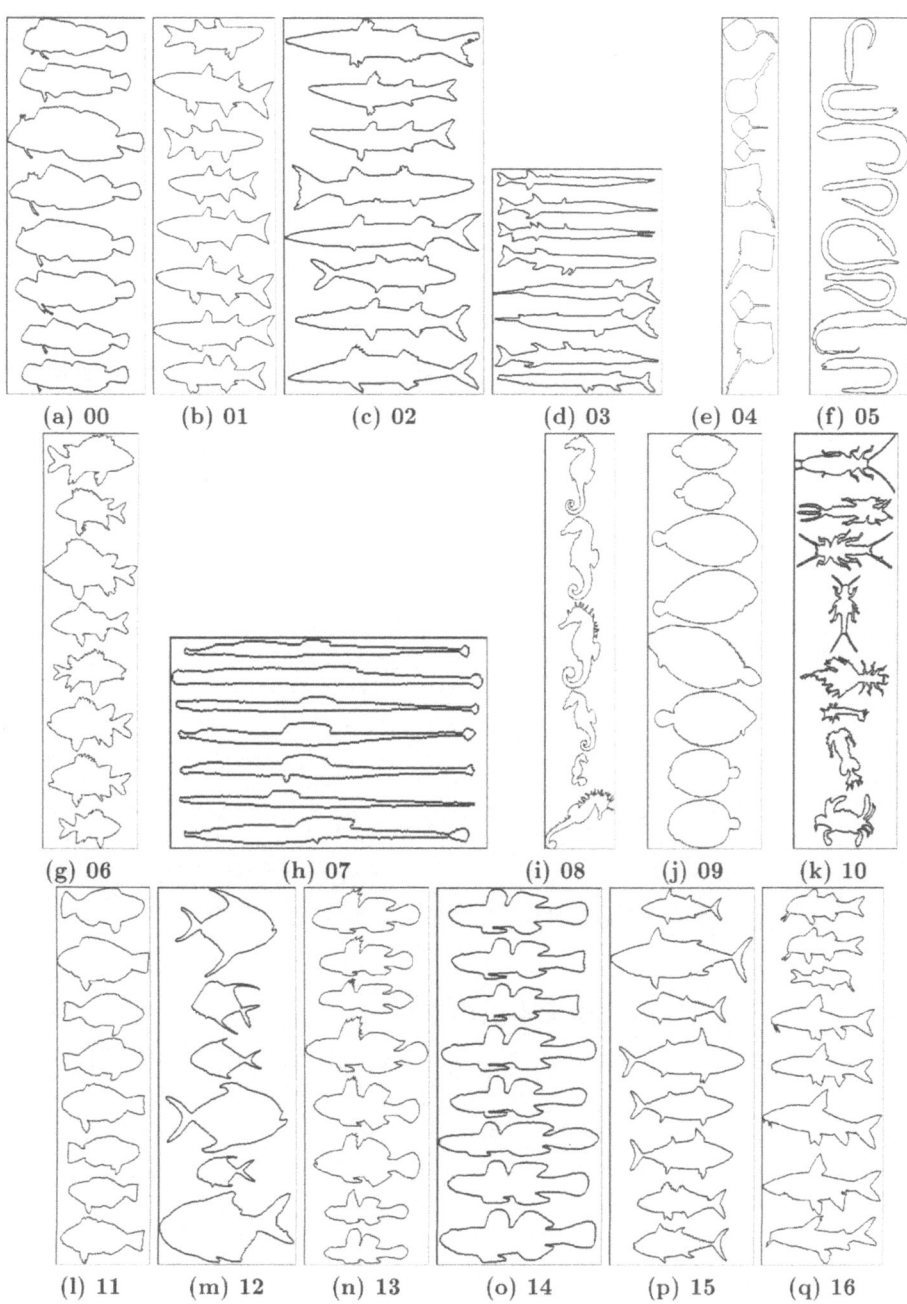

(a) 00 (b) 01 (c) 02 (d) 03 (e) 04 (f) 05

(g) 06 (h) 07 (i) 08 (j) 09 (k) 10

(l) 11 (m) 12 (n) 13 (o) 14 (p) 15 (q) 16

Figure 3.3. Classified database used for objective evaluation.

multiply it by 100 and let the result be the performance measure of the system for that particular object (note that this measure is similar to the concept of *Recall* [90] and that it depends on the total number of objects retrieved from the database).

- Repeat the previous steps for all members of class one. Determine the performance of the system for class one by averaging the performance measures of all members of this class.

- Determine the performance measures for all classes, repeating the above steps.

- Finally, find the performance measure of the system for the whole classified database by averaging the performance measures of all classes.

Using this method, we measured the performance of the system using different approaches and compared them in the following section.

5.   Results for Original Method

In this section the results of our experiments are presented and discussed. We start with the original CSS matching which uses the maxima of the CSS image to represent the shape. It will then be compared to a modified version which includes the mirror-image extension. These are presented in subsection 5.1. In the first step of the new matching algorithm, we also create more nodes. As a result, the chance for a better normalisation in the starting point and change in orientation increases. We present the results of our experiments on using the CSS representation with the global parameters in shape similarity retrieval in subsection 5.2.

5.1.   CSS without Global Parameters

This is the first version of our system and is called the *reference method*. We modified the original CSS matching algorithm [190] by considering additional nodes and solving the mirror-image problem. Two examples are presented in Figs. 3.4(a) and 3.4(b). In both examples, the input query has appeared as the first output of the system. In 3.4(a), the seventh output is not as similar to the query as the other outputs. The dissimilarity between the two shapes is mainly due to the global appearance of them. We can deal with this problem by using global parameters. This argument also applies to the seventh output of Fig. 3.4(b).

The results of objective evaluation for this approach is presented in the first row of table 3.1. The result for group 04 is 100%. This means

G	00	01	02	03	04	05	06	07	08	09	10
RM	88	86	59	31	100	100	78	33	72	95	31
WM	78	53	44	30	98	100	44	35	69	95	25
Dif	-10	-33	-15	-1	-2	0	-34	2	-3	0	-6

G	11	12	13	14	15	16	T
RM	100	78	89	100	86	59	76
WM	61	78	69	86	72	52	64
Dif	-39	0	-20	-14	-14	-7	-12

Table 3.1. Effects of not using the mirror-image consideration. WM and RM stand for Reference Method and Without mirror-image respectively. The third row shows the difference in percents caused by cancelling the mirror-image consideration.

that whenever one of the members of this group is used as the query, all other members appear in the first 15 outputs of the system. The same results are achieved for groups 05, 11 and 14. Apart from groups 03, 07 and 10, the results for other groups are more or less acceptable. Groups 03 and 07 include shapes with shallow concavities [10, 8], and group 10 consists of unusual shapes which are not quite similar to each other. However, considering objects in other groups, we believe that these objects can be grouped together.

Note that for this test, we have not used any global parameters except the aspect ratio of the CSS images.

5.1.1 CSS without Mirror-Image

The problem of mirror-image in CSS matching was explained earlier. Here we repeat the queries of Figs. 3.4a and 3.4b by disabling the mirror-image option. The results are presented in Figs. 3.4c and 3.4d respectively. It is observed that the orientation of outputs is the same as that of the input queries in these cases. Some relevant models like the first three outputs of Fig. 3.4b have not appeared in Fig. 3.4d. The same comment applies to the fifth and the sixth output of Fig. 3.4a. The problem with the eighth output of this example is related to its shallow concavities described in section 7.

The results of the objective evaluation is presented in the second row of table 3.1. A dramatic drop in performance measure is observed when we remove the modifications. This drop is more considerable for groups such as 06 and 11 which include shapes with different orientations.

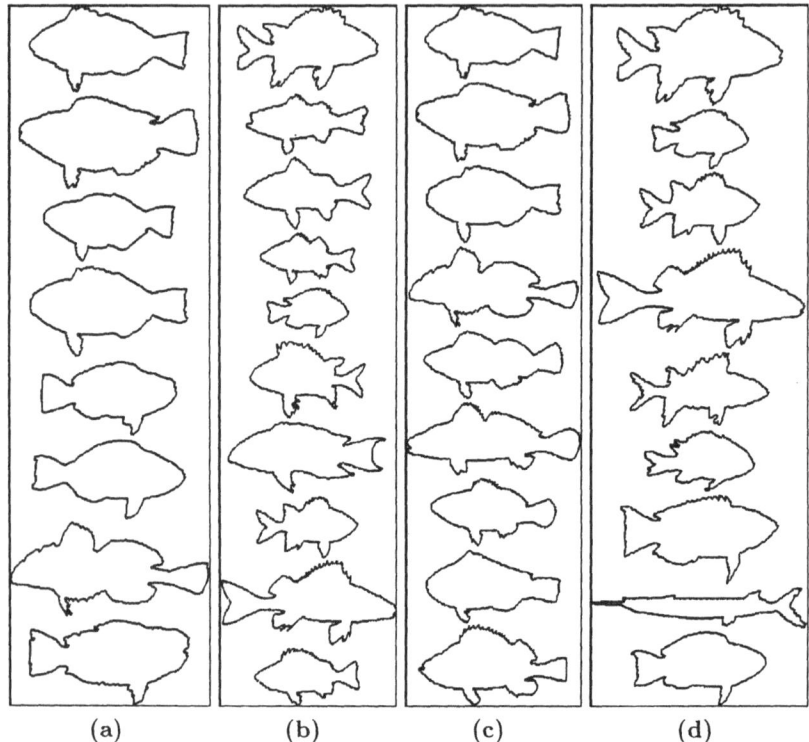

(a)	(b)	(c)	(d)

Figure 3.4. a) and b) correspond to the reference method. c) and d) are without mirror-image consideration.

5.2. CSS with Global Parameters

Eccentricity and circularity contain considerable information about the global shape of an object. When this information is used in conjunction with the maxima of the CSS image, which contain more local information about the shape, the best results are achieved.

We study the effects of these parameters in two stages. In stage one, we observe the improvement caused by each of these parameters. Then we study the effects of using them together. Using circularity alone may increase the performance measure of the method by up to 7% which is achieved by setting α_{ct} to any value in the wide range $[0.28 - 0.46]$. This is shown in Fig. 3.5(a). Note that we have not used eccentricity in this experiment. Also note that when we choose a value for α_{ct}, it is fixed for all images of the database.

Since the aspect ratio of the CSS image was used with the reference method, it is also used in the experiments of this section. This makes the comparison easier and more meaningful. The total performance measure

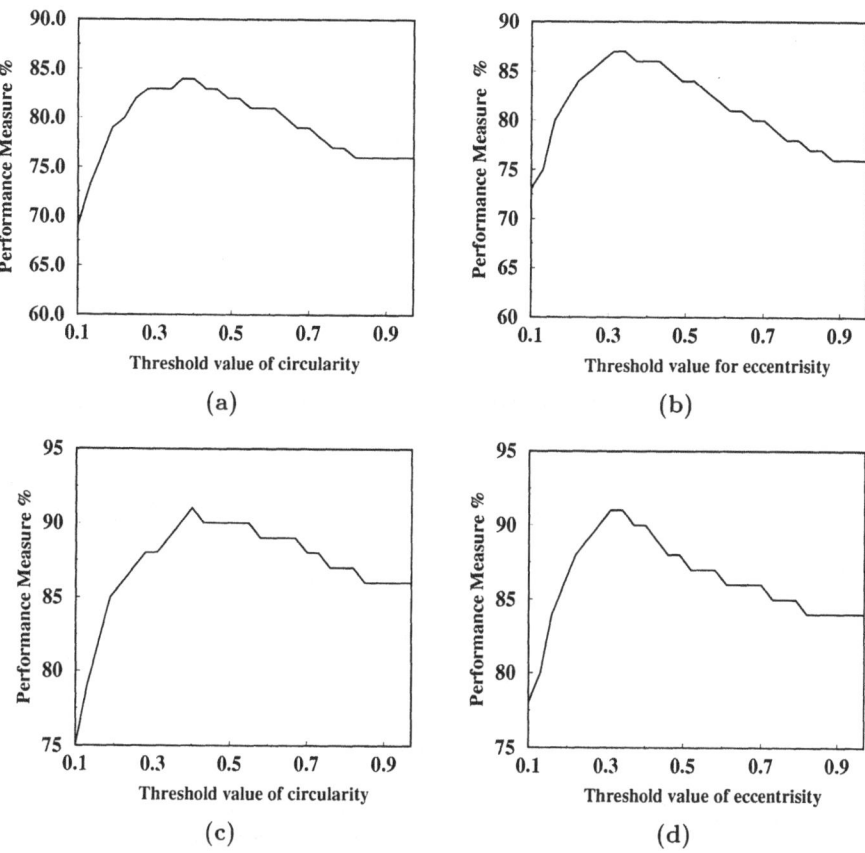

Figure 3.5. Effects of using global parameters. a) circularity alone, $\alpha_{et} = 1$. b) eccentricity alone, $\alpha_{ct} = 1$. c) Both but α_{et} is fixed at 0.33. d) Both but α_{ct} is fixed at 0.4. Note: For all cases α_{at} is fixed at 0.25.

for the CSS matching plus aspect ratio as the global parameter was 76%, as shown in table 3.1.

The better results are achieved when we use eccentricity alone. Any value for α_{et} in the range [0.28 – 0.42] leads to 10% improvement in the performance measure which is quite considerable. The total performance measure is 86%. Fig. 3.5(b) represents the results of this experiment.

Now we choose α_{et} as 0.33 and study the performance of the system by changing the value of α_{ct}. This time the increase in the performance measure is 14% when we choose α_{ct} in the range [0.37 – 0.55]. This is shown in the plot of Fig. 3.5(c). Note that for the values in the range [34% – 67%], this figure will be 13%.

If α_{ct} is fixed as 0.4, and α_{et} is chosen and fixed at any value in the range [0.28 – 0.4], then the same result is obtained, i.e. 14% increase is achieved. This is shown in Fig. 3.5(d).

We can conclude that by using these three global parameters, the performance measure of the system will be more than 90% for a wide range of threshold values.

6.   Comparison to Other Methods

In this section we compare our method to two other popular methods: Fourier Descriptors and Moment Invariants.

6.1.   Fourier Descriptors

A closed curve $\Gamma(t) = (x(t), y(t))$, can be considered as a complex periodic function of t, where $-\infty \leq t \leq +\infty$. This function can then be sampled by N equidistant points. The Discrete Fourier Transform of $\Gamma(t)$ is then defined as:

$$F_k = \sum_{n=0}^{N-1} e^{-j2\pi nk/N} \Gamma_n$$

where

$$-N/2 + 1 \leq k \leq N/2 - 1$$

and Γ_n is the nth sample of $\Gamma(t)$.

As stated in [315], F_1 always has the highest magnitude among F_k's (F_0 is not considered), provided that the contour is traced in the counterclockwise manner and the contour does not cross itself. We will observe that this may not be true in some special cases.

Provided that the starting point, position, orientation and scale have been chosen properly, the degree of similarity between two sets of FDs can be measured by the sum of Euclidean distances of corresponding components. This will be proportional to the sum of Euclidean distances of the contour points. A method is suggested in [315] to normalise the FDs. All contours of the database are turned so that the orientation and the starting points of all similar shapes become similar and therefore the Euclidean distance is minimised.

Applying the inverse Fourier transform, we can recreate a contour from its FDs. In particular, F_1 and F_{-1} create an ellipse which is quite similar to the outline of the contour without any details. Each pair such as F_i and F_{-i} will create a particular ellipse which is traced i times. The superposition of these ellipses recreates the contour.

Using 20 pairs of FDs of every contour of our database, we employed the method explained in [315] to normalise them. We then used the

Euclidean distance between the FDs of the input query and those of the models to find the most similar shapes.

A number of examples are given in Fig. 3.6. We can learn the following points from these examples.

- In the first three examples, the input query can be covered by an ellipse which is presented by F_1 and F_{-1}, and therefore the results are globally similar to the input query.

- The main reason for the good results of the second example is that the input query is symmetrical. This is not true for Figs. 3.6(d) and 3.6(e), and therefore, the results of these examples are not very good.

- The problem of mirror-image is not considered in the method. Looking at group 06, we realise that there are several other shapes similar to the mirror-image of the input query of the third example which have not appeared as outputs.

- For the input of Fig. 3.6(e), F_1 does not have the largest magnitude as claimed in [315], and therefore the method fails to apply a proper normalisation.

The same conclusions can be made by considering the results of the objective evaluation presented in table 3.2, where they are compared to the CSS with global parameters.

We have very good results for groups 00, 02 and 07. On the other hand, very poor results are observed for groups 05, 06, 08, 10, and 11. In group 00, the orientations and the starting points are the same for all contours. This is a result of using the same algorithm for extracting these contours from the original color images. Shapes in group 02 are also symmetric and shapes in group 07 are unique in terms of the largest ellipse. For more complex shapes like groups 08 and shapes with different orientations like groups 06 and 11, we observe a poor performance. For some of the contours of group 05, F_1 does not have the largest magnitude component and therefore we come across very poor results.

In conclusion, the FD method introduced by [315] is simple, quite fast, easy to implement and contains useful global information about the input shape. On the other hand, lack of local support and ambiguities in starting point and orientation are the most important shortcomings of the method. Moreover, we observed that [315] have not considered the problem of mirror-image and their assumption about F_1 as the component with the largest magnitude among FDs is not always correct.

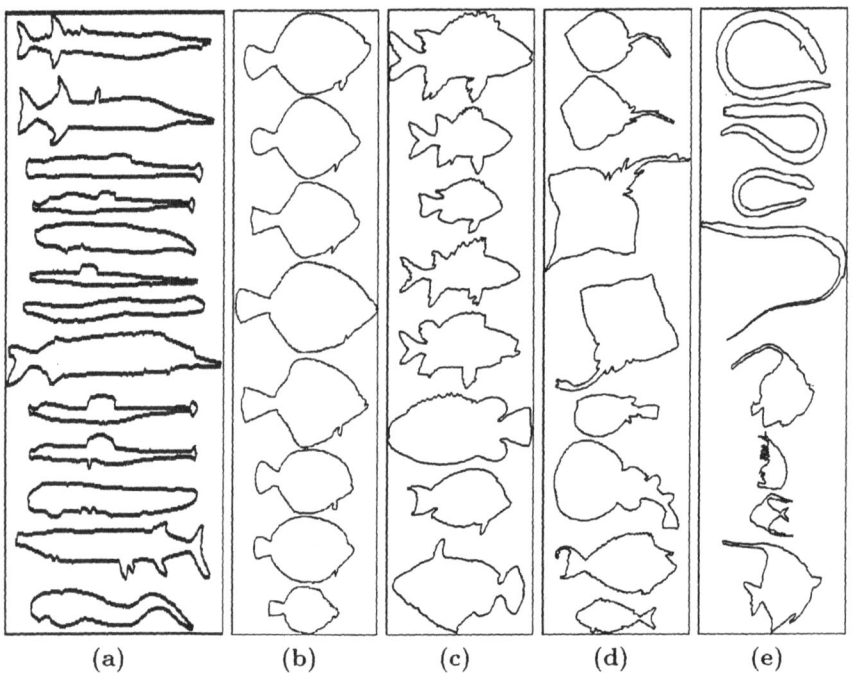

| (a) | (b) | (c) | (d) | (e) |

Figure 3.6.   Examples of query results based on Fourier Descriptors method

G	00	01	02	03	04	05	06	07	08	09	10
CSS	89	86	81	95	100	100	91	98	75	92	81
FD	100	84	100	75	78	42	80	100	36	98	13
Dif	-11	2	-19	20	22	68	9	-2	39	-6	68

G	11	12	13	14	15	16	T
CSS	100	94	89	100	95	69	91
FD	70	78	98	73	97	48	75
Dif	30	16	-9	27	-2	21	16

Table 3.2.   FD results in comparison to CSS + global parameters.

6.2.   Moment Invariants

The geometric moment of order pq of the boundary points of a shape is defined as follows.

$$\mu_{pq} = \sum_x \sum_y x^p y^q$$

The same definition also applies to the points of the solid shape. The original *moment invariants* introduced by [118] are seven functions of μ_{pq} where p and q are between 0 and 3:

$$M_1 = (\mu_{20} + \mu_{02})$$
$$M_2 = (\mu_{20} - \mu_{02})^2 + 4\mu_{11}^2$$
$$M_3 = (\mu_{30} - 3\mu_{12})^2 + (3\mu_{21} - \mu_{03})^2$$
$$M_4 = (\mu_{30} + \mu_{12})^2 + (\mu_{21} + \mu_{03})^2$$
$$M_5 = (\mu_{30} - 3\mu_{21})(\mu_{30} + \mu_{12})$$
$$M_6 = (\mu_{20} - \mu_{02})\left[(\mu_{30} + \mu_{12})^2 - (\mu_{21} + \mu_{03})^2\right]$$
$$M_7 = (3\mu_{21} - \mu_{03})(\mu_{30} + \mu_{12})$$

Functions M_1 to M_7, are invariant to rotation and reflection as well as a combination of them. Scale invariance can be achieved by normalising the functions using *radius of gyration* which is defined as follows.

$$r = (\mu_{20} + \mu_{02})^{\frac{1}{2}}.$$

This parameter is proportional to the size of object boundary. A set of six normalised moment invariants is obtained as follows.

$$M'_2 = \frac{M_2}{r^4}$$
$$M'_3 = \frac{M_3}{r^6}$$
$$M'_4 = \frac{M_4}{r^6}$$
$$M'_5 = \frac{M_5}{r^{12}}$$
$$M'_6 = \frac{M_6}{r^8}$$
$$M'_7 = \frac{M_7}{r^{12}}$$

We use the latter to represent the boundary as well as solid shape and experimentally compare this method to our proposed method. Each object is then represented by a 12 dimensional feature vector, including two sets of moment invariants, one from object boundary and the other from solid shape. The Euclidean distance is used to measure the similarity between different shapes.

The results of objective evaluation is presented in table 3.3. The best results are for groups 02, 03, 07. These are globally different from the other groups.

G	00	01	02	03	04	05	06	07	08	09	10
CSS	89	86	81	95	100	100	91	98	75	92	81
MI	34	78	91	100	88	41	72	86	47	53	23
Dif	55	8	-10	-5	12	59	19	12	28	39	58

G	11	12	13	14	15	16	T
CSS	100	94	89	100	95	69	91
MI	91	53	45	72	73	58	65
Dif	9	41	43	28	22	11	26

Table 3.3. Moment invariants results compared to CSS + global parameters.

7. The Problem of Shallow Concavities

Gaussian smoothing with resampling converges to heat diffusion which is a form of curvature deformation [142, 222]:

$$\frac{\partial \mathbf{\Gamma}}{\partial t} = \kappa \mathbf{N} \quad given \quad \mathbf{\Gamma}(s,0) = \mathbf{\Gamma}_0(s). \tag{3.1}$$

where κ denotes curvature, $t = \frac{\sigma^2}{2}$, and σ is the standard deviation of the kernel. In this equation, $\mathbf{\Gamma}_0(s)$ represents the original boundary of the shape. $\mathbf{\Gamma}$ is the evolved curve, and the equation defines its evolution as t (and in fact σ) increases. This equation clearly shows that the rate of movement of a point on a segment is proportional to the value of curvature at that point.

Now consider a shallow segment with two curvature zero crossings at end-points. As σ increases, the zero crossing points approach each other *very slowly* due to low values of curvature. As a result, the relevant CSS contour turns out to be quite large. For a curved deep segment, the rate of movement can be large. However, in order to convert a deep segment to a straight line, each point must experience a large movement. As a result, the maximum of the relevant contour in the CSS image will have a large σ coordinate. Therefore, deep and shallow concavities may create the same large contours in the CSS image.

An example of a shape with shallow concavities is shown in Figure 3.7(a) where the associated CSS image is shown in Figure 3.7(b). After removing the noise in the first stages ($\sigma \leq 3$) the major segments of the shape with two curvature zero crossings at the end-points will appear. For $\sigma = 8$, the whole shape is very close to a convex shape and one may believe that it is converted to a convex curve soon after this stage.

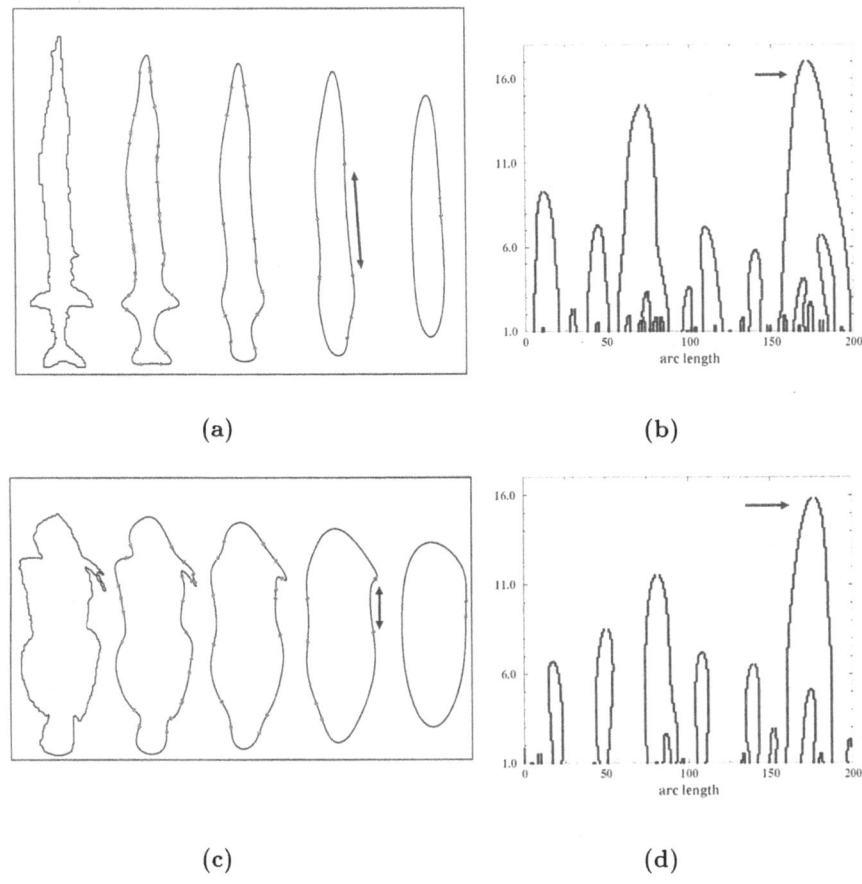

Figure 3.7. a) A boundary with shallow concavities and b) its CSS image. c) A boundary with curved concavities and d) its CSS image. The marked contours relate to the marked concavities.

However, due to shallow segments and especially the marked one, this does not happen as long as σ is less than 16.6.

An example of a shape with curved concavities and its CSS image are shown in Figures 3.7(c) and 3.7(d). The marked segment is still curved at $\sigma = 8$, however the zero-crossings join at $\sigma = 15.5$ which is even lower than the previous example.

Considering Figures 3.7(b) and 3.7(d), we observe that the two sets of CSS maxima are almost similar, where one set relates to shallow and the other relates to curved segments. It is not surprising that the shape of Figure 3.7(c) comes out as the third most similar shape according to the CSS matching, when the shape of Figure 3.7(a) is used as a query.

In the following subsections, we introduce three different solutions to resolve this problem. The results of objective evaluation, presented in section 8, reveal the advantages and disadvantages of each approach.

7.1.   Height Adjusted Method

The basic idea of this method is to select a new maximum for each CSS contour. For every contour, the new maximum is lower than the actual one. The difference depends on the shape of the concavity. For shallow concavities the difference will be large, whereas for curved concavities it will be negligible. The new maximum will be defined by the level of smoothing at which the curve segment between the two zero crossings is converted to a straight line. We consider a segment as a straight line if the absolute values of curvatures of all its points fall below a threshold.

In this subsection we explain all aspects of this approach. This includes the algorithms for determining and extracting the new maximum of each contour and introducing a new global parameter (subsection 7.1.3) which is used to reject dissimilar shapes prior to the matching algorithm. The matching algorithm for this method was explained in section 2, the only difference is that we use the new set of maxima to represent the shape.

7.1.1   New Maxima

In principle, we modify the representation such that for each contour we find another point instead of the maximum to represent the relevant segment. This point, called new maximum, will be almost the same as the actual maximum for deep concavities and will be lower for shallow segments. The new maximum will be defined by the level of smoothing where the curve segment between the two zero crossings is converted to a straight line. We consider a segment as a straight line if the absolute value of curvature of all its points fall below a threshold.

To clarify the issue, we have plotted curvature versus arc length in Figure 3.8. These are for smoothed version ($\sigma = 10$) of the objects of Figure 3.7. At this level, there are two concavities left on each smoothed contour. We have marked them with numbers 1 and 2 on the plots and the curves. Considering the shape with shallow concavities, we observe that for concavity number 2 the absolute value of curvature remains near zero between the two curvature zero crossings of each concavity. It can be assumed that at this level, there are no actual curvature zero crossings in this part of the shape and the segment has been converted to a straight line. Therefore, the maximum of the relevant CSS contour can be chosen around $\sigma = 10$ instead of $\sigma = 16.6$. To determine the exact value of the new maximum for every contour of the CSS image,

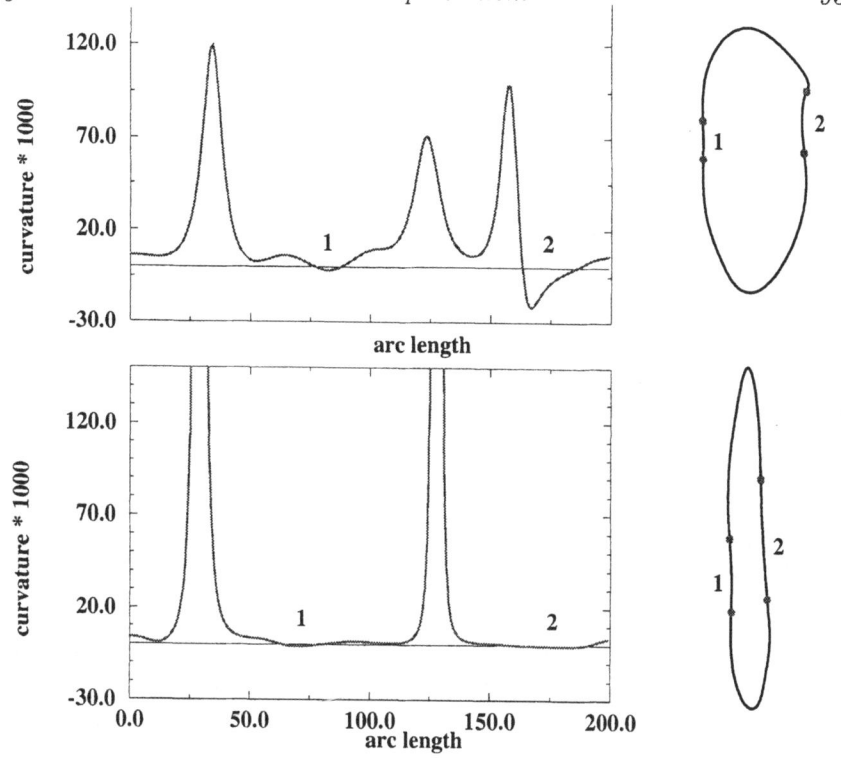

Figure 3.8. The curvature between the two curvature zero crossings is near zero for shallow concavities at earlier levels of smoothing. Compare the concavities labelled by 2 and the related parts on the curvature plots. Plots are for $\sigma=10$.

we look for the level of smoothing where *the absolute value of maximum curvature* in the region between the two zero crossings falls below a certain threshold. The value of this threshold can be determined by studying the results of objective evaluation on a classified subset of the database. It can be shown that the system is not sensitive to this value [10]. Figure 3.9 shows the new maxima of the CSS images of Figure 3.7. Note that the sets of maxima are not very similar.

The absolute value of curvature is not robust with respect to scale. Therefore, we have to normalise the perimeter of the boundaries so that all boundaries have identical perimeter. To normalise the perimeter, we first calculate the average perimeter of the objects in our database. We then calculate the scale factor for each object of the database as follows:

$$S_i = \frac{P_{av}}{P_i}$$

where S_i is the scale factor for the ith object and P_i and P_{av} refer to the ith object and the average perimeter respectively. To obtain a new

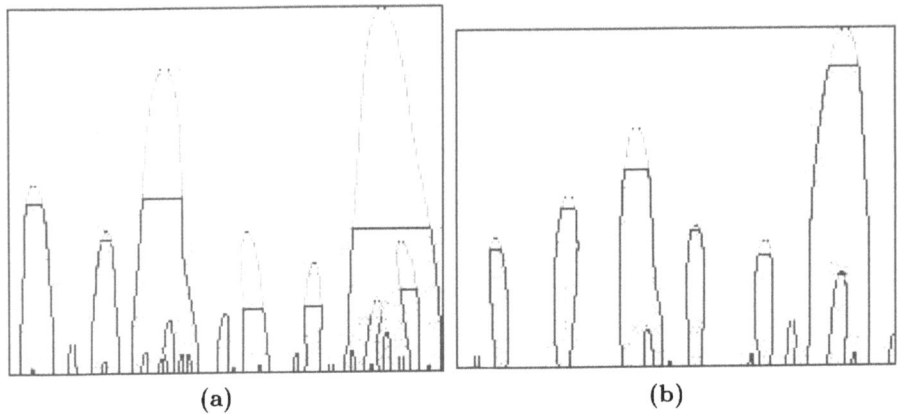

(a)   (b)

Figure 3.9. a) The new maxima are significantly different for shallow concavities. b) No considerable changes for curved concavities.

perimeter for the ith object, we multiply the x and y coordinates of its boundary points by the relevant scale factor, S_i.

7.1.2   Shape Segmentation

As mentioned earlier, at every stage of smoothing, each concavity or convexity is represented by a pair of curvature zero crossings. However, during the process of smoothing, it is not possible to determine whether two successive zero crossings on a smoothed curve are related to each other or they belong to different concavities.

Therefore, the only way to segment the curve and track each segment during the evolution is using the CSS image itself. After extracting the maximum of a contour in the CSS image, we follow each of the two branches of this contour, and find the locations of curvature zero crossings related to this segment of the curve at each level of smoothing. For example, by using this method and tracking the two branches of the largest contour in the CSS image of Figure 3.7(d), it is determined that the relevant pair of zero crossings for $\sigma = 6$ occur at $160th$ and $184th$ samples of the curve. We then examine the absolute value of curvature between these points at different levels to find the level in which it falls below a certain threshold.

We can summarise our algorithm for locating the new maximum of a contour of the CSS image as follows:

1 Determine the maximum of the largest contour and start the process from this level.

2 For the next lower scale, find the locations of zero crossings by following the two branches of the contour.

3 For the current level of smoothing, find the absolute value of curvature at every point between the two zero crossings and determine the maximum curvature on this segment.

4 If the maximum curvature is above a certain threshold, set the new maximum as the middle point of this segment, otherwise go to Step 2.

5 Exit if there are no maxima left, otherwise determine the next largest maximum of the CSS image and repeat Steps 2 to 5.

7.1.3   A New Global Parameter

We introduce a new global parameter for a closed planar curve as:

$$D = \sum_{i=1}^{n} d_i$$

where d_i is the difference between the old and new maximum of the ith contour of the CSS image of the curve. The parameter will be near zero for curved shapes and will be large for shapes with shallow concavities. We have used this parameter for indexing. In response to a query, α_D is calculated for every model in the database as follows:

$$\alpha_D = \frac{\mid D_i - D_m \mid}{max(D_i, D_m)}$$

where D_i and D_m represent the new parameter for the image and the model respectively. If α_D is above a threshold, α_{DT}, the model is rejected without applying the CSS matching. This simple test narrows down the range of searching. A more sophisticated way of using this parameter is explained in subsection 8.1.

7.2.   Average Curvature Method

In this approach, each segment of a shape is represented by the relevant actual maximum of the CSS image contour as well as the average curvature on the segment at a certain level of scale. This level is equivalent to $\frac{1}{6}$ of the height of the CSS image. Note that if the height of a contour is less than this level, it is considered as noise and is not used for representation. Therefore, an object boundary will be represented by a set of 3-D points. Each point consists of u and σ coordinates of the maximum of a CSS image contour, and the normalised average curvature on the relevant segment as the z coordinate.

7.2.1   Normalization of Average Curvature

Normalization of curvature is applied in two stages. Since curvature is not scale invariant, in the first stage, we re-scale all shapes of the database so that they all have the same perimeter. In the second stage, we should re-scale the curvature so that the absolute value of the third dimension of each point falls in a suitable range in comparison to the other dimensions. Since we use the Euclidean distance later in the matching algorithm, if the absolute value of average curvature is too low, then its contribution to the final matching value would be too small.

After performing a sequence of experiments, we concluded that by leaving the scale-factor of average curvature at 1.0 and limiting the maximum value of average curvature to 0.1, the best results can be achieved. We studied the distribution of average curvature over all segments of all objects of our database to achieve this result.

7.2.2   Matching Algorithm

The matching algorithm of section 2 must be slightly modified to accommodate the third dimension. For example if the difference between the average curvature of a maximum of the model and its nearest maximum of the image is higher than a threshold, they are considered unmatched maxima. For all unmatched maxima, the length of the projection over the $\sigma - z$ plane is added to the matching value. For matched maxima, the Euclidean distance between the points are used as the cost of the match. The complete matching algorithm for this approach can be found in [228].

7.3.   Mean-Distance Method

This approach is almost the same as the previous one. The only difference is that instead of the average curvature, we use another parameter to describe the relevant segment between the two curvature zero crossings. To calculate the parameter, a straight line is drawn between the two end-points. The average distance to this line over all points of the segment is then calculated. This value is then normalised by dividing it by the distance between the two end-points. From Figure 3.10 this parameter can be calculated as follows.

$$Mean - dist = \frac{\frac{1}{n}\sum_{i=1}^{n} d_i}{d}$$

where d_i is the distance of ith point of the segment from the line, d is the length of straight line between the two curvature zero crossings and n is the number of points on the segment.

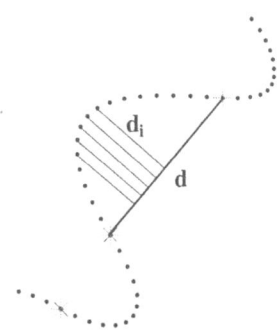

Figure 3.10. The Mean distance parameter

It is obvious that for shallow concavities, this parameter will be near zero and for deep ones, it is significantly larger.

8. Performance Evaluation and Experimental Results

Here we present an objective evaluation which involves a small classified subset of our database. There are 17 classes in this database, each consisting of about 8 objects. The whole database was shown in Figure 3.3. The performance of the system was evaluated using the procedure described in section 4. Using that method, we measured the performance of the system using different approaches. The original CSS matching which uses the actual maxima of the CSS image as the shape representation is used as a *reference method*. The results of the reference method is then compared to the results of more sophisticated versions of the CSS representation introduced in previous sections.

8.1. Height Adjusted CSS image

This method was explained in section 7.1. We introduce a new maximum for each contour of the CSS image which is lower than the actual one. The new maximum is equal to the level of smoothing where the maximum of curvature on the relevant segment falls below a certain threshold. For shallow segments the difference between the new and old maxima is significant while for deep segments it is negligible.

A new global parameter, used for indexing, indicates whether the shape contains shallow segments or not. If we do not use this parameter, the improvement in the results of objective evaluation is about 5%, which is very good. The improvement is significant for shapes with

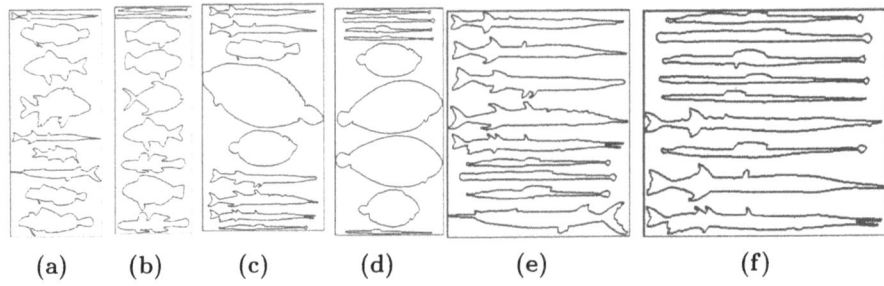

(a)   (b)   (c)   (d)   (e)   (f)

Figure 3.11. Two examples of different methods, a) and b) reference; c) and d) height adjusted e) and f) Enhanced height adjusted.

shallow concavities (groups 3 and 7). Using the new global parameter increases the performance measure of the system for shapes with shallow concavities. This increase may be even higher if we set different thresholds for different types of shapes. More details and final results with another 3% increase in the performance measure are presented in this section.

8.1.1   Without Global Parameter

If we use the new set of maxima as shape representation and do not change the matching algorithm, a significant improvement is achieved for shapes with shallow concavities. Two examples are shown in Figure 3.11. Figures 3.11(a) and 3.11(b) are related to the reference method while Figures 3.11(c) and 3.11(d) are related to the new method respectively. In all of these examples, the first output is the same as the input query. Figures 3.11(e) and 3.11(f) are related to a more sophisticated method with the global parameter which is described later.

Due to significant changes in the maxima of shapes with shallow concavities, and small changes in the maxima of shapes with deep segments, it may happen that two new sets of maxima become similar, while the actual contours are not. This can be observed in the third to fifth outputs of Figure 3.11(c) and fifth to eighth outputs of Figure 3.11(d). This problem can easily be solved by using a method to distinguish between the two types of shapes prior to the application of the matching algorithm. However, we can observe a considerable improvement in these two examples. This is also confirmed by the results of objective evaluation given in table 3.4. The results for groups 3 and 7 have been improved by 33% and 53% respectively.

G	00	01	02	03	04	05	06	07	08	09	10
RM	88	86	59	31	100	100	78	33	72	95	31
HA	89	89	59	64	100	100	81	86	75	98	26
Dif	1	3	0	33	0	0	3	53	3	3	-5

G	11	12	13	14	15	16	T
RM	100	78	89	100	86	59	76
HA	100	78	95	100	77	55	81
Dif	0	0	6	0	-9	-4	5

Table 3.4. Height adjusted without new global parameter. RM and HA stand for Reference Method and height adjusted respectively. The third row shows the improvement in percents for each group.

G	00	01	02	03	04	05	06	07	08	09	10
RM	88	86	59	31	100	100	78	33	72	95	31
NHA	88	89	68	100	100	97	81	96	75	100	29
Dif	0	3	11	69	0	-3	3	63	3	5	-2

G	11	12	13	14	15	16	T
RM	100	78	89	100	86	59	76
NHA	100	78	95	100	78	55	84
Dif	0	0	6	0	-8	-4	8

Table 3.5. Height adjusted with new global parameter.

8.1.2 With Global Parameter

Using the new global parameter, introduced in subsection 7.1.3, will result in further improvement to the performance measure.

The appropriate threshold value, α_{DT} can be determined by studying the results of objective evaluation for different values of α_{DT}. As Figure 3.12(a) shows, for small values of α_{DT}, the performance measure is low. This is due to the rejection of a large number of good candidates prior to CSS matching. As α_{DT} increases, better results are achieved. For the best results, this parameter should be chosen around 0.80. This will lead to about 1% increase in the performance measure in comparison to the same method without the new global parameter.

Now consider Figures 3.12(b) and 3.12(c). These plots show the performance measure of the system versus α_{DT} for two different groups. As

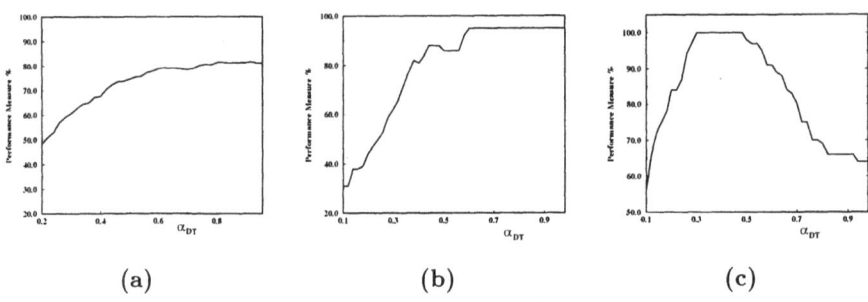

(a)   (b)   (c)

Figure 3.12. a) The performance measure of height adjusted method versus α_{DT}, the threshold value of the global parameter. b) and c) The performance measure versus α_{DT}, for groups 13 and 03.

Figure 3.12(b) shows, for group 13 which contains shapes with curved segments, the performance measure increases as α_{DT} increases. For groups such as 03, the best results are achieved by considering lower values like 0.4 for α_{DT}. This indicates that different values of α_{DT} should be chosen during the query processing. We observed that the most impressive results may be achieved by considering different values for α_{DT} based on the type of the input query.

The type of input can be determined by this simple algorithm:

- Find the maximum drop, d_{imax}, among the new maxima of the input.

- Calculate the percentage of the drop by dividing d_{imax} by the value of the relevant maximum. If this is more than a threshold, consider the input as a shape with shallow segments and choose the relevant value for α_{DT}. Otherwise, choose α_{DT} as determined by the user.

The examples of Figures 3.11(a) and 3.11(b) are repeated in Figures 3.11(e) and 3.11(f), using the new modification. The results are dramatically improved. It should also be pointed out that the volume of the rejected candidates based on global parameters has increased from 23% to 88% for the query of Figure 3.11(e). For the other example it remains unchanged at 89%.

The results of the objective evaluation is presented in table 3.5, where α_{DT} is chosen as 0.42 and 0.92 for shapes with shallow and shapes with curved segments respectively. As a result of using the new method, the performance measure for almost all groups especially group 3 are increased.

G	00	01	02	03	04	05	06	07	08	09	10
RM	88	86	59	31	100	100	78	33	72	95	31
A-C	67	88	72	52	100	100	72	71	72	95	50
Dif	-21	2	13	21	0	0	-6	38	0	0	19

G	11	12	13	14	15	16	T
RM	100	78	89	100	86	59	76
A-C	100	83	78	98	80	59	79
Dif	0	5	-11	-2	-6	0	3

Table 3.6. Average curvature results

8.2. Average Curvature Method

In this approach the third dimension is the normalised average curvature between the two end points of a segment. The method was explained in section 7.2. Using this method, a very good improvement is observed for shapes with shallow concavities as the average curvature is near zero on the shallow segments. However, since curvature can be quite high at sharp corners, average curvature on these segments is unpredictable. For example, in Figure 3.14, two segments are associated with tails of animals. Both segments include sharp corners. In all of these cases, the average curvature on one of the segments is more than double the other one. This is the reason an upper limit was used for curvature.

The results are presented in table 3.6. A good increase in performance measure of the system is observed for groups containing shapes with shallow concavities (2, 3 and 7). At the same time, there are drops for groups 0, 13 and 15. The drop in group 0 can be disregarded if we believe that this group can be considered similar to groups 11 and 14, as in most cases, in response to a query from group 0, the incorrect answers belong to these groups.

Three examples are given in Figure 3.13, where (a), (b) and (c) are the results of queries to our reference method. The results for the same queries to the average curvature method are presented in (d), (e) and (f) respectively. In Figure 3.13(a), there are six shapes from the same group as the input. This figure is five for 3.13(d). However, the last output of Figure 3.13(a) is an irrelevant selection, while all outputs of Figure 3.13(d) seem to be relevant. The same facts are observed in the remaining examples.

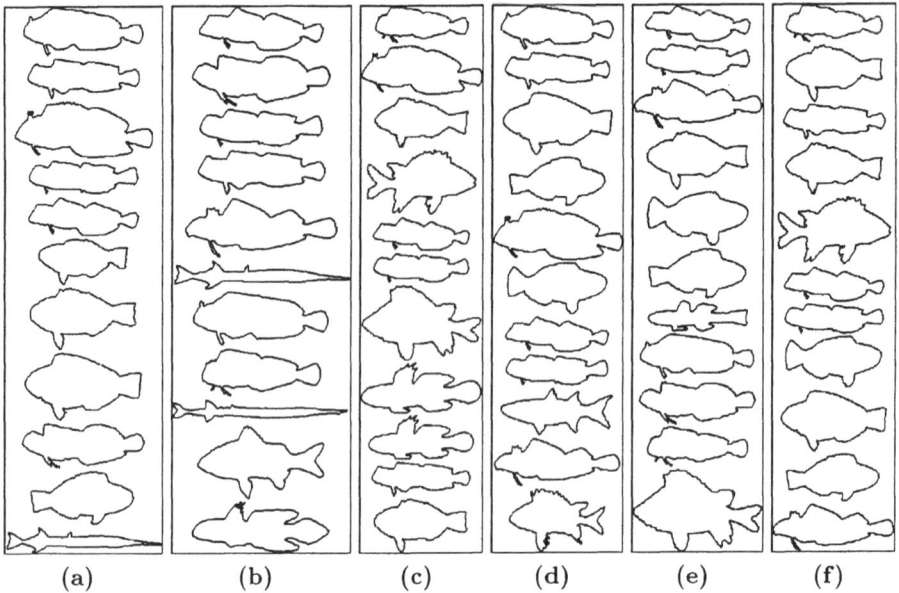

(a)   (b)   (c)   (d)   (e)   (f)

Figure 3.13. Example of using average curvature method in comparison to the reference method. The results of reference method is presented in *a*, *b* and *c*, the results of improved method are presented in *d*, *e* and *f* respectively.

8.3.   Mean-Distance Method

This method was explained in section 7.3. For each segment of the shape, a straight line is drawn between the two end points. The average distance to this line over all points of the segment is then calculated. This value is then normalised by dividing it by the distance of the two end points. It is then considered as the third dimension of the point which represents the segment. The other two dimensions are u and σ coordinates of the corresponding maximum in the CSS image.

Using this method, we observe a 2% improvement in the total result of the objective evaluation. The results are presented in table 3.7, where they are compared to the results of the reference CSS matching without global parameters. A substantial improvement is observed in groups 3 and 7 which include shapes with shallow concavities. The drop in the performance measure of the system for group 0 is also considerable. This can be explained by noting the fact that for the shapes of this group, although the overall appearances are similar, different segments have different shapes. This is shown in Figure 3.14. The three shapes appear similar, but looking at the segments between the pairs of curvature zero crossings marked by +, we realize that these segments are quite different in terms of the new parameter.

G	00	01	02	03	04	05	06	07	08	09	10
RM	88	86	59	31	100	100	78	33	72	95	31
M-d	78	88	63	41	100	100	86	69	72	95	31
Dif	-10	2	4	10	0	0	8	34	0	0	0

G	11	12	13	14	15	16	T
RM	100	78	89	100	86	59	76
M-d	100	78	81	100	81	63	78
Dif	0	0	-6	0	-5	4	2

Table 3.7. Results of using Mean-Distance method in comparison to the reference method.

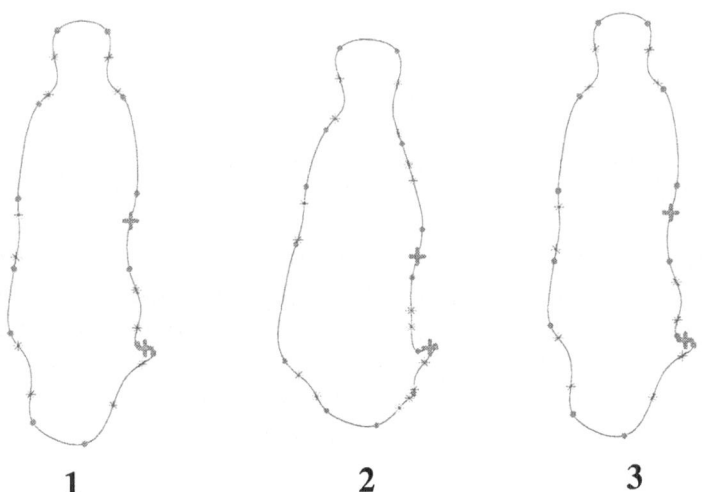

1   2   3

Figure 3.14. The segments between two curvature zero crossings, marked by +, are not similar in terms of Mean-Distance parameter.

9. Application to Chrysanthemum Leaf Classification

In this section, we address the problem of 2D shape representation and matching in presence of self-intersection for large image databases. This may occur when part of an object is hidden behind another part and results in a darker section in the gray level image of the object. The boundary contour of the object must include the boundary of this part which is entirely inside the outline of the object.

We study the effects of contour self-intersection on the Curvature Scale Space image. When there is no self-intersection, the CSS image contains several arch shape contours, each related to a concavity or a convexity of the shape. Self-intersections create contours with minima as well as maxima in the CSS image. An efficient shape representation method has been introduced in this chapter which describes a shape using the maxima as well as the minima of its CSS contours. This is a natural generalisation of the conventional method which only includes the maxima of the CSS image contours. The conventional matching algorithm has also been modified to accommodate the new information about the minima. The method has been successfully used in a real world application to find, for an unknown leaf, similar classes from a database of classified leaf images representing different varieties of Chrysanthemum [5]. For many classes of leaves, self-intersection is inevitable during the scanning of the image.

9.1.   The Problem of Leaf Classification

In Britain, plant breeders who develop a new variety of plant are granted exclusive right to sell that variety for a period of time. One of the requirements imposed by current Plant Breeders Rights legislation is the distinctness of the new varieties. They should be different in at least one characteristic from all existing varieties. The distinctness test is carried out by the National Institute for Agricultural Botany (NIAB). There are over 3000 registered varieties, each represented by ten leaf images, and NIAB receives about 300 new applications to be tested each year. The distinctness tests and leaf classification are currently carried out by NIAB experts based on a number of heuristic features. These features have not been well defined yet.

The main aim of our work has been to ease the process of test and classification by finding the most similar varieties to the input leaf image. We randomly selected a subset of the classified NIAB leaf images to create our prototype database. It consists of 120 leaf images from 12 different varieties of Chrysanthemum. Every image contains one object on a simple background. The 8-bit grey-scale images have been scanned at 400dpi.

Five members of three classes of leaf images are shown in Figure 3.15. Considering these images, one can easily appreciate that the problem of automatic classification of leaf images is a difficult task:

- *Overlaps* of some adjacent parts of leaves are sometimes unavoidable. They create major differences among the boundary contours of similar leaves.

- The between-class similarity is considerable, while the within-class similarity is not adequate. Therefore, misclassification can happen frequently.

- The texture of leaves is rather similar, and texture features such as the parameters derived from the co-occurance matrix are not useful to classify the leaf images. A complex and more sophisticated texture analysis may help, but it will be much more time consuming in comparison to the present system.

- Even in our prototype database the number of classes is notably large, while the number of samples in each class is quite small.

As a result, it is almost impossible to classify these images automatically. However, as the results of our experiments show, it is possible to find the most similar classes to an input image and help the user make the final decision.

9.2. The Problem of Self-Intersection

Fig. 3.16 shows how an intersection occurs. The actual boundary of the object of Fig. 3.16(a) is partly hidden as a result of self-occlusion. In order to extract the boundary of the object, one may ignore the hidden parts and extract the outline of the object as shown in Fig. 3.16(b), where some information is missing. This is a trade-off which reduces the complexity of segmentation. A simple thresholding and a contour tracing algorithm recovers the boundary of the object. The actual boundary, as shown in Fig. 3.16(c) includes three points of self-intersection which need to be recovered interactively. During the process of contour tracing, the user should help the system follow the contours inside the object rather than its boundary.

This contour must finally be represented by appropriate shape descriptors. The CSS image of the contour provides a good source of information which has been used to describe the shape. In section 9.4, we explain how the CSS image of a self-crossing contour is constructed and how useful information is recovered from this image.

9.3. Image Segmentation

The aim of this stage is to recover the boundary of a leaf, taking into account the self-intersection parts. Using a gray level histogram, we find

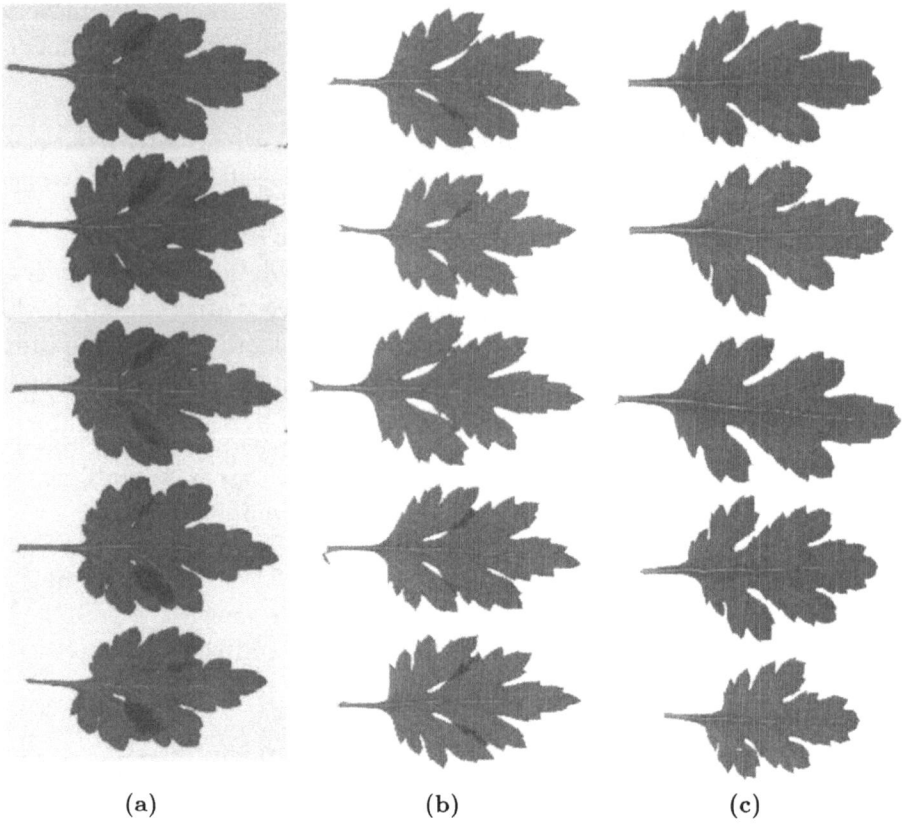

(a)　　　　　　　　　(b)　　　　　　　　　(c)

Figure 3.15. Three classes of images. Due to intraclass similarity and interclass dissimilarity, misclassification is likely to happen.

the best threshold and then separate the object from the background automatically. We then use a simple contour tracing method to extract the boundary of object (see Figure 3.17).

If there were no overlaps, this method would extract the actual boundary of the object. To extract the boundary of an object in presence of an overlap, we employ an interactive method. The gray level image is first segmented into three regions, namely background, overlapping parts of the object, and other parts of the object. The contour tracing method starts from an arbitrary point of the boundary, indicated by the user. The system traces the boundary between the background and the object until it reaches an intersection point, indicated by the user. The system then changes the tracing direction and goes inside the object to trace the boundary between the darker and the lighter parts of the object. This process continues until the whole boundary of the object (including the

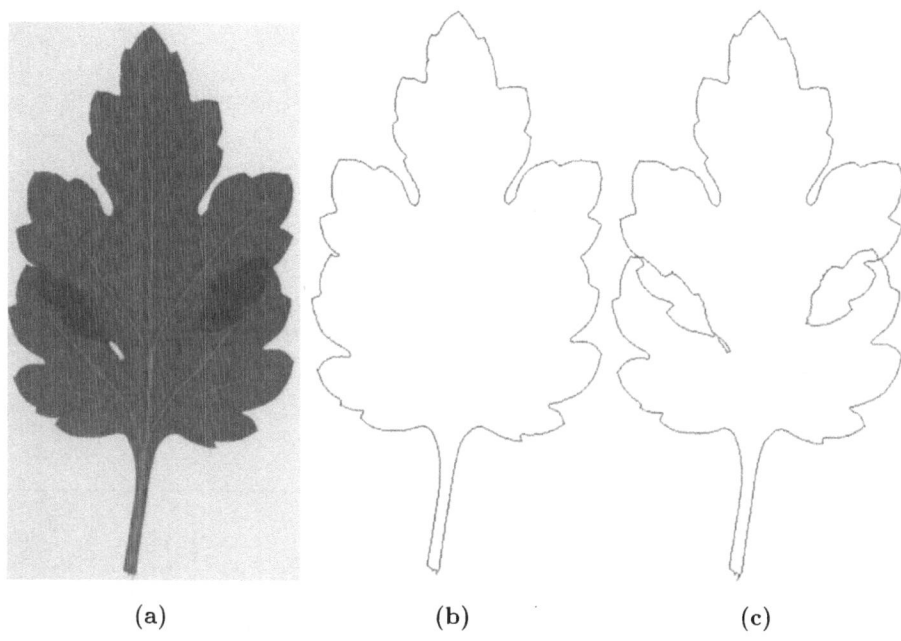

(a) (b) (c)

Figure 3.16. An example of self-intersection. a) Gray level image. b) The boundary of object without considering self-intersection. c) The actual boundary of the object.

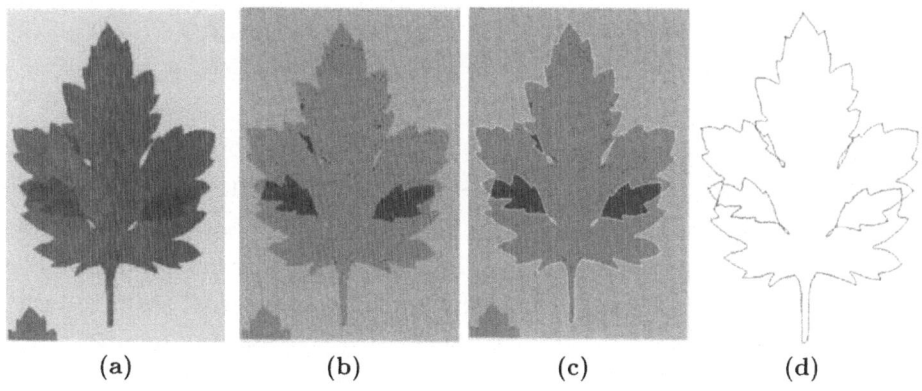

(a) (b) (c) (d)

Figure 3.17. Image segmentation; a) Original image, b) Multi-level thresholding c)Interactive contour tracing, d)Object boundary

internal segments) is extracted. User help is needed at all stages as the internal segments do not show a regular pattern. For example they may create one or two loops, and they may or may not share a border with the background of the image.

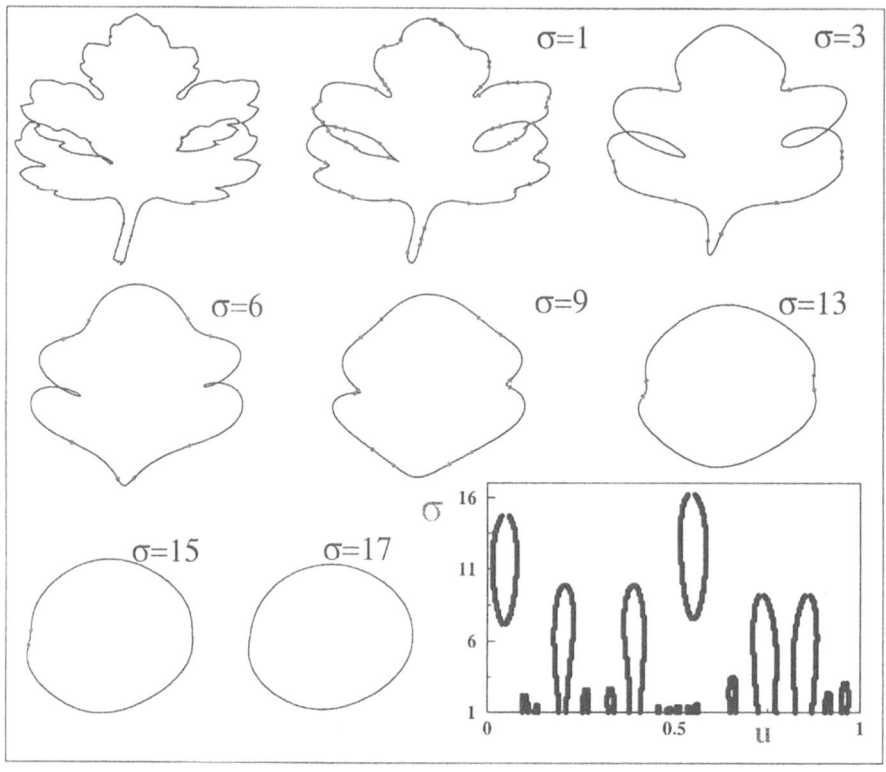

Figure 3.18.   Evolution of a shape with self-intersection

9.4.   CSS Image of Self-Crossing Boundaries

An example of the evolution of such a shape is shown in Fig. 3.18. The original shape includes self-intersections and is seen at top left. For $\sigma = 1$, there are some inflection points inside the internal segments due to small concavities on those segments which disappear in very early stages and before $\sigma = 3$. When $\sigma = 3$, there are no inflection points left on the internal segments. However, the internal loop gradually vanishes and a concavity appears in its place. This concavity, in turn, creates a contour in the CSS image which obviously does not start from $\sigma = 1$. The two branches of this contour are created from the moment that the internal loop vanishes and the concavity is born. A minimum is then created in the CSS image at the relevant σ. Clearly the height of this minimum is proportional to the size of the internal loop.

The location of a CSS minimum in the CSS image conveys information about the internal loop of the shape. The horizontal coordinate of a

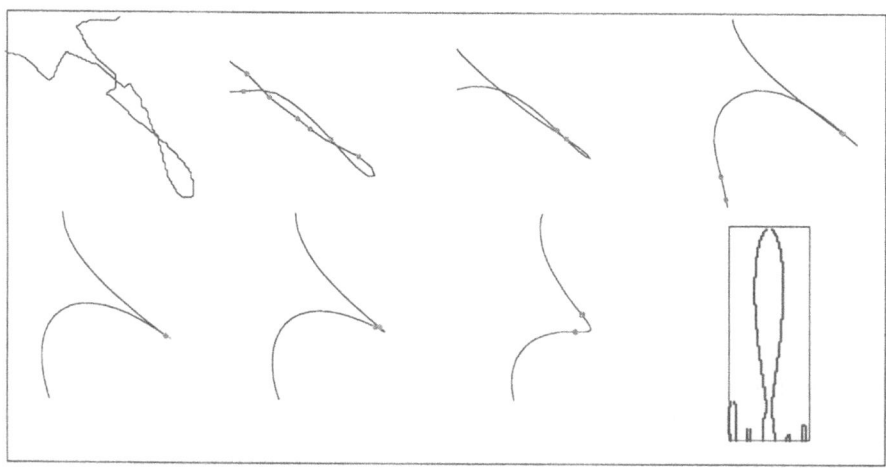

Figure 3.19. Sometimes self-intersection does not create a minimum in the CSS image

minimum reflects the position of the loop on the shape, while the vertical coordinate indicates the size of the loop.

We discovered that while we might expect a minimum to appear in the CSS image for every internal segment, this is not always the case in practice. In fact, if the size of the internal loop is small, the minimum is expected to appear in early stages when the inflection points inside the loop have just disappeared. As a result, the maximum of the CSS contour created by those small features joins the minimum of the CSS contour created when the loop vanishes. This can clearly be seen in Fig 3.19. The original segment of the shape is seen in top left. As σ increases, all inflection points on the segment disappear except for one pair. When the loop disappears, this pair disappears but a new pair of inflection points is born. The result has been shown in the corresponding contour of the CSS image in the lower-right part of the Figure. The thin part of the contour corresponds to the moment when this event takes place.

It should be noted that even if a minimum is created in such situations, due to its small height, its effect is not considerable.

9.5. Recovering Maxima and Minima of CSS Contours

We represent every image in the database with the locations of its CSS contour extrema. For example, in Figure 3.18 there are six maxima and two minima. Therefore, the shape will be represented by 8

Figure 3.20. To find a CSS maximum, we start from the top and scan each row of the CSS image, looking for a pair of black points with a small gap.

pairs of integer numbers. The locations of extrema are not readily available and must be extracted from the image. The CSS contours are usually connected everywhere except sometimes in a neighbourhood of their maxima as seen in Figure 3.20. We find the peaks of both branches of a contour in the CSS image and consider the middle point of the line segment joining the pair as a maximum of the CSS image. Starting from the top, each row of the CSS image is scanned for a black pixel. When found, the search continues in the same row to find another one in the neighbourhood. After finding a maximum of a contour, both branches of the contour are marked and at the same time, search for a possible minimum of the contour begins.

An explanation of our algorithm follows.

1   Start by scanning the second row. If a zero-crossing (black) point is found examine its neighbouring points. If there is no zero-crossing neighbour at the row just above and there is just one zero-crossing neighbour at the following row, go to step 3, otherwise go to step 2.

2   Scan the remaining points of the current row, and start scanning the next row if it is not the last one to be scanned. If a candidate is found go to step 3, and if this is the last row to be scanned , stop.

3   Scan the same row to find another zero-crossing as described in step 1, within a reasonable distance. If the next candidate is not found, mark the first one and go to step 2. If it is found, carry out the following:

- Consider the middle point of the line segment joining the pair as a maximum.

- Mark (delete) all zero-crossings on both branches of the corresponding contour whose maximum has just been found.

- When marking a branch, look for a minimum. A minimum is a point with no black neighbor immediately below.

- If a minimum is found, go to step 2 immediately. Otherwise mark the contour down to the last row to be scanned and then go to step 2.

Note that usually the last few rows of a CSS image represent some information about the existing noise on the actual image contour, so the last row to be scanned in step 2 has $\frac{1}{6}$ the height of the CSS image. Also note that if a candidate in step 2 is in the first few columns, its corresponding point may exist in the last few columns of the same row and vice versa. In this case the search for the matching candidate must include the relevant interval.

9.6.   Matching CSS minima

A straight forward approach to accounting for the minima is to match the two sets of minima, one from the image and the other from the model exactly the same way as we do with two sets of maxima. In this approach, the two sets of maxima are first matched and the corresponding matching value is determined. Then the two sets of minima are matched and the resulting matching value is added to the former to produce the final matching value for the two CSS images. In this approach, the shift parameter which is used to compensate for the effect of change in orientation, may be different for the two parts of matching. In another approach, one may match the maxima and obtain the best shift parameter which is then used for matching the minima. The third approach is to match maxima and minima simultaneously. In other words, when two maxima are matched, their possible corresponding minima are also matched and the Euclidean distance between the minima is added to the Euclidean distance between the maxima. We examined these approaches and discovered that the first approach leads to the best results which are presented in the following section.

9.7.   Experiments and Results

In order to test our method, we used a database of classified leaf images representing different varieties of Chrysanthemum. For many classes of leaves, self-intersection is inevitable during the scanning of the image.

We tested our method on a prototype database of 120 leaf images from 12 different varieties, both with and without considering the self-intersection. The task was to find out whether an unknown leaf belongs to one of the existing varieties or whether it represents a new variety. The system finds the most similar varieties to the input and allows the user to make the final decision. The results indicated a promising performance of the new approach and its superiority over the conventional method.

To reject the dissimilar images based on the global parameters, we first calculate α_e, α_c and α_a as follows:

$$\alpha_e = \frac{|\ e_i - e_m\ |}{max(e_i, e_m)} \qquad \alpha_c = \frac{|\ c_i - c_m\ |}{max(c_i, c_m)} \qquad \alpha_a = \frac{|\ a_i - a_m\ |}{max(a_i, a_m)}$$

where e and c represent the eccentricity and circularity of the boundary and a represents the aspect ratio of the CSS image, while i and m stand for image and model respectively.

According to their definition, α_e, α_c and α_a are between zero and one. We need to choose a threshold for each of these parameters so that if one of them is above the relevant threshold, the corresponding model is rejected. For our system we chose these threshold values:

$$\alpha_{et} = 0.125 \quad \alpha_{ct} = 0.25 \quad \alpha_{at} = 0.30.$$

The CSS matching is then applied to the surviving models to find the leafs most similar to the input query and also make a decision about the class of the input leaf. Each image consisted of just one leaf on a uniform background. The system software was developed using the C language under Unix operating system. The response rate of the system was less than one second for each user query.

To evaluate the method, we considered every image in the database as an input and in each case, asked the system to identify the variety of the input, based on the first k similar images. Obviously, the first output of the system is identical to the input, but the system does not consider it in classification. In fact, we first pull each sample out of the database and classify it based on the remaining classified samples. The best varieties are then selected based on the number of their samples in the best k similar samples. The output of the system is the name of the first 3 classes.

We first used an automatic segmentation method to recover the outline of the objects without considering the self-intersection (see Fig. 3.16). The resulting representations included only the maxima of the CSS images. We also used some global parameters to reject dissimilar candidates prior to CSS matching [9]. The results for different values of k, the number of observed outputs for each query, has been presented in table 3.8. As mentioned earlier, in response to a query, the system returns the top 3 classes which are most similar to the input. The first row of table 3.8 shows the success rate of the system to identify the correct class as its first choice is 75.8% if the judgement is based on the first 5 outputs of the system. This figure is 77.5% for $k = 6$ and so on. The

	$k = 5$	$k = 6$	$k = 7$	$k = 8$	$k = 9$	$k = 10$
1	75.8%	77.5%	78.3%	75.0%	72.5%	70.0%
<= 2	90.0%	88.3%	86.6%	87.5%	88.3%	89.1%
<= 3	91.6%	94.1%	94.0%	94.1%	95.8%	93.2%

Table 3.8. Results of evaluation for different values of k, initial method

	$k = 5$	$k = 6$	$k = 7$	$k = 8$	$k = 9$	$k = 10$
1	81.0%	85.0%	81.7%	79.2%	80.0%	78.3%
<= 2	94.3%	94.2%	93.4%	92.5%	91.7%	92.5%
<= 3	97.6%	97.5%	98.4%	97.5%	95.8%	95.8%

Table 3.9. Results of evaluation for different values of k, improved method.

second row shows the success rate of the system to identify the correct class as its first or second choice, and similarly for the third row.

As this table shows, for this particular database, one may obtain good results. However, as shown in table 3.9 even better results may be achieved by including the self-intersections in the process. It should be noted that the performance of the system is not sensitive to the value of k, specially when we consider the last row of this table. Overall, the superiority of the improved method over the conventional one is seen in these two tables.

10. Conclusions

This chapter addressed the problem of shape similarity retrieval in image database systems. We used the maxima of the CSS image image to represent boundaries of objects. We then observed and explained the problems of this representation with regard to shapes with shallow concavities. In order to achieve a robust shape representation, the Curvature Scale Space image was modified. Three different approaches were introduced and the results were observed on a database of marine creatures. An objective evaluation method which used a classified subset of the SQUID database was also explained and utilized. The results indicated that among the methods proposed in this chapter, the height adjusted method offers the best performance on shapes with shallow concavities. Moreover, it does not negatively affect the good results on shapes with curved segments. The overall increase of 8% in the total performance measure is very promising. This chapter also discussed an application of our system to Chrysanthemum leaf classification. The system demonstrated encouraging results.

Chapter 4

EXTENSIONS OF THE CURVATURE SCALE SPACE IMAGE, AND BEHAVIOR UNDER AFFINE TRANSFORMS/NON-RIGID DEFORMATIONS

The maxima of the CSS image have been used to represent 2-D shapes in different applications. The representation has shown robustness under the similarity transformations. scaling, orientation changes, translation and even noise can be easily handled by the representation and its associated matching algorithm. In this chapter, we examine the robustness of the representation under general affine transforms.

The SQUID database consists of 1100 images of marine creatures. The contours in this database demonstrate a great range of shape variation. A database of 5000 contours has been constructed using 500 real object boundaries and 4500 contours which are the affine transformed versions of real objects. The CSS representation is then used to retrieve similar shapes from this prototype database. The results provide substantial evidence of stability of the the CSS image and its contour maxima under affine transforms.

The method is also evaluated objectively through a large classified database and its performance is compared with the performance of two popular methods, namely Fourier Descriptors and Moment Invariants.

1. Introduction

Considerable amount of information exists in 2-D boundaries of objects which enables us to recognise objects without using further information. However, despite great effort [165], the problem of shape representation in computer vision and image processing is still a very difficult one, and remains so in shape similarity retrieval [67, 179, 237, 243, 270, 279].

A shape is often defined by the x and y coordinates of its boundary points which is subject to change if the distance between camera and object changes or the object is rotated in front of a fixed camera or the origin is altered. As a result, a shape representation must be robust under similarity transformation which includes scaling, changes in orientation and translation. If the camera is allowed to change its viewpoint with respect to the object, the resulting boundary of the object will be deformed. The deformation can be approximated by general affine transforms.

A number of shape representations are suggested to recognise shapes even under affine transformation. Some of them are the extensions of well-known methods such as Fourier Descriptors [20] and Moment Invariants [82, 336, 119]. The methods are then tested on a small number of objects. In both methods, the basic idea is to use a parametrisation which is robust with respect to affine transformation. The arc length representation is not transformed linearly under general affine transform and therefore is replaced by *affine length* [101, 62]. In all cases, the methods are tested on a small number of objects and therefore the results are not reliable. Moreover, almost the same results can be achieved with the conventional methods without modifications [20, 336].

Affine invariant scale space is introduced in [272] and [16]. It generalises the definition of curvature and introduces *affine curvature* (see also [61]). This curve evolution method is proven to have similar properties to curvature evolution [142, 286, 141], as well as being affine-invariant. However, an explicit shape representation has yet to be introduced based on the theory of affine invariant scale space [60]. The prospective shape representation might be computationally complex as the definition of the affine curvature involves higher order derivatives.

We have used the maxima of the CSS image to represent shapes of boundaries in similarity retrieval applications [208, 210]. The representation has been proven to be robust under similarity transformation which includes translation, scaling and changes in orientation. In this chapter, we examine the robustness of the representation under general affine transforms.

The CSS representation finds its roots in curve evolution [220, 17], curvature deformation and heat equation. In fact, the *resampled* curvature scale space [222] implements curvature deformation [142]. This is carried out by convolving each coordinate of a closed planar curve, with a Gaussian function at different levels of scale. At each stage and before being convolved by a larger width Gaussian, the curve is represented in terms of arc length parameter. In *regular* curvature scale space [222] the resampling is not applied. As a result, the process is not equivalent to curvature deformation anymore. However, the implementation is carried out much faster and the representation has shown good performance in object recognition [190], and shape similarity retrieval [9, 210].

The CSS image representation employs the arc length parametrisation which is not affine invariant. As a result, we expect some deviation in the maxima of the CSS image under general affine transformation. It has been shown that affine invariance can only be achieved by an affine invariant parametrisation. Indeed, affine length has been used by a number of authors [20, 82, 272, 336]. We also examine the utility of using affine length instead of arc length to parametrise the curve prior to computing its CSS image [1].

The SQUID database consists of 1100 images of marine creatures. The contours in this database demonstrate a great range of shape variation. A database of 5000 contours has been constructed using 500 real object boundaries and 4500 contours which are the affine transformed versions of real objects.

The following is the organisation of the remainder of this chapter. Section 2 discusses the behaviour of the CSS image under affine transforms. Section 3 is about affine transforms and the procedure we follow to create our large databases. Section 4 is on affine length, and section 5 is on affine curvature. Section 6 discusses some implementation issues related to the integration of affine length and affine curvature into the CSS image. Section 7 presents the experimental results, and section 8 presents a comparison to other methods. Concluding remarks are given in section 9.

2. CSS Image under Affine Transforms

We show that the curvature zero crossing points are preserved under affine transformation. As a result, the overall configuration of the CSS image is also preserved. This can also be verified by recalling the fact that the number and orders of the shape segments remain unchanged under general affine transformation. We then verify these facts through several examples.

The general affine transformation can be represented mathematically with the following equation.

$$\begin{cases} x_a(u) = ax(u) + by(u) + e \\ y_a(u) = cx(u) + dy(u) + f \end{cases} \tag{4.1}$$

where $x_a(u)$ and $y_a(u)$ represent the coordinates of the transformed shape. There are six degrees of freedom in this transformation. Translation and non-uniform scaling are each represented by two degrees of freedom, while change in orientation needs just one parameter. The remaining parameter is related to *shear*. The general affine transformation includes all of these transformations.

THEOREM 4.1 *If Γ_a is the transformed version of a planar curve Γ under general affine transformation, there is a one-to-one correspondence between the curvature zero crossings of Γ and Γ_a.*

Proof. Recall the formula for curvature on the original contour:

$$\kappa(u) = \frac{\dot{x}\ddot{y} - \dot{y}\ddot{x}}{(\dot{x}(u)^2 + \dot{y}(u)^2)^{3/2}} \tag{4.2}$$

In order to derive a formula for curvature on the affine-transformed contour, note that from equation (4.1) we have:

$$\dot{x}_a(u) = a\dot{x}(u) + b\dot{y}(u)$$
$$\dot{y}_a(u) = c\dot{x}(u) + d\dot{y}(u)$$

Similarly,

$$\ddot{x}_a(u) = a\ddot{x}(u) + b\ddot{y}(u)$$
$$\ddot{y}_a(u) = c\ddot{x}(u) + d\ddot{y}(u)$$

And therefore,

$$\kappa_a(u) = \frac{(a\dot{x}(u) + b\dot{y}(u))(c\ddot{x}(u) + d\ddot{y}(u))}{((a\dot{x}(u) + b\dot{y}(u))^2 + (c\dot{x}(u) + d\dot{y}(u))^2)^{3/2}}$$
$$- \frac{(a\ddot{x}(u) + b\ddot{y}(u))(c\dot{x}(u) + d\dot{y}(u))}{((a\dot{x}(u) + b\dot{y}(u))^2 + (c\dot{x}(u) + d\dot{y}(u))^2)^{3/2}}$$

Hence,

$$\kappa_a(u) = \frac{(ad - bc)(\dot{x}\ddot{y} - \dot{y}\ddot{x})}{((a\dot{x}(u) + b\dot{y}(u))^2 + (c\dot{x}(u) + d\dot{y}(u))^2)^{3/2}} \tag{4.3}$$

Now compare the numerators of equations (4.2) and (4.3) and observe that the curvature zero-crossings of the original and the transformed shapes have a one-to-one correspondence.   ⋈

As this theorem shows, in all cases, even those with severe deformations, the number of the CSS contours are preserved. The order of shape segments is also preserved under general affine transforms. As a result, the number and order of the CSS contours are preserved. We observed that even the heights of the corresponding CSS contours are not significantly affected. However, small changes in the locations of maxima are inevitable. As a result of the changes in the CSS maxima, the matching value between a shape and its transformed version is not zero.

3. Affine Transforms and Affine Databases

The affine transformed version of a shape can be represented by equation 4.1, where $x_a(u)$ and $y_a(u)$ represent the coordinates of the transformed shape [2]. Translation is represented by e and f, while rotation, scaling and shear are reflected in the remaining four parameters. Since the CSS image is invariant with respect to translation, we do not consider e and f in our studies. Therefore matrix A defined as:

$$A = \begin{pmatrix} a & b \\ c & d \end{pmatrix}$$

can be assumed to represent an affine transform without loss of generality.

Now if we choose

$$a = d$$

$$c = -b$$

and set e and f to zero, then the transformation is equivalent to a similarity transform. Any deviation from these constraints will cause some affine deformation.

We introduce γ as a measure of the degree of affine deformation, where each parameter such as a, can be replaced by either $(1 - \gamma)a$ or $(1 + \gamma)a$. In particular, if we choose $\gamma = 0.5$, we can create affine transformed versions of our existing shape contours. Five different sets of affine parameters are constructed in this case as follows.

$$A_1 = \begin{pmatrix} 20 & 10 \\ -10 & 20 \end{pmatrix}$$

$$A_2 = \begin{pmatrix} 30 & 15 \\ -5 & 10 \end{pmatrix}$$

$$A_3 = \begin{pmatrix} 10 & 5 \\ -15 & 30 \end{pmatrix}$$

$$A_4 = \begin{pmatrix} 30 & 5 \\ -15 & 10 \end{pmatrix}$$

$$A_5 = \begin{pmatrix} 10 & 15 \\ -5 & 30 \end{pmatrix}$$

where A_1 represents a similarity transform and the others represent four different affine transformations. By choosing larger values for γ (such as 0.7), we can create more affine-deformed shapes. Figure 4.1 shows an

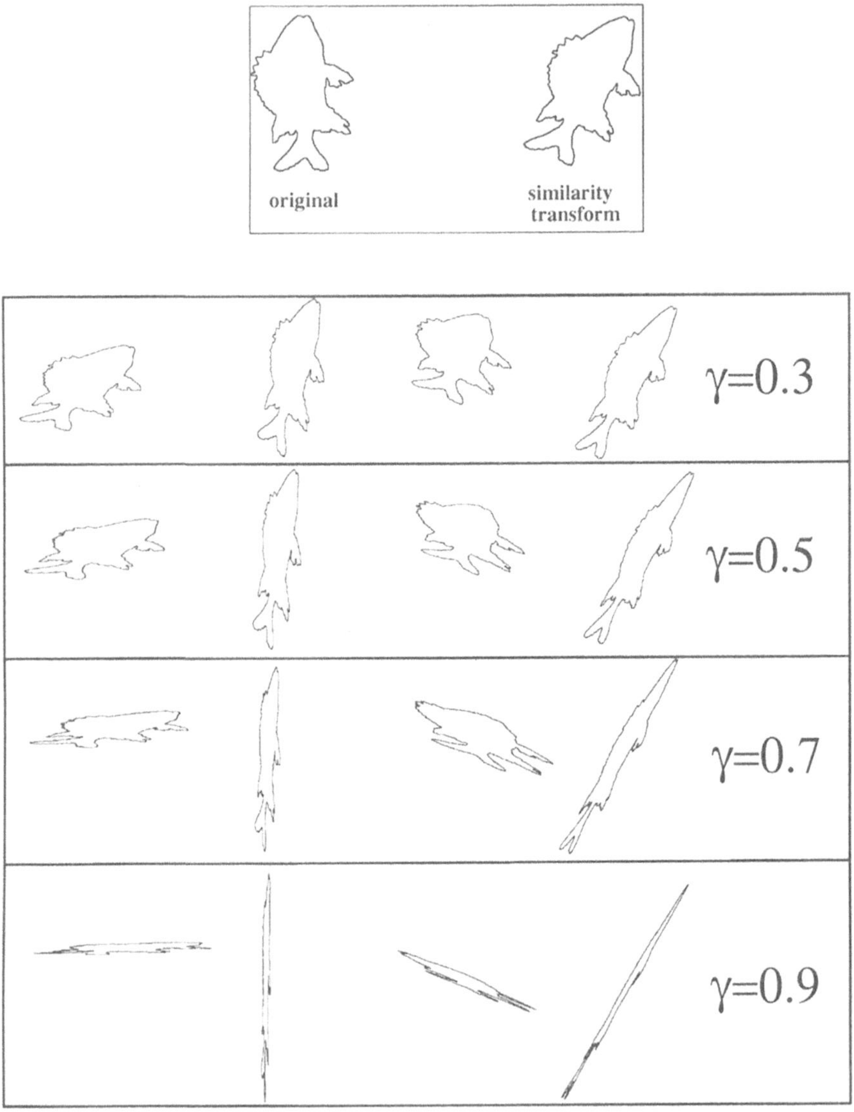

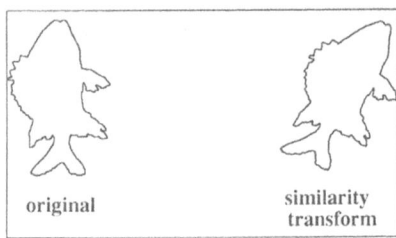

original

similarity
transform

γ=0.3

γ=0.5

γ=0.7

γ=0.9

Figure 4.1.   Affine transformation can be considered as a change in camera viewpoint. The original is in top left, top right shows a similarity transform with $a = 20, b = 10, c = -10$ and $d = 20$. The first through fourth rows show four different groups of affine-transformed shapes with $\gamma = 0.3$, $\gamma = 0.5$, $\gamma = 0.7$ and $\gamma = 0.9$ respectively. Note that we have re-sized the original shape so that comparison is possible.

example illustrating how a shape is deformed under each of these transformations. Four different values, 0.3, 0.5, 0.7 and 0.9 are considered for γ.

By choosing four different values for γ, we created four different databases, each consisting of 5500 shapes out of which 1100 were our original database of marine creatures and 4400 were the affine transformed versions of the shapes. Different experiments were carried out on these databases to investigate the performance of the CSS representation under affine transformation [203, 202].

Now we consider several examples of CSS images under affine transforms. The shapes of Figure 4.1 have been shown in Figure 4.2 together with their corresponding CSS images. The effects of change in orientation have not been reflected in the CSS images of this figure so that comparisons can be made more easily. All CSS contours experience the same circular shift as a result of change in orientation which is easily detected during the matching algorithm.

As this figure shows, in all cases, even those with severe deformations, the numbers and orders of the CSS contours are preserved. Even the heights of the corresponding CSS contours are not significantly affected. However, small changes in the locations of maxima are inevitable. Note for example that the largest maxima in the CSS images of the left column are different from the largest maxima of the other columns. Note also the last row of this figure which shows the shapes and their corresponding CSS images for $\gamma = 0.9$. The deformations of shapes are noticeably severe, however, even small contours of the CSS images are preserved and they are quite similar to the CSS image of the original shape. This demonstrates the robustness of the CSS image under affine transformation. As a result of the changes in the CSS maxima, the matching value between a shape and its transformed version is not zero. We have already used the CSS representation in shape similarity retrieval [11] and learned that a matching value between zero and 0.5 reflects a very good and between 0.5 and 1.0 a reasonably good measure of similarity between the two shapes. Therefore, we expect that in response to a query, the system should be able to identify the transformed versions of the query and retrieve them from the database.

In order to verify this expectation, we measured the matching value corresponding to each shape and its transformed versions. Starting with 1100 shapes in the SQUID database and 4 transformed versions for each shape, we obtained 4400 values, each related to an original shape and one of its transformed versions. We observed that for $\gamma = 0.3$, in 98.3% of cases the matching value was less than 0.5. This can be seen in Figure 4.3(a), where an accumulative histogram of these values is presented

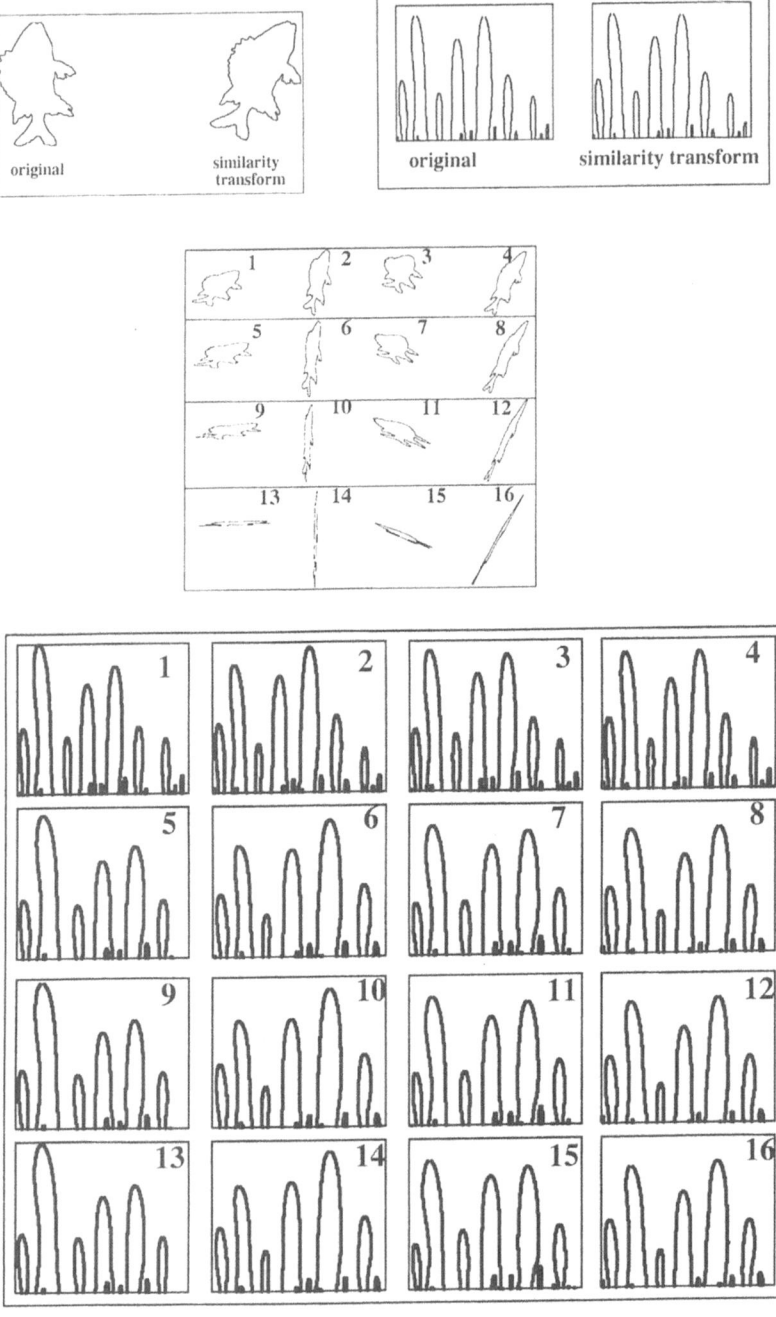

Figure 4.2. Affine transformed shapes and their associated CSS images. Above: similarity transform does not change the CSS image. Below: affine transform causes minor changes in the CSS images. Note that the CSS images are shifted to compensate the effects of orientation changes.

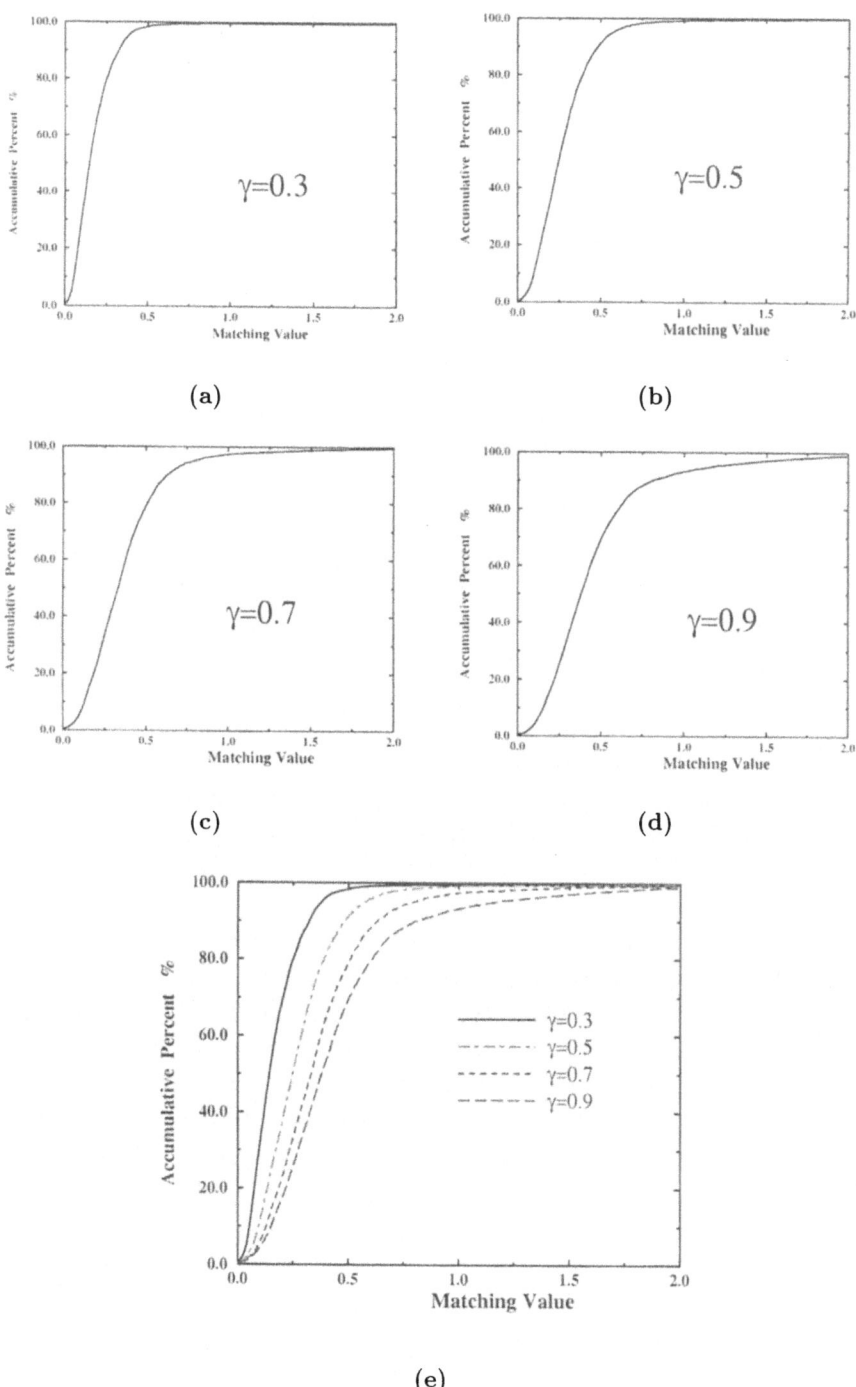

Figure 4.3. Matching value between a shape and its affine-transformed versions is small.

	$\gamma = 0.3$	$\gamma = 0.5$	$\gamma = 0.7$	$\gamma = 0.9$
n= 10	99.6%	90%	77%	69%
n= 20	98.6%	95%	90%	77%
n= 30	100%	98%	94%	88%
n= 40	100%	99%	96%	90%

Table 4.1. The average success rate for retrieval of the affine transformed versions of a shape among the first n outputs.

for $\gamma = 0.3$. A similar histogram is presented in Figure 4.3(b) where $\gamma = 0.5$ indicates the degree of deviation in affine transform parameters. In this case, although the deformation is considerable, the changes in CSS maxima and the corresponding matching values are not significant. In 91.1% of cases, the matching value is less than 0.5. This figure is 80.0% and 70.6% for $\gamma = 0.7$ and $\gamma = 0.9$ presented in Figures 4.3(c) and (d) respectively.

As Figure 4.3(a) shows, for $\gamma = 0.3$, the matching value between a shape and its transformed versions is always less than 1.0. For $\gamma = 0.5$, in 99.1% of cases the matching value is less than 1.0. This figure is 97.3% and 93.2% for $\gamma = 0.7$ and $\gamma = 0.9$, respectively.

In conclusion, this experiment demonstrates the robustness of the CSS maxima under affine transformation, even in presence of sever deformations.

In order to evaluate the performance of the CSS representation under affine transformation, we carried out another experiment using the SQUID database. Every original shape was selected as the input query and the first n outputs of the system were observed. We found out that for $\gamma = 0.3$, almost all transformed shapes appear in the first 10 outputs of the system. For $\gamma = 0.5$, on average 95% of transformed shapes of the input query are among the first 20 shapes retrieved by the system as shapes similar to the input query. This figure is 90% if we look at the first 10 outputs. The results are summarised in table 4.1. As this table shows, for $\gamma = 0.9$, and $n = 10$, the success rate of the system is about 70%, which may seem a bit low at first glance, however, this result means that if the database consists of 1100 original shapes and 4400 transformed shapes (a total of 5500) and one of the original shapes is considered as the input query, almost 3 of its 4 transformed shapes appear in the first 10 outputs of the system. This result appears to be much more interesting when we consider the huge deformation in the transformed shapes in the case of $\gamma = 0.9$ in Figure 4.1.

Although the general affine transform in equation (4.1) contains translation, scaling, change in orientation and shear, it is possible to apply only one of these transforamtions at a time. scaling, change in orientation and shear are represented by the following matrices:

$$A_{scaling} = \begin{pmatrix} S_x & 0 \\ 0 & S_y \end{pmatrix}$$

$$A_{rotation} = \begin{pmatrix} cos\theta & -sin\theta \\ sin\theta & cos\theta \end{pmatrix}$$

$$A_{shear} = \begin{pmatrix} 1 & k \\ 0 & 1 \end{pmatrix}$$

If S_x is equal to S_y, $A_{scaling}$ represents a uniform scaling. A shape is not deformed under rotation, uniform scaling and translation. However, non-uniform scaling and shear contribute to the shape deformation under general affine transformation. We examine the performance of the CSS representation under shear transform. The measure of shape deformation depends on parameter k, *shear ratio*, in matrix A_{shear}. In the present form of the matrix A_{shear}, x-axis is referred to as the *shear axis* since the shape is pulled toward this direction.

Figure 4.4 shows the effects of affine transformation on shape deformation. In this figure, shear ratio is selected as $k = 1$. In order to achieve different shear axes, we have changed the orientation of the original shape prior to applying the pure shear transformation. The values of θ range from 20° to 180°, with 20° intervals. As this figure shows, the deformation is severe for $k = 1.0$. For larger values of k (such as 1.5 and 2.0), the deformation is much more severe.

In order to create different databases, we chose different values for shear ratio, 1.0, 2.0 and 3.0. We then applied the transformation to a database of 500 original object contours from the SQUID database. From every original contour, we obtain 9 transformed shapes corresponding to different values of θ. Therefore, each database consists of 500 original and 4500 transformed shapes. We then carried out a series of experiments on these databases to verify the robustness of the CSS image representation under affine transformations. Section 7 describes those experiments.

4. Affine Length

In order to construct the CSS image of a planar curve, the curve is first parametrized by its normalized arc length parameter. The normalized

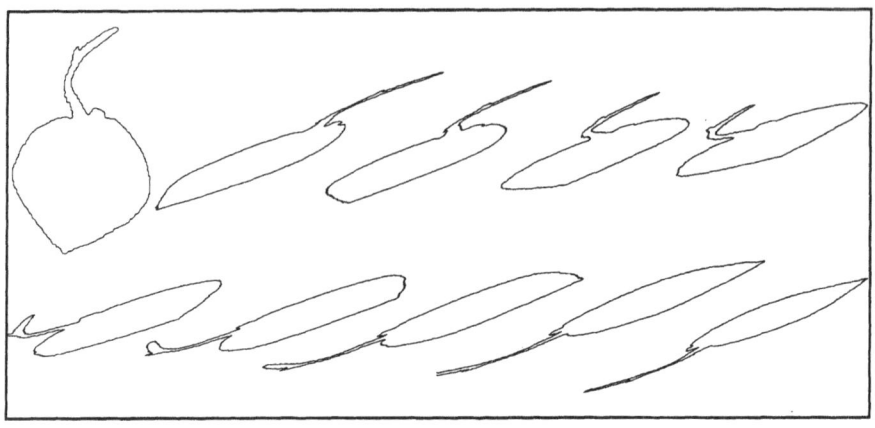

Figure 4.4.   The deformation of shapes is considerable even with $k = 1$ in shear transform. The original shape is presented in top left. Others represent transformation with $k = 1$ and $\theta = 20°, 40°, ..., 160°, 180°$.

arc length parameter can be calculated using the following formula:

$$s(u) = \frac{\int_0^u (\dot{x}^2 + \dot{y}^2)^{\frac{1}{2}}}{\int_0^1 (\dot{x}^2 + \dot{y}^2)^{\frac{1}{2}}}$$

It is obvious that s is not preserved under the transformation described by equation (4.1). As a result, the normalised arc lengths of the corresponding points on the original and the transformed shapes differ. In order to achieve an affine invariant parametrization, arc length is usually replaced by affine length which has the following definition:

$$\tau(u) = \frac{\int_0^u (\dot{x}\ddot{y} - \ddot{x}\dot{y})^{\frac{1}{3}}}{\int_0^1 (\dot{x}\ddot{y} - \ddot{x}\dot{y})^{\frac{1}{3}}}$$

The main disadvantage of affine length is that its computation requires higher order derivatives. However we can parametrise the curve using this formula.

The CSS image can be constructed using affine length instead of arc length [205]. In the regular CSS image, only the initial representation is affine length and re-parametrization is not applied as the curve is smoothed. In section 7, we present the results of our experiments which confirm the advantage of using affine length over arc length.

5.   Affine Curvature

In computer vision and image processing, we tend to look for those descriptors of a shape which are invariant under specific transformations. For example, curvature is invariant under similarity transforms (rotation, translation and scaling). It remains unchanged under rotation and translation; and its value under uniform scaling is a linear function of scale factor. However, under general affine transforms, the change in curvature is not a linear function of the transformation matrix. *affine curvature* has been defined as an alternative for curvature which changes linearly under affine transforms.

The definition of affine curvature is based on *affine length* parametrisation which is a replacement for arc length parametrization as follows:

$$\tau(u) = \int_0^u [\dot{x}(t)\ddot{y}(t) - \ddot{x}(t)\dot{y}(t)]^{\frac{1}{3}} dt \qquad (4.4)$$

Affine curvature is then defined as follows:

$$\mu(\tau) = x^{''}(\tau)y^{'''}(\tau) - x^{'''}(\tau)y^{''}(\tau). \qquad (4.5)$$

Note the similarity of equation (4.5) and the numerator of (4.2). It is more convenient to compute affine curvature if it is expressed as a function of an arbitrary parameter. In order to express μ as a function of an arbitrary parameter u, we first determine the second and third derivatives of x and y:

$$x^{''}(\tau) = \frac{d^2 x}{du^2} \left(\frac{du}{d\tau}\right)^2 + \frac{dx}{du}\frac{d^2 u}{d\tau^2}$$

and

$$x^{'''}(\tau) = \frac{d^3 x}{du^3}\left(\frac{du}{d\tau}\right)^3 + 3\frac{d^2 x}{du^2}\frac{du}{d\tau}\frac{d^2 u}{d\tau^2} + \frac{d^3 u}{d\tau^3}\frac{dx}{du} \qquad (4.6)$$

The same formulae are derived for $y^{''}(\tau)$ and $y^{'''}(\tau)$ by replacing x with y.

From equation (4.4):

$$\frac{d\tau}{du} = (\dot{x}\ddot{y} - \ddot{x}\dot{y})^{\frac{1}{3}}$$

and

$$\frac{du}{d\tau} = \frac{1}{(\dot{x}\ddot{y} - \ddot{x}\dot{y})^{\frac{1}{3}}} \qquad (4.7)$$

$$\frac{d^2 u}{d\tau^2} = \frac{d}{d\tau}\left(\frac{du}{d\tau}\right) = \cdots = \frac{\tilde{x}\dot{y} - \dot{x}\tilde{y}}{3(\dot{x}\ddot{y} - \ddot{x}\dot{y})^{\frac{5}{3}}} \qquad (4.8)$$

where $\dot{x}$, $\ddot{x}$ and $\tilde{x}$ denote the first, second and third derivatives of x with respect to u, respectively. Derivatives of y are defined similarly. Note also that:

$$\frac{d^3u}{d\tau^3} = \frac{d}{d\tau}\left(\frac{d^2u}{d\tau^2}\right) = \cdots = \frac{3(\hat{x}\dot{y} - \dot{x}\hat{y} - \ddot{x}\tilde{y} + \tilde{x}\ddot{y})(\dot{x}\ddot{y} - \ddot{x}\dot{y}) - 5(\dot{x}\tilde{y} - \tilde{x}\dot{y})^2}{9(\dot{x}\ddot{y} - \ddot{x}\dot{y})^3}$$

(4.9)

where $\hat{x} = \frac{d^4x}{du^4}$ and $\hat{y} = \frac{d^4y}{du^4}$. By substituting equations (4.7), (4.8) and (4.9) in equation (4.6), and the results in (4.5), the final formula for affine curvature as a function of an arbitrary parameter u is obtained:

$$\mu(u) = \frac{\ddot{x}\tilde{y} - \tilde{x}\ddot{y}}{(\dot{x}\ddot{y} - \ddot{x}\dot{y})^{\frac{5}{3}}} - \frac{3(\hat{x}\dot{y} - \dot{x}\hat{y} - \ddot{x}\tilde{y} + \tilde{x}\ddot{y})(\dot{x}\ddot{y} - \ddot{x}\dot{y}) - 5(\dot{x}\tilde{y} - \tilde{x}\dot{y})^2}{9(\dot{x}\ddot{y} - \ddot{x}\dot{y})^{\frac{8}{3}}}$$

(4.10)

Equation (4.10) presents an explicit formula for affine curvature as a function of an arbitrary parameter u [204]. The following facts can be observed about this equation:

- The formula is rather complicated and involves up to fourth derivatives of x and y. This can be problematic in digital implementation of the formula, where the precise approximation of high order derivatives is a difficult task. These derivatives tend to cause instability and involve larger errors.

- Unlike regular curvature, affine curvature does not have a straight forward physical interpretation.

- Affine curvature is not defined at inflection points. The denominator in equation (4.10) (without a power), is the same as the numerator of equation 4.2 without a power. As a result, whenever conventional curvature goes to zero, affine curvature goes to infinity.

- The main advantage of this descriptor is expected to be its invariance under general affine transforms. In fact, it is not absolute invariant: the value of affine curvature does not remain constant under affine transformation. However, it is a linear function of the determinant of matrix A. More precisely, by substituting for x and y in equation (4.10) with x_a and y_a from equation (4.1), we obtain the following result:

$$\mu_a = \mu \times \frac{1}{[det(A)]^{\frac{2}{3}}}$$

where μ and μ_a denote affine curvature on the original and affine transformed curve respectively and *det* denotes determinant. The difference between the affine curvature of the original and affine transformed shape is represented by a scale factor. This suggests that affine curvature can be used in conjunction with the maxima of the CSS image to improve the performance of shape similarity retrieval under affine transforms.

6.   Implementation Issues

In this section, we explain how affine length and affine curvature are used to enrich the CSS representation.

6.1.   Affine Length

As mentioned earlier, in order to construct the CSS image using arc length parametrization, we first re-sample the curve by 200 equally distant points. The resulting x and y coordinates of these points are then convolved with Gaussian filters to smooth the curve. The same procedure is followed in the case of affine length parametrisation. The only difference is that the distance between two points is defined by the following formula:

$$\int_0^u (\dot{x}\ddot{y} - \ddot{x}\dot{y})^{\frac{1}{3}}$$

Apart from this, there are no other changes and the whole process including the extraction of the CSS maxima and the matching algorithm will be the same.

6.2.   Affine Curvature

To construct the CSS image of a digital curve, its curvature zero-crossing points should be determined at different levels of smoothing. In Gaussian smoothing, there will be two curvature zero-crossings on every concave or convex part of the shape and as the curve becomes smoother these points approach each other and create a contour in the corresponding CSS image. When the segment is filled, the two points join and represent the maximum of the relevant contour. The height of this contour then reflects the depth and size of the concavity or convexity.

We examined the value of affine curvature on each segment of the shape at different levels of scale. For each segment affine curvature approaches to $-\infty$ at end-points, as they are curvature zero crossings. As a result, affine curvature always has a maximum on the segment. We examined the maximum value of affine curvature on each segment at different levels of scale and found out that this value increases as the curve becomes smoother. For each segment of the shape, we then found the level of smoothing where the maximum value of affine curvature exceeds a threshold value. We denote this value by y_{ac}. The sizes of all curves of the database were adjusted to have the same perimeter prior to the procedure of CSS image construction and affine curvature computation.

In the new representation, each segment is represented by a three dimensional vector $(x_m, y_m, y_m - y_{ac})$, where (x_m, y_m) indicate the location of the maximum of its relevant CSS contour.

A matching algorithm which compares two such representations and assigns a match value as the similarity measure of the two corresponding shapes can be found in [12].

7.   Experiments and Results

This section describes experiments on the CSS image when incorporating affine length and affine curvature.

7.1.   Affine Length

We examined the performance of the representations through two different experiments. The first one was performed on the databases of section 3. Every original shape was selected as the input query and the first n outputs of the system were observed to see if the transformed versions of the query are retrieved by the system. Both regular and resampled CSS representations with arc length and affine length parametrisation were examined. The results indicated that using affine length parametrisation instead of arc length improves the performance of both representations.

Considering each original (not affine-transformed) shape as an input query, we observed the first n outputs of the system and determined m, the number of outputs which are the affine-transformed versions of the input. The success rate for a particular input is calculated as follows:

$$Success\ rate\ for\ an\ input\ query = \frac{m}{m_{max}} \times 100 \qquad (4.11)$$

where m_{max} is the maximum possible value of m. Note that m_{max} is equal to n if $n \leq 10$; if not, m_{max} is equal to 10. The success rate of the system for the whole database will be the average of the success rates for each input query.

It is obvious that the success rate is a function of n, the number of observed outputs. If n is large, it is more likely that all 10 affine transformed versions of the input query appear among the outputs. On the other hand; the first few outputs of the system almost always are

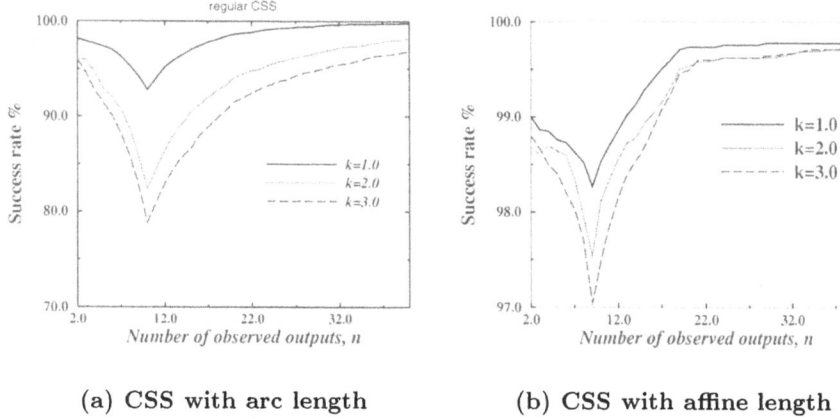

(a) CSS with arc length   (b) CSS with affine length

Figure 4.5. Identifying transformed versions of the input query. k is the shear ratio and represents the measure of deformation.

the transformed versions of the input. As a result; if n is small, then the success rate is large. The success rate is also a function of k, the shear ratio. The larger values of k result in more deformation and perhaps lower success rate.

We chose different values for n, ranging from 2 to 40, and in each case found the average success rate of the system for *all* 500 original shapes. The same experiment was carried out on four different CSS representations, including regular and resampled CSS images with arc or affine length parametrization.

The results are presented in Figure 4.5(a) and 4.5(b). Each Figure includes three curves associated with three values of k, the shear ratio. Each curve shows the average success rate for a particular type of the CSS representation and for different values of n, the number of observed outputs.

Starting from 4.5(a), we observe that the regular CSS image shows good results. For example, with $k = 1.0$ and in spite of severe deformation, more than 93% of outputs are always the affine transformed versions of the input query. This figure drops to 80% as k increases to 3.0 but it is still reasonably large.

With affine length parameterisation, the method shows much better results. Almost all affine transformed versions of an input query appear among the first outputs of the system. The results are also robust with respect to k, the shear ratio.

In conclusion we observe the following:

- Regular CSS image is almost robust with respect to affine transforms.

- Since the transformation is applied mathematically, the effects of pre-processing noise has not been considered. In real world applications [3], when the object boundaries must be extracted from images taken from different camera viewpoints, noise can affect the object boundaries dramatically. However, we expect that using affine length instead of arc length improves the performance of the method even in presence of such noise.

In the second experiment, we selected 76 shapes and classified them into 10 different groups as presented in Figure 4.6. We produced 9 transformed shapes from each original shape. As a result, a group with 8 members had 80 members after adding the affine transformed shapes. The whole database then had 760 shapes in 10 different groups.

In order to assign a performance measure to the method using this classified database, we chose every member of each group as the input query and asked the system to find the n most similar shapes from the database. We then observed the number of outputs, m, which are from the same group as the input. The *success rate* of the system is described by equation (4.11), where n is the number of observed outputs and m_{max} is the maximum possible value of m. Note that m_{max} is equal to n if n is less than the number of group members; if not, m_{max} is equal to the number of group members. The success rate of the system for the whole database is the average of the success rates for each input query. We chose different values for n and for each case, computed the success rate of the methods. The results are presented in Figure 4.7(a) and 4.7(b). We observe that both arc length and affine length parameterizations lead to good results. However, the results for CSS with affine length are better and are not sensitive to the measure of affine deformation.

7.2.   Affine Curvature

The results are presented in Figure 4.8, where the performance of the new technique is compared with the conventional method. The shear ratio, k has been chosen as 1.0, and as a result the deformation is severe. We observe that for lower values of n ($n < 10$), almost all outputs are in the same group as the input query and the success rate is high.

As Figure 4.8 shows, the performance of the new method is better than the conventional one. However, the difference is not very significant. This modest improvement in performance should be weighed against the added complexity of computation of affine curvature in user applications.

8.   Comparison to Other Methods

Fourier Descriptors [315] and Moment Invariants [72, 118] have been widely used as shape descriptors in similarity transform environments.

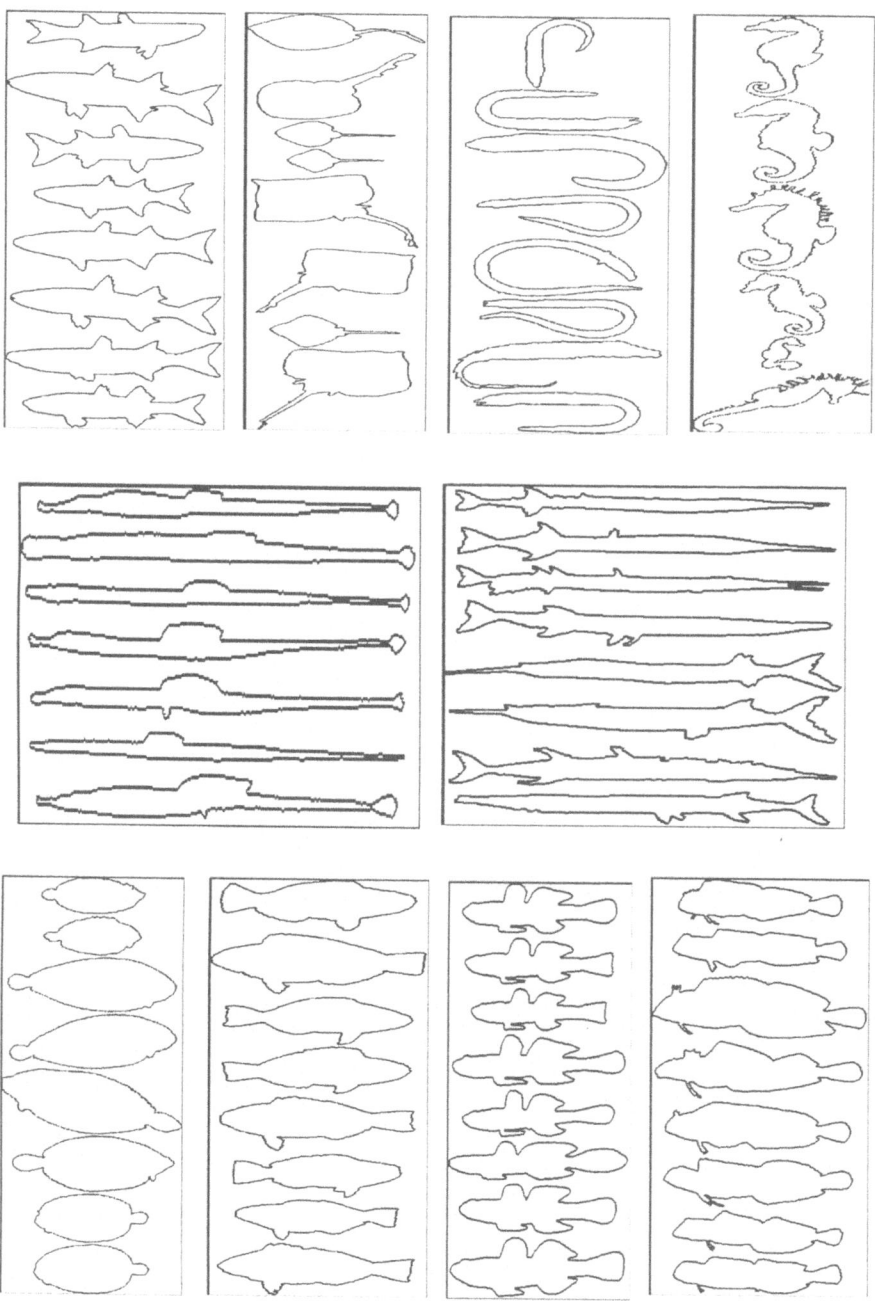

Figure 4.6. Classified database used for objective evaluation consists of 10 different classes. Note that for each original shape, nine affine transformed versions are also generated. Therefore, the actual size of the database is 10 times larger than the size of this database.

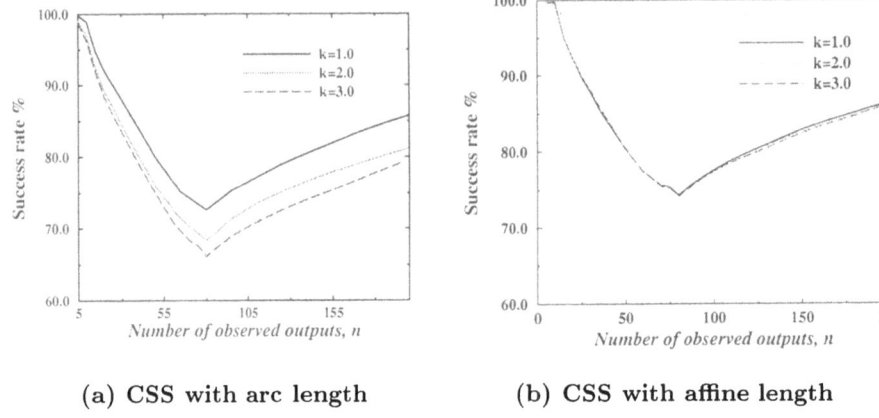

(a) **CSS with arc length**   (b) **CSS with affine length**

Figure 4.7. The results of objective evaluation for CSS with arc length and affine length parametrizations.

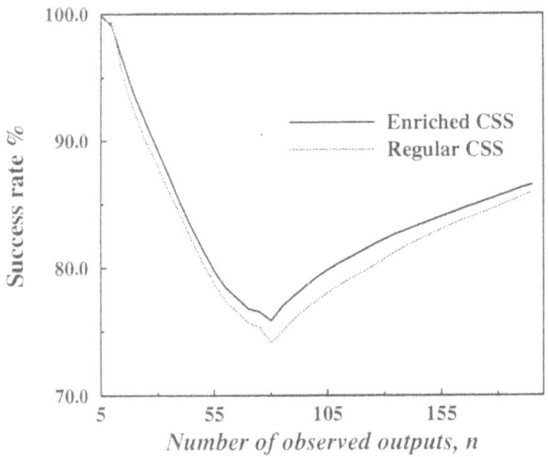

Figure 4.8. Comparison between the performance of the regular and affine curvature enriched CSS

Both methods represent the global appearance of shapes in their most important components. For example, the largest magnitude component of Fourier Descriptors represents the dimensions of the ellipse best fitted to the shape. Since affine transformation changes the global appearance of a shape, it is expected that the performance of these methods is negatively affected under the transformation.

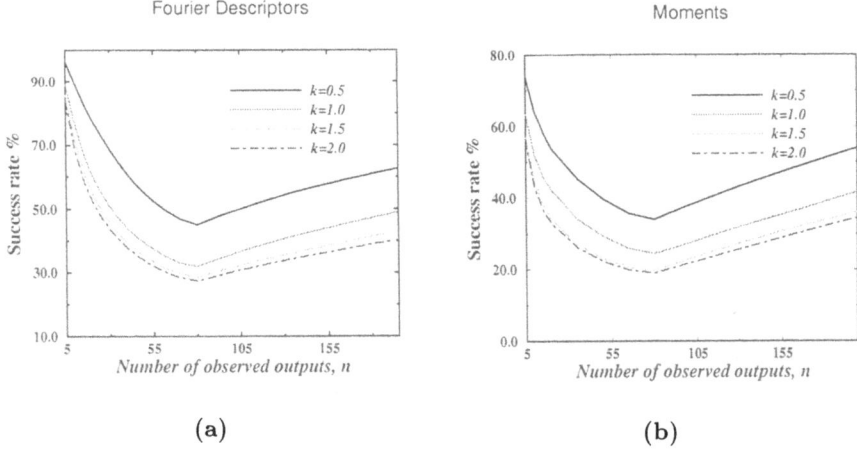

Figure 4.9. The results of objective evaluation for other methods.

The modified versions of these methods have been introduced to deal with affine transformation [20, 82, 336]. They are used in object recognition applications and clustering a small number of shapes. However, it has been shown that the improvements in modified versions are not very significant in comparison to the conventional versions. Considering the fact that the implementation of the modified versions is not a straight forward task, we decided to examine the conventional versions of these methods and compare the results to the results of our method. We can observe that the difference between the performance measure of our method and the performance measure of each of these methods is very large. Even if we allow $10 - 15\%$ improvement for the modified versions of the methods, their performance are still well behind the performance of the CSS representation.

We implemented the method described in [315] to represent a shape with its normalised Fourier Descriptors. Every original shape and its transformed versions were represented by their first 20 components. The Euclidean distance was used to measure the similarity between the two representations.

The results for different values of k are presented in Figure 4.9(a). The classified database shown in figure 4.6 was also used for this experiment. The minimum of the performance measure for different values of k is around 30%, compared to 70% for the CSS method. At the same time, the slope of the plots after the minimum points is not high which indicates that most of the missing models are not ranked even among the first 200 outputs of the system.

For Moment Invariants, each object is represented by a 12-dimensional feature vector, including two sets of normalised Moment Invariants [72], one from object boundary and the other from a solid silhouette. The Euclidean distance is used to measure the similarity between different shapes. The results are presented in Figure 4.9(b). Comparing this plot with the plots of Figure 4.7 for the same shear ratio, a large difference between the performance of the CSS representation and this method is observed.

9.   Concluding Remarks

The maxima of CSS image have been used to represent closed planar curves in shape similarity retrieval under affine transforms. In conventional forms, arc length parametrisation is used to resample the curve at first stage of smoothing. We also examined the utility of using affine length instead of arc length to parametrise the curve prior to computing its CSS image. In different sections of this chapter, we reviewed the background of the representations as well as parametrisations. We then carried out a number of experiments to compare the performances of our shape similarity systems using different approaches.

We constructed a database of 5000 contours using 500 real object boundaries of marine creatures from the SQUID database and 4500 contours which are the affine transformed versions of those boundaries.

We observed that the performance of the CSS representation in shape similarity retrieval under affine transforms is promising and further improves by using affine length parametrisation as well as affine curvature. It is believed that the method can also be used to represent and recognise free-form 3-D objects using a small number of silhouette contours obtained from different view points [3]. Since the CSS representation is robust under affine transform, it can be used to represent each silhouette contour.

In conclusion, the CSS representation and its associated matching algorithm can be used for the purpose of shape similarity retrieval in an affine transformed environment and also for 3-D free-form object recognition.

Chapter 5

FREE-FORM 3-D OBJECT RETRIEVAL FROM ARBITRARY VIEWPOINTS

This chapter addresses the problem of retrieval of free-form 3-D objects using arbitrary views of those objects as queries. It also addresses the issue of automatic selection of the best and the optimum number of views for each object in multi-view 3-D object representation and retrieval. The object boundary in each view is a 2-D shape and is represented by the maxima of its CSS image contours. After representing each object by an optimal number of views, an unknown object is recognised using a single image taken from an arbitrary viewpoint using the same shape representation.

The chapter also presents a novel method for fusion of results from multiple shape descriptors. The utilisation of this method for multi-view 3-D object representation and retrieval has been explored. The object boundary of each view is represented effectively using the CSS technique, Moment Invariants, and Fourier Descriptors. It has been shown that the results obtained from the fusion method are superior to the results obtained from any single technique.

The method has been tested on a collection of 3-D objects consisting of 15 aircrafts of different shapes. Each object has been modelled using an optimal number of silhouette contours obtained from different view points. This number varies depending on the complexity of the object and the measure of expected accuracy. A comprehensive analysis of the performance of the system has been given in this chapter as the number of views varies.

A complete system for free-form 3-D object retrieval using partially occluded or noisy views of those objects has also been presented in this chapter. Substantial experiments show that this system is also robust and reliable for retrieval despite background clutter, noise and partial occlusion of the objects in query images.

1.   Introduction

There are two major methods of data acquisition in 3-D object representation. In laser-based systems, range data is produced which essentially records the distances between the camera and different points of the object. In image-based systems, a CCD camera is used to produce 2-D images from the object. In both methods, the 3-D object is finally represented by a number of features extracted either from the range data or from 2-D images. While the range data provides more accurate information about the surface of the 3-D object, the laser-based systems are more expensive and the related recognition methods are more time-consuming. The computational cost of a method has recently become much more important in search and retrieval from large databases. Although hierarchical methods, and different indexing techniques have been introduced to narrow down the search space in such applications, the need for rapid matching methods to find the best matches among the remaining candidates still exists.

A large number of 3-D object representation methods have been introduced in the literature. They can be categorised based on the data acquisition techniques or the type of descriptors they extract from this data to represent the 3-D object. Note that some methods impose specific restrictions on the classes of geometrical objects that can be handled.

In [150], parallel lines and ellipses are used to describe different viewpoints of an object. A strategy is suggested to recognise an object from an unknown viewpoint. The method is based on earlier works by Brooks [44] which has been modified later on by others [94, 111]. In [280], a model is established from a large number of viewpoints taken from a video sequence. The input of the system is also a video sequence of an unknown object. The system first builds a 2-D representation of the object. If the representation matches one of the objects of the database, it is modified based on new information extracted from the new sequence. Otherwise the object is recognised as a new one and its representation is stored. The number of views may be reduced with a more sophisticated preprocessing [41]. Representation with multiple views and recognition using a single view were further proposed in [71]. The number of viewpoints used to represent a complex object was down from about 2000 in [280] to 20 − 30 as reported in [71]. Shape [317, 77] and color [239] features have also been used in 3-D object representation.

Multi-view representations have not yet successfully dealt with the following issues [4, 215]:

- What is the optimal number of views?

- How to select the optimal views?

In this chapter, we propose a method for automatic selection of optimal views of a free-form 3-D object. In order to represent an object efficiently, we eliminate similar views and select a relatively small number of views using an optimization algorithm. This number varies from 5 to 25 depending on the complexity of the object and the measure of expected accuracy. To identify an unknown object from a single viewpoint, its representation is matched with all images of the database and the best matches are retrieved and displayed.

In order to represent each view of the object, we need a contour shape descriptor. Since the camera is allowed to change its viewpoint with respect to the object, the resulting boundary of the object may be deformed. The deformation can be approximated by an affine transformation and therefore the descriptor must be affine invariant. A number of shape representations have been proposed to recognize shapes even under affine transformations. Some of them are the extensions of well-known methods such as Fourier Descriptors [20] and Moment Invariants [82, 336]. Affine invariant scale space was introduced in [272]. It generalizes the definition of curvature to introduce *affine curvature*. This curve evolution method was proven to have properties similar to curvature evolution [141, 142], as well as being affine-invariant. However, an explicit shape representation has yet to be introduced based on the theory of affine invariant scale space [60]. The prospective shape representation might be computationally complex as the definition of affine curvature involves higher order derivatives. A number of shape representation techniques are based on level-set methods [171][172] and volumetric diffusion [145, 146]. These representations suffer from inefficiency and lack of robustness with respect to occlusion. Other techniques based on curve evolution [286] are more suitable for applications other than shape representations.

In our system, each view of an object is represented by the locations of the maxima of its CSS image. The representation has already been used to represent shapes of boundaries in similarity retrieval applications [11] and proven to be robust under general affine transforms [3].

In this chapter we also explore the fusion of results from multiple shape descriptors for automatic selection of optimal views in multi-view 3-D object recognition. In both view selection and recognition stages, the results of three different descriptors are combined to achieve the best possible performance. Since each descriptor captures some features of the contour, a combination of well-selected shape descriptors will result in better representation of the silhouette as well as the 3-D object. Our optimal view selection method is independent of shape descriptors. In fact, it can also be used to select optimal views based on other features such as colour and texture. Using each shape descriptor, the view selection method returns a set of views. These sets are combined at the next stage to find the final set of optimal views. To identify an unknown object from a single viewpoint, its representation is matched to the representations of all images of the database, and the best matches are retrieved and displayed. In our experiment with a collection of 15 toy aircrafts of different shapes, we observed that the results obtained from the fusion method are superior to the results obtained from any single technique.

Finally, a complete system for free-form 3-D object retrieval using partially occluded or noisy views of those objects has also been presented in this chapter. Substantial experiments show that this system is also robust and reliable for retrieval despite background clutter, noise and partial occlusion of the objects in query images.

The following is the organization of the remainder of this chapter. Section 2 is on multi-view free-form object representation and retrieval. Section 3 discusses robust automatic selection of optimal views of free-form 3-D objects. Section 4 presents a system for free-form 3-D object retrieval with occlusion from arbitrary viewpoints. Section 5 contains the concluding remarks.

2.   Multi-View 3-D Object Representation and Retrieval

The encouraging results of the previous chapter suggest that it may be possible to model a 3-D object using a small number of silhouette contours obtained from different viewpoints [6]. In fact the main difficulties in implementing this idea include occlusion, transformations and segmentation.

occlusion occurs when some parts of objects are hidden behind other parts. As a result, 2-D boundary contours of those objects are severely affected. The CSS representation can deal with minor occlusions as they affect the shape locally and only one or two CSS maxima may be affected. However, major occlusions must be dealt with by increasing the number of viewpoints used to represent the object. The optimum number of viewpoints very much depends on the geometry of each individual object and varies from one to another. Since each view is represented by a small

Figure 5.1. Three examples from our collection. In our first experiment, these four standard views are used to recognise a randomly grabbed view.

size feature vector, and the matching process is very fast, this method can be used in retrieval applications dealing with a large number of objects, each represented by a number of viewpoints.

affine transformation is another phenomenon which occurs in multi-view object representation and it is important for the features extracted from the images to be affine invariant. In the previous chapter we showed that the CSS representation is robust under general affine transforms.

Segmentation of images is necessary to extract the boundary of objects prior to CSS computation. This is not always an easy task and can not always be carried out automatically. Interactive methods such as active contours [37] may be used in the case of complicated scenes.

Our experiments were carried out on a collection of 18 toy aircrafts. A video sequence was prepared for each object using a 3CCD digital video camera. Color images were then grabbed from those video sequences. In order to avoid a complicated pre-processing stage, the objects were placed on a simple blue background and the scene was illuminated by three different lights to minimize shadows.

From the video sequence, a number of images of each object were grabbed from different views. The views consisted of four *standard* views, namely top, side, front and back; as seen in Figure 5.1. The top view of all objects of our collection is presented in Figure 5.2.

A number of *arbitrary* views of each object were also grabbed. From the arbitrary views, three images for each object were randomly selected and each of them was separately used as an input query. After pre-

Figure 5.2.   The top views of all objects in our collection. Note that the input queries are viewed from different angles (see Figure 5.3).

processing and extracting the object boundary, the system computed its CSS image and extracted the contour maxima of the CSS image. The resulting maxima were then matched to each model of the database to find the most similar objects. The first two outputs of the system were then observed to find out whether they included one of the standard views of the input query. Due to differences between the standard views and the arbitrary view, recognition was a difficult task.

Examples of random views which are used as input queries are shown in Figure 5.3. It can be seen from this figure that apart from affine transformation, the inputs may also suffer from self-occlusion. Also they may

Figure 5.3. The random views of some of the objects.

include some additional parts which are occluded when the objects are observed from standard views. Moreover, since the camera moves during the video recording, additional noise is sometimes created in some images. An example is shown on the left hand side of Figure 5.4. Since CSS computation involves smoothing, this noise is removed in early stages of the evolution process and does not affect the CSS image. Since we do not use a sophisticated segmentation method, wherever the background is similar to the object in terms of color, or in the case of inevitable shadows, the boundary of the object is not extracted correctly. This can be seen in the example presented on right hand side of Figure 5.4. As long as this distortion is not very severe, the CSS representation can handle it.

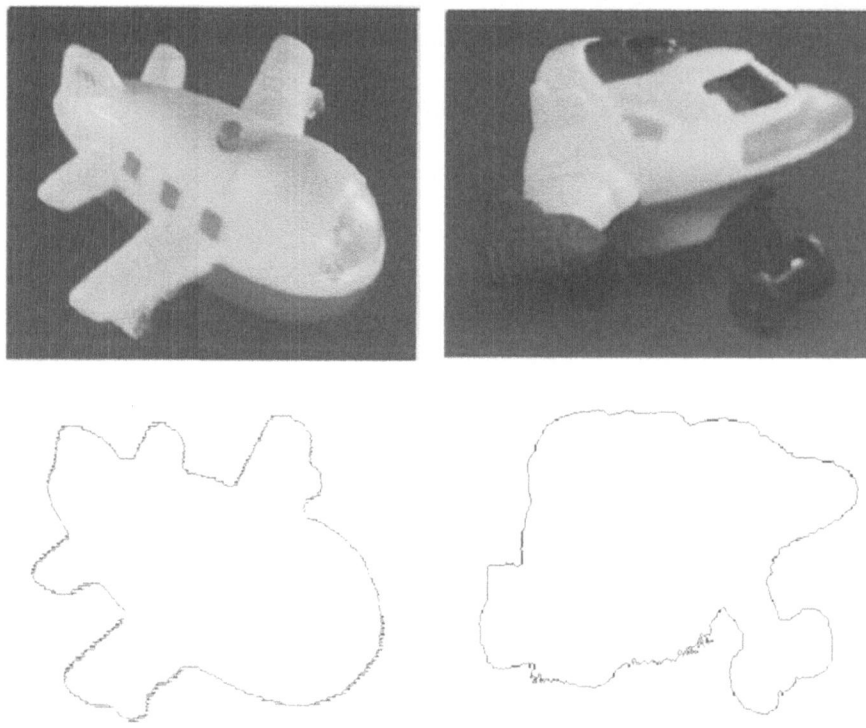

Figure 5.4.   Left: noise as a result of moving camera. Right: noise as a result of poor segmentation.

In order to deal with the problem of self-occlusion, the number of standard views should be increased. Since CSS matching is very fast, the speed of the system does not suffer as a result of adding more views. In our second experiment, we used five additional standard views and modelled each object using nine silhouette contours.

As mentioned earlier, three images from each object were selected and used as input queries. In response to a query, the system returns the most similar objects according to CSS matching. Three examples are shown in Figure 5.6 where nine views of each object are used in the database (see figure 5.5). For each example, the input query is shown on the left prior to system results. Note that in all three examples, the first output of the system is the same as the input object.

By applying the system to all input queries and using nine views per object, we observed that in 93% of cases, at least one of the first two outputs of the system was another view of the input object while in 80% of cases the first output also represents the same object. These figures

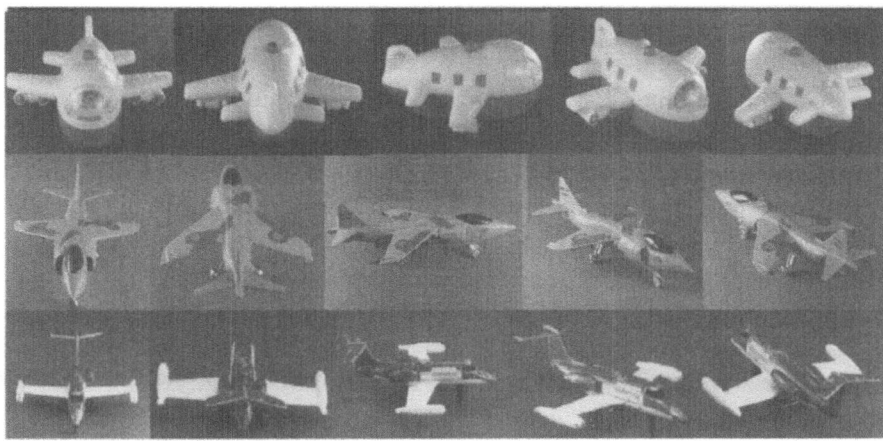

Figure 5.5. The additional five standard views, used in conjunction with four standard views of Figure 5.1 in our second experiment.

Figure 5.6. Three examples of query results. In each example, the image presented on left hand side is the input query, followed by the first four outputs of the system.

were 87% and 69% respectively when we used four views to represent each object.

The results indicate that by representing an object with several views, recognition based on the first output of the system is feasible. At the same time, in the image database approach, where the user verifies the output of the system, the number of standard views can be reduced.

3.   Robust Automatic Selection of Optimal Views

This section addresses the issue of automatic selection of the best and the optimum number of views for each object in multi-view 3-D object representation and recognition. The object boundary in each view is considered as a 2-D shape and is represented by the maxima of its CSS image contours. After representing each object by an optimal number of views, an unknown object is recognised by a single image taken from an arbitrary viewpoint using the same shape representation.

This section also presents a novel method for fusion of results from multiple shape descriptors. The utilization of this method for multi-view 3-D object representation and recognition has been explored. The object boundary of each view is represented effectively using the CSS technique, Moment Invariants, and Fourier Descriptors. It has been shown that the results obtained from the fusion method are superior to the results obtained from any single technique.

The method has been tested on a collection of 3-D objects consisting of 15 aircrafts of different shapes. Each object has been modelled using an optimal number of silhouette contours obtained from different view-points. This number varies depending on the complexity of the object and the measure of expected accuracy. A comprehensive analysis of the performance of the system has been given as the number of views varies. Since each view is represented by a small size feature vector, and the matching process is very fast, this method can be used in retrieval applications dealing with a large number of objects, each represented by a number of viewpoints.

Affine transformation is another phenomenon which occurs in multi-view object representation. It is important for the features extracted from the images to be affine invariant. It has been shown [3] that the CSS representation is robust under general affine transforms.

Segmentation of the images is necessary to extract the boundary of objects prior to CSS computation. This is not always an easy task and can not always be carried out automatically. Interactive methods [19] may be used in the case of complicated scenes. Our experiments were carried out on a collection of 15 toy aircrafts. A video sequence was prepared for each object using a 3CCD digital video camera. Color images were then grabbed from the video sequences. In order to avoid a complicated pre-processing stage and segmentation errors, the objects were put on a simple stand and illuminated by two lamps against a dark background. One view of each object of our database is represented in Fig. 5.17.

From the video sequence, a number of images from each object were grabbed from different views. On average, 53 images per object were grabbed at this stage. Fig. 5.7 shows all views of a particular object. In order to eliminate similar views and achieve an efficient representation, we introduced an algorithm for automatic selection of optimal views as explained in the following subsection.

3.1. Optimal View Selection

When an object is pictured from a large number of viewpoints, it is likely that some of the resulting images are similar and convey no additional information. As a result, an algorithm is required to identify the optimal number of images needed to represent an object. Using an arbitrary descriptor and its associated matching method, one can measure the similarity between two different images grabbed from a single object. The algorithm which selects the most suitable views according to an arbitrary representation is as follows:

1 Obtain many views of the object from different viewpoints. The jth object will have n_j views:

$$V_1(o_j), V_2(o_j), ..., V_{n_j}(o_j)$$

2 Segment the images obtained in step 1 to recover the boundary contours. For each boundary, compute the descriptors to obtain:

$$DES(V_1(o_j)), DES(V_2(o_j)), ..., DES(V_{n_j}(o_j))$$

3 Select a threshold value t which will be used to define which views are similar. If the matching cost between two views is less than t, they are marked as similar:

$$if \ (M\_cost(DES(V_j), DES(V_k)) \le t)$$

$$\Rightarrow Sim(V_j, V_k) = TRUE$$

$$else \quad Sim(V_j, V_k) = FALSE$$

4 Calculate the matching cost between each representation, obtained from a contour in step 1, and all other representations, obtained from other contours. If the matching cost is less than t, declare the two views as similar. Assign a rank r to each view defined as the number of views that are similar to it.

$$r(V_j) = size - of\{V_k|\ Sim(V_j, V_k) = TRUE\}$$

Figure 5.7. An object in our database is initially represented by about 50 views. Some of these views are discarded during the automatic view selection process.

5 Create a sorted list L of all views. Sort the views in L according to their decreasing rank values. Each view will have a pointer to other views similar to it.

$$L = \{V_j, V_k, ..., V_l | \; r(V_j) \geq r(v_k) \geq ... \geq r(V_l)\}$$

6 Start from the top of L and place the first view in the set C of characteristic views. Remove all views similar to the first view from L to obtain a reduced list.

7 Move down the reduced list L and repeat the procedure in 6 until the end of L is reached.

At the end of this process, the set C will contain the full set of characteristic views of the input object determined automatically.

Note that the algorithm involves the matching of each view of an object to all other views of the same object, and it must be applied to all objects. In the case of large databases, the process may be time consuming; however, it is carried out off-line and has no effects on the recognition stage which compares the input query to all selected views of different objects.

The result of this process for the object of Fig. 5.7 is presented in Fig. 5.8 using the CSS descriptor.

3.2. Combining Optimal Views

The optimal view selection process is independent of the type of descriptors. As a result, one can use the algorithm to combine several descriptors to achieve better 3-D representation and recognition.

We used the three formerly described shape representations separately to determine the relevant sets of optimal views. To combine the resulting sets, we selected those views which were present in the majority of sets.

In another approach, one may select those views which are present in at least one set. We noticed that in this case the resulting final set of optimal views is large and therefore the method loses its efficiency.

Note that it is also possible to combine the representations at the view selection stage and obtain only one set of optimal views from this stage. In that case, $M\_cost(DES(V_j))$ in step 3 will be the weighted sum of matching values obtained from each representation as discussed below.

3.3. Recognition Experiment

After selecting the optimal views of each object, the corresponding representations, consisting of the locations of maxima of their CSS image contours, Moment Invariants and Fourier Descriptors were stored

Figure 5.8. For the example of Fig 5.7, 19 views are selected in the case of $t = 0.3$ for the CSS descriptor. Note that most of the similar views have been discarded.

in separate files. A pointer to the original object was also stored with each representation. The process of optimal view selection and matching is first applied separately to each shape representation method. In the second experiment three methods were combined in order to obtain improved results.

In order to test the performance of the method, we recorded another video sequence for each object. A set of 10 test images were then randomly grabbed from each sequence (see Figure 5.9). After extracting the boundaries of the objects, the relevant descriptors were used to find similar images from the database produced by the automatic view selection algorithm.

In response to a query, three different values were assigned to each model of the database. They were the matching values associated to CSS, Moment Invariants and Fourier Descriptors. Using these values separately, we obtained the results of the retrieval system and observed that the CSS representation performed better that the other two descriptors.

In order to obtain even better results, we combined the three matching values and assigned a single matching value to each model of the

Figure 5.9. Test images for the object of Figs. 5.7 and 5.8.

database, as follows.

$$M = \alpha_c m_c + \alpha_f m_f + \alpha_m m_m$$

where M was the final matching value and m_c, m_f and m_m were the three values corresponding to each method. The weights α_c, α_f and α_m controlled the contribution of each method in the final similarity measure. Note that setting these values as equal did not necessarily lead to equal contribution of each method as the matching values of different descriptors were not in the same range. As a result, the weights were determined statistically. In order to do this, we used each representation separately and added up the matching values assigned to the first two outputs of the system. We then found the average of this over the whole query set. The corresponding weight for each method would then be inversely proportional to this average. For example, in the case of CSS we have:

$$\alpha_c = \frac{1}{\frac{1}{P}\sum_{j=1}^{P}\left(m_{cj}(0) + m_{cj}(1)\right)}$$

where P was the number of queries.

Each test image was then used as an input query and the system was asked to retrieve all similar images based on the multiple representation and multiple matching method. We observed that the results obtained from the fusion method were superior to the results obtained from any single technique. We also observed that in almost all cases, recognition was possible as the first output of the system was another view of the same object. The results are presented in the following subsection.

3.4. Results

We start this subsection by presenting a number of query results in Fig. 5.10. Each example is presented in one row, starting with the input

Figure 5.10.   Query results: each object has been represented on average by 20 views.

query on the left hand side followed by the outputs of the system for that query. In these examples, the CSS representation has been used and each object is presented by about 20 views on average.

n_a	33	26	22.5	19.5
MI- (%)	86	86	77	69
FD- (%)	94	92	90	79
CSS- (%)	97	96	93	91
MUL-(%)	99	98	96	94

Table 5.1. The success rate, S.R. and the average number of images per object, n_a, for different descriptors. Note that the multiple descriptor in the last row always gives better results.

As seen in these examples, in response to a query, in addition to the closest view to the input query, other similar views are also retrieved. It is also notable that due to the movement of the camera, the size of the object may also change even if it is viewed from the same viewpoint.

The key parameter which most affects the results is n_j, the number of images used to represent the jth object. This parameter varies from an object to another and therefore we use its average n_a as follows.

$$n_a = \frac{1}{M} \sum_{j=1}^{M} n_j$$

where M is the number of objects. Both n_j and n_a are controlled by the threshold value t in the view selection algorithm. We used different values of t for each descriptor and obtained different values for n_a, as shown in Figure 5.11. The results for different descriptors when used separately as well as the results of the multiple representation are presented in table 5.1.

In response to a query, the system retrieves those views of all models of the database which are similar to the input query. However, only if the first output of the system is a view of the same object as the input, we consider it as a *success* for the system. The success rate in table 5.1 refers to the average success of the system over all 150 input queries. For example, if we use Moment Invariant only, and in the optimal view selection process the selection of the threshold t leads to $n_a = 33$, in 86% of cases, the system is able to recognize the input object correctly. This figure is higher for Fourier Descriptors and even higher for CSS. However, the best performance, almost 100%, can be achieved by using multiple shape descriptors.

It is also clear that the higher values of n_a result in higher success rates, as shown in Figure 5.12. We can conclude from table 5.1 that even if an object is represented by a relatively small number of views, a

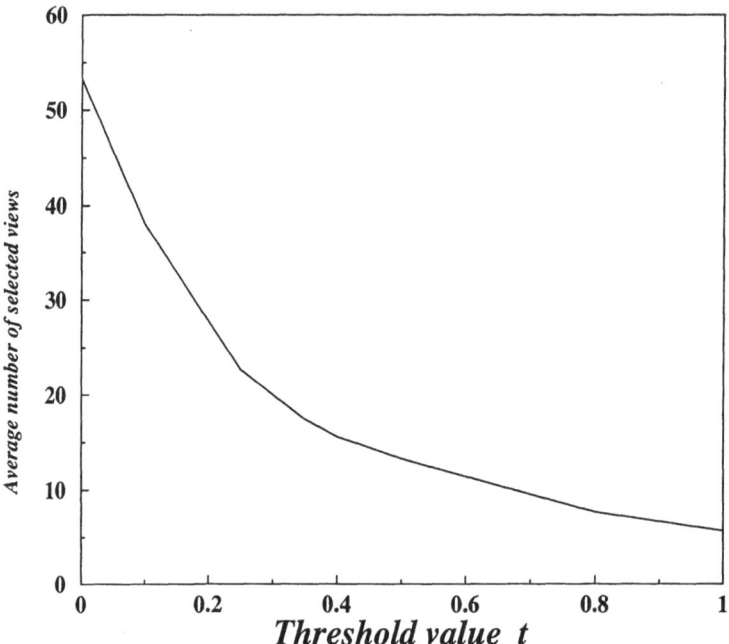

Figure 5.11. The average number of views per object is directly related to the threshold value, t. See subsection 3.1

good recognition rate can be achieved by the multiple shape descriptors method.

4.   Free-form 3-D Object Retrieval with Occlusion from Arbitrary Viewpoints

This section describes a novel approach to multi-view 3-D object representation and recognition with occlusion which can be used in image database retrieval applications [206]. 3-D objects are recognised using a relatively small number of images taken from different views. An unknown object is then recognized using a single image taken from an arbitrary viewpoint. A multi-scale edge-based technique is used for robust segmentation. Input image edges include shadow boundaries as well as internal object edges. Furthermore, object boundary edges are usually broken. Each local segment is part of an edge contour having only two curvature zero-crossings at its end-points. The segmentation is carried out using the CSS technique. Each segment is described by a number of features which are used to narrow down the search space during the recognition process. In response to an input query, geometric

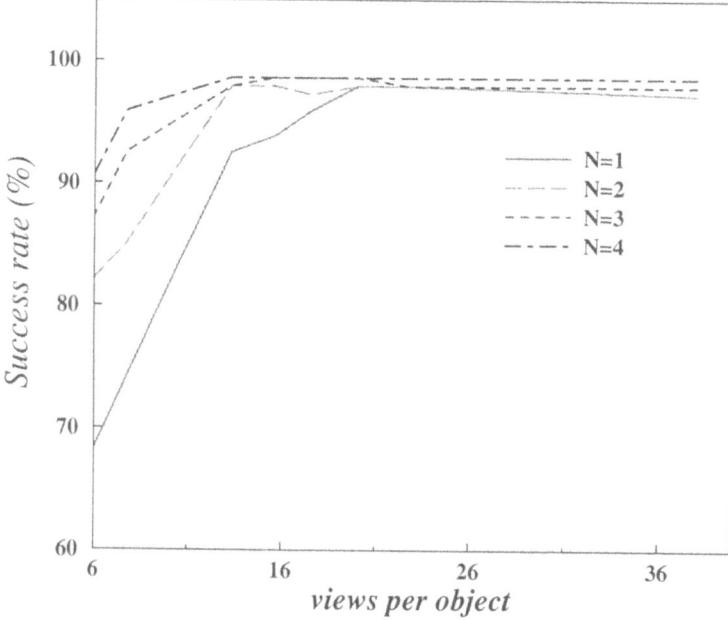

Figure 5.12. The success rate increases as the number of views per object increases. N is the number of observed outputs.

hashing is first used to find the number of matched segments between the input image and every model image. Those models with larger numbers of matched segments are then passed to the verification stage. Two different approaches are suggested for verification. In the first approach, the harmony among the transformation parameters obtained from each pair of matched segments is examined to assign a match value to each model candidate and select the best match. In the second approach, the distance between the input query edge contours and the corresponding model contours is measured. The measurement is then optimized and used as the match value. The process is initialized by the transformation parameters obtained from a pair of matched segments and is then optimized by an efficient and robust algorithm to determine the best transformation parameters between the query and each model candidate. The methods have been tested on a collection of 3-D objects consisting of 15 aircrafts of different shapes.

4.1.   System Overview

In most shape-based methods, segmentation is necessary to extract the boundary of the object. This is not always an easy task and can not always be carried out automatically. Our multi-scale segmentation method [194], however, uses edge contours of the image to extract local features. Using edge contours rather than the complete boundary contour enables us to deal with the problem of occlusion more effectively. The recognition of an occluded object is carried out through the visible parts of the object as the edge segments of these parts remain intact.

Canny edge detector [45] is used to extract the edge contours of the image. Curvature zero-crossings of each contour are then used to divide it into different segments. Each segment has only two curvature zero-crossing as its endpoints. The image is initially represented by a number of geometric features extracted from these segments. For each feature, a hash table is produced which re-arranges all segments of all models of the database based on the value of that feature.

In response to a query, each feature of a segment is separately indexed to the corresponding hash table to vote for a number of segments of the database. Those segments which are voted for by the majority of features will then be selected and their associated *models* are voted for. As a result, each segment of the input query votes for a number of models. At the end of this stage, when all segments of the input query are indexed to the hash tables, those models with highest votes are passed to the verification stage.

In the verification stage, each candidate is examined against the input query. For each pair of matched segments, the transformation parameters which transform image segment to the model segment are computed and are used as initial estimation for registration of all image points into the model points. The registration is then optimized by successively re-estimating the transformation parameters based on the corresponding edge points. At the end of this process the distance between the image and the candidate is measured. Note that each pair of matched segments provides a distance measure, the final distance between the model and the image is the minimum possible distance obtained from the matched segments.

A faster approach in this stage is to obtain the transformation parameters corresponding to each pair of matched segments and examine them. If these parameters are similar, we can conclude that the model is the transformed version of the input. To implement the idea, each set of transformation parameters is mapped to a point in the parameter space. We then look for significant clusters in this space and assign a match value to the model using the properties of these clusters.

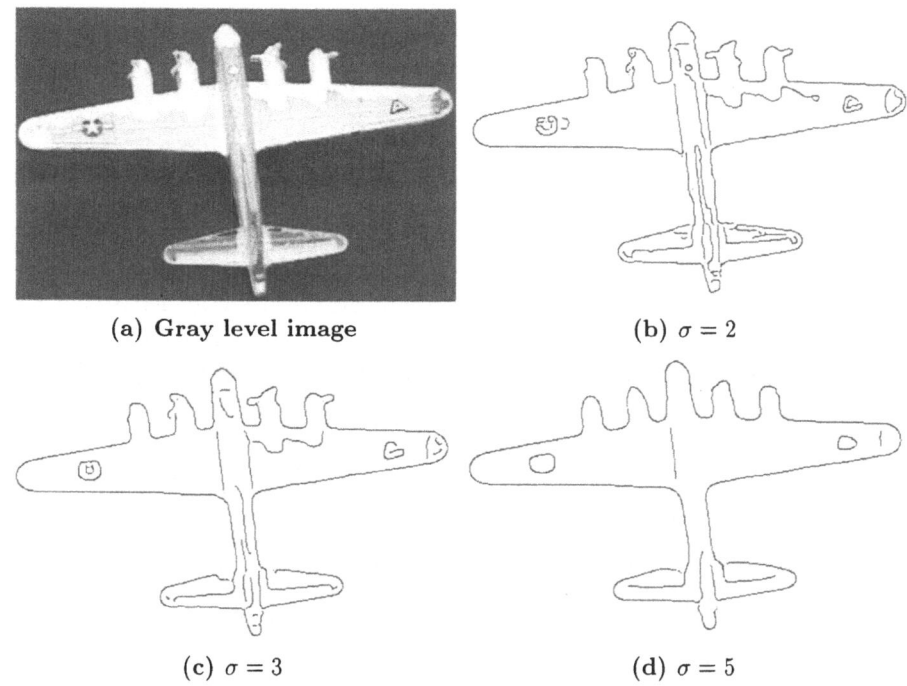

(a) Gray level image (b) $\sigma = 2$

(c) $\sigma = 3$ (d) $\sigma = 5$

Figure 5.13. Canny edge detector with different levels of smoothing.

4.2. Feature Extraction

Feature extraction consists of three stages, multi-scale edge detection, multi-scale contour segmentation and feature computation. In this section, we explain these stages in more detail.

4.2.1 Multi-Scale Edge Detection

Canny edge detector [45], smoothes the image by a Gaussian filter prior to edge detection to remove noise. The level of smoothing which is indicated by σ, the filter width, affects the output of the edge detector and any later process which may be applied to the edge contours. We minimize these effects by using different levels of smoothing at this stage. Three different binary images corresponding to three levels of smoothing are produced from the input gray level image at this stage. As seen in Figure 5.13, at a low level such as $\sigma = 2$, the output is rather noisy, and edges tend to be cut more frequently, but more details of the image edges are preserved. At a higher level such as $\sigma = 5$, however, longer edges are observed but some details disappear. We believe that by taking into account both lower and higher levels of smoothing more information can be extracted from the image which is used later on in the matching stage.

From each binary image the edge contours are extracted. The x and y coordinates of each edge contour is stored separately in the same file. The very small contours, which do not convey significant information are eliminated at this stage.

4.2.2   Multi-Scale Contour Segmentation

At this stage, each edge contour is divided into smaller segments by means of curvature zero-crossings. Each segment has only two curvature zero-crossings at its endpoints. Edge contours are smoothed using the CSS technique to remove the remaining noise. CSS also helps to compute curvature and find curvature zero-crossings [7].

The number of segments on an edge contour depends on the scale of smoothing. While using a small scale may result in a larger number of small segments, a large scale removes small features and retains the global appearance of the contour. Figure 5.14 shows how σ, the width of Gaussian filter which indicates the scale of smoothing, affects the contour segmentation. For small values such as $\sigma = 2$, the number of curvature zero-crossings is very large due to noise. As a result, many small segments are produced, which do not convey useful information. As σ gradually increases, the number of curvature zero-crossings decreases and a good segmentation can be achieved. When σ is very large, as seen in Figure 5.14(e), the contour is no longer similar to the original contour and most of the details disappear.

In conclusion, it is very difficult to determine a particular scale at which the contour can be segmented properly. Therefore, the solution is to use a multi-scale approach. We start with an initial scale which is proportional to the length of the contour. At this scale, the contour is segmented and the end-points of the segments are stored. Then we increase σ gradually and segment the contour at each new scale. As σ increases, the number of curvature zero-crossings decreases. As a result, new segments may be discovered and stored. These segments are generally larger than the initial segments. The process ends when σ is sufficiently large in comparison to the length of the contour. The following algorithm may be used to segment a contour at different levels of scale.

- Select an initial scale, based on the length of the contour.

- Smooth and segment the contour at the current scale. Compare each segment with all selected segments from previous scales. If both its endpoints are close to one of the selected segments, remove it, otherwise add it to the list of selected segments.

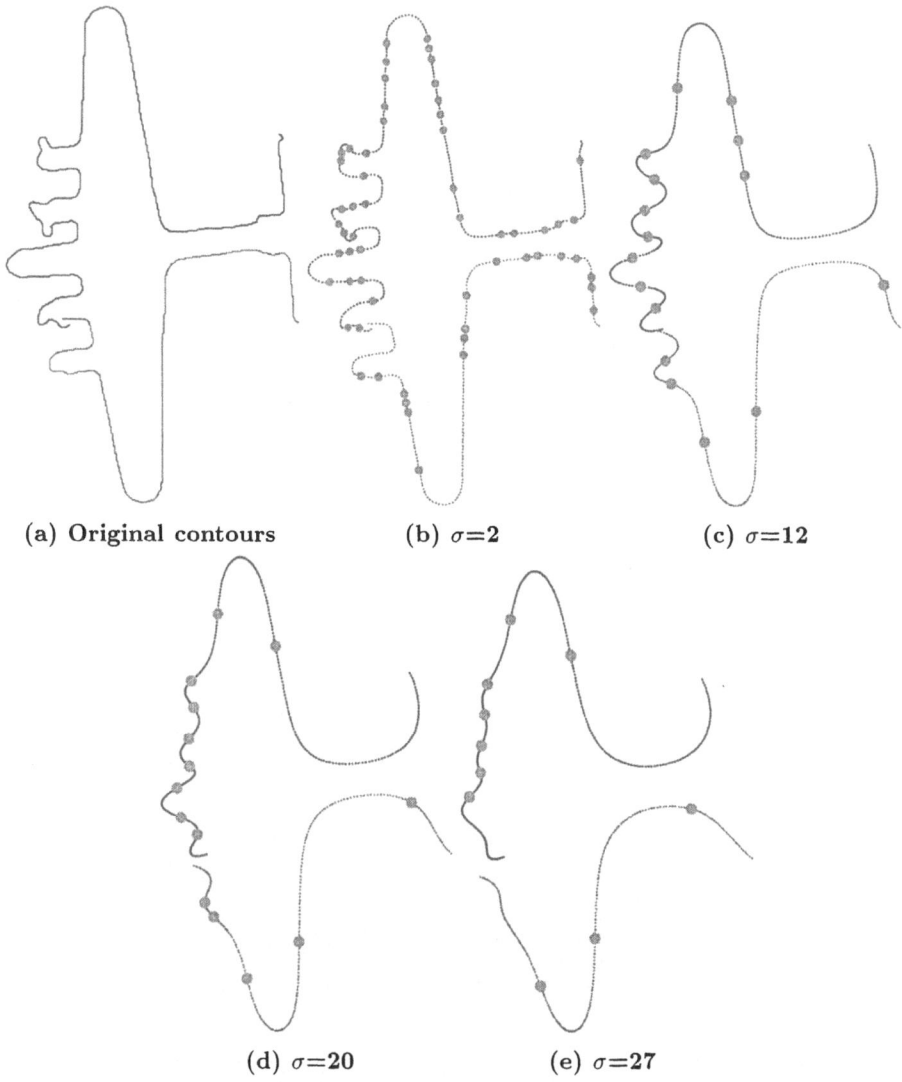

(a) Original contours   (b) $\sigma=2$ (c) $\sigma=12$

(d) $\sigma=20$ (e) $\sigma=27$

Figure 5.14. Segmenting edge contours at different levels of smoothing.

- Increase the value of σ by 0.2, if $\sigma \times 7$ is larger than the size of the contour stop, otherwise go to previous step.

The maximum value of σ is chosen such that the size of the Gaussian filter is not larger than the size of the contour. In our experiments we choose $\sigma \times 7$ as the size of this truncated filter.

At lower levels of scale, and particularly in straight parts of the contour, the number of curvature zero-crossings is relatively large which gives rise to many small segments on the contour. These small segments do not convey useful information and therefore are discarded from the final segment list. In general, if the size of a segment is less than a certain threshold, it will be discarded. As σ increases, some of the curvature zero-crossings disappear and the small segments are joined to create larger segments which are then included in the list.

4.2.3   Segment Features

After segmenting a contour, each segment should be described effectively by a number of features. It is desirable for these features to be invariant under affine or at least similarity transforms. Since the number of affine invariant features are limited and they usually involve higher order derivatives, we used a number of features which are invariant under similarity transforms.

The first feature of a segment indicates the convexity or concavity of the segment. Between two successive curvature zero-crossings, the sign of curvature does not change. For convex segments, curvature is positive, whereas for concave segments it remains negative.

The second feature is normalized average curvature. Since curvature is not scale invariant, average curvature is not invariant under similarity transform. Curvature is inversely proportional to the length of the segment. Therefore, we normalize average curvature by multiplying it by the length of the segment.

There is always a local curvature extremum between the two curvature zero-crossings which conveys some information about the shape of the segment. To take this parameter into account, our third feature is obtained by dividing the second feature by the absolute value of this extremum.

The fourth feature is based on affine curvature [204]. The average affine curvature is divided by the maximum affine curvature of the segment. Since affine curvature is not defined at curvature zero crossings, only those points of the segment which are sufficiently far from the endpoints are considered in this computation.

The fifth feature of a segment is the normalized average distance to the endpoints line. To calculate this parameter, a straight line is drawn between the two endpoints. The average distance to this line over all points of the segment is then calculated. This value is then normalized by dividing it by the distance between the two endpoints. From Figure 5.15 this parameter can be calculated as follows:

$$Mean - dist = \frac{\frac{1}{n}\sum_{i=1}^{n} d_i}{d}$$

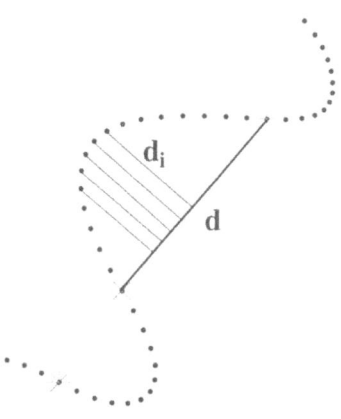

Figure 5.15. The Mean-dist parameter.

where d_i is the distance of ith point of the segment from the line, d is the length of the straight line between the two curvature zero-crossings and n is the number of points on the segment.

4.3.   Matching

After segmenting all edge contours of a model, and computing all features of each segment, the results for all models are separately stored in a file. A record for a model includes the number of edge contours; and for each contour the number of its segments; and for each segment, the locations of its end-points, and the values of its features.

This information is analyzed to generate a number of hash tables, one for each feature. To generate the hash table for a feature value such as f, a value for Δ is first determined as:

$$\Delta = (f_{max} - f_{min})/n$$

where n is the desired number of intervals. A pointer to a segment is then associated to the kth interval, if the following statement is true for its feature value, f:

$$f_{min} + (k - 1) \times \Delta \leq f < f_{min} + k \times \Delta \qquad k = 1, 2, ..., n$$

Pointers to all segments of all models are then stored in n groups. It is obvious that the number of segments may differ from one group to another. The difference is particularly large for marginal groups (k close to 1 or n) as the maximum and minimum of the feature values for some segments may be unusually higher or lower than average. To deal with this problem, before applying the above mentioned algorithm, the margins for the first and the last group are determined so that each of these groups accommodates $\frac{1}{n}$th of the whole number of segments.

In response to a query which is in the form of an image taken from an arbitrary view of an unknown object, all segments of the image are indexed to these tables separately to find the best candidates as described in the following subsection.

4.3.1   Indexing

This is the first step in the matching process. Each feature value of every segment of the input query is examined against all segments of all models of the database using the hash tables just described. When a feature value of an input segment falls in the range of the kth interval of the relevant hash table, all segments of that interval are primarily selected and transfered to a list associated to that feature. For each input segment, when the lists for all features are prepared, those segments which are in the majority of lists are then selected and their associated models are voted for. At the end of this stage, all models are ranked according to their votes. The first m models are then selected for the verification stage.

It is also possible to consider more than one interval of a hash table in this process. In this case, instead of the kth interval, from $(k-i)$th to $(k+j)$th intervals are considered where $f \times (1-\delta)$ falls in the $(k-i)$th interval and $f \times (1+\delta)$ falls in the $(k+j)$th interval and δ normally has a value less than 0.1. We noticed that this approach is more effective as the user can control the procedure by changing the value of δ.

4.3.2   Verification through Registration

At the indexing stage, local features are used to select the best candidates. The verification stage is designed to globally compare these candidates to the input query to find the best match. To do this, we estimate the best transformation parameters which map models edge point to image edge points. After mapping, the distance between the two sets is computed and considered as the measure of similarity between the image and the model.

The process starts with a pair of matched segments, one from the image and the other from the model. We find the best parameters which map the model segment to the image segment and use them to map the whole model edge points to the image edge points. The distance between the corresponding points is then computed. This process is repeated for

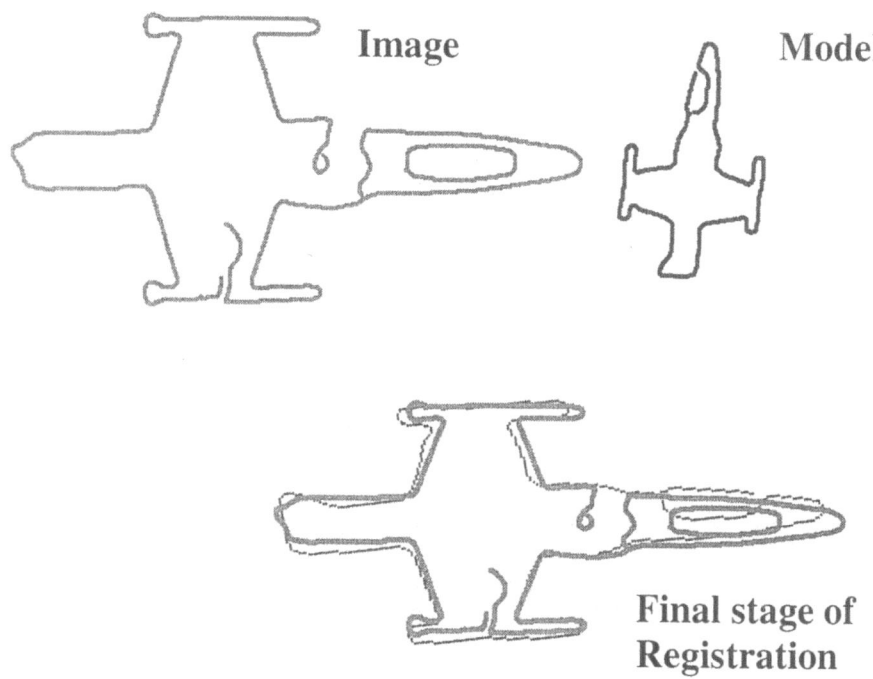

Figure 5.16. The process of registration. Observe how the model is transfered to match the image. Note also that the edge contours are different in the model and the image.

all pairs of matched segments to obtain a distance measure for each of them. The minimum distance measure is then selected to represent the similarity measure between the model and the image.

Let (x_j, y_j) be a point of the image segment and (ξ_j, ψ_j) be the corresponding point of the model segment. They are related to each other through the following equations:

$$\begin{cases} x_j = a\xi_j + b\psi_j + c \\ \\ y_j = -b\xi_j + a\psi_j + d \end{cases} \tag{5.1}$$

where a and b represent rotation and scaling, while c and d represent translation. Note that to avoid complexity, we have only considered similarity transforms.

The *least-squares estimation* method is used to estimate the values of a, b, c and d. Let Γ be the difference between the model segment and the image segment defined by:

$$\Gamma = \sum_{j=1}^{\eta} (x_j^t - x_j^c)^2 + (y_j^t - y_j^c)^2$$

where (x_j^c, y_j^c) is the corresponding point on the image segment to transformed model segment point (x_j^t, y_j^t) and η is the number of points on the model segment. Since η is usually different from the number of image segment points, an adjustment is needed to find the corresponding point. Using equation (5.1) to eliminate x_j^t and y_j^t yields:

$$\Gamma = \sum_{j=1}^{\eta} (a\xi_j + b\psi_j + c - x_j^c)^2 + (-b\xi_j + a\psi_j + d - y_j{}^c)^2.$$

The partial derivatives of Γ with respect to a, b, c and d should be set equal to zero. The result is a linear system of four equations in four unknowns which is solved to obtain estimates of a, b, c and d:

$$a = \frac{\sum \xi_j x_j^c + \sum \psi_j y_j^c - \frac{1}{\eta} \sum x_j^c \sum \xi_j - \frac{1}{\eta} \sum y_j^c \sum \psi_j}{\sum \xi_j^2 + \sum \psi_j^2 - \frac{1}{\eta} \sum \xi_j \sum \xi_j - \frac{1}{\eta} \sum \psi_j \sum \psi_j},$$

$$b = \frac{\sum \psi_j x_j^c - \sum \xi_j y_j^c + \frac{1}{\eta} \sum y_j^c \sum \xi_j - \frac{1}{\eta} \sum x_j^c \sum \psi_j}{\sum \xi_j^2 + \sum \psi_j^2 - \frac{1}{\eta} \sum \xi_j \sum \xi_j - \frac{1}{\eta} \sum \psi_j \sum \psi_j},$$

$$c = \frac{\sum x_j^c - a \sum \xi_j - b \sum \psi_j}{\eta},$$

$$d = \frac{\sum y_j^c + b \sum \xi_j - a \sum \psi_j}{\eta}.$$

Once an estimate of the transformation parameters is available, it is possible to map the model edge points to the image edge points to measure the image-model distance. Note that although the parameters are estimated based on a pair of segments, the distance measurement takes into account all image and model edge points.

The following procedure is used to determine the image-model distance.

- Apply the transformation to the model, using the parameters obtained from a pair of matched segments.

- Set the distant measure, $D = 0$.

- Find the closest image edge point to each point of the transformed model, (x_j^t, y_j^t) and store it in (x_j^c, y_j^c). Add the distance between the two points to the distance measure.

- Consider model points and their corresponding image points and apply the same method as described above to estimate the transformation parameters. Note that η will be the number of model edge points and not the number of points on the segment.

- Repeat the above steps using the new transformation parameters as long as the new distance measure is less than the old one. Otherwise, consider the old distance measure as the final one and stop.

This procedure is followed for each pair of matched segments. The final distance measure between the input and the model is the lowest distance measure obtained from one of the matched segments.

An example has been shown in Fig. 5.16. Note that although the model and the image look similar, the details of edge contours are quite different. The transformation of the model edges to the image edges has been shown in this figure. Note how the size and orientation of the model change as it is transformed to the image. At the final stage, the model is very close to the image and the distance between the corresponding points has been minimized.

4.3.3 Verification through Clustering

This method is faster than the previous one. To assign a matching value to each of the candidates, the transformation parameters which transform the model segment to the image segment are computed for each matched segment. To make the process even faster, only the end-points are used to estimate these parameters. If (x_1, y_1) and (x_2, y_2) are the end-points of the image segment and (x_1', y_1') and (x_2', y_2') are the corresponding endpoints of the model segment, they are related to each other through the following equations:

$$
\begin{cases}
x_1' = ax_1 + by_1 + c \\[2mm]
y_1' = -bx_1 + ay_1 + d \\[2mm]
x_2' = ax_2 + by_2 + c \\[2mm]
y_2' = -bx_2 + ay_2 + d
\end{cases}
\tag{5.2}
$$

where a and b represent rotation and scaling, while c and d represent translation. Note that again, we have only considered similarity transforms. We can solve this linear system of four equations in four unknowns to obtain estimates of a, b, c and d as follows:

$$a = \frac{(y_2 - y_1)(y_2' - y_1') + (x_2 - x_1)(x_2' - x_1')}{(y_2 - y_1)^2 + (x_2 - x_1)^2}$$

$$b = \frac{x_1' - x_2'}{y_1 - y_2} + a\frac{x_2 - x_1}{y_1 - y_2}$$

$$c = x_1' - ax_1 - by_1$$

$$d = y_1' + bx_1 - ay_1$$

Using these equations, we estimate the transformation parameters for each pair of matched segments. If the query and the model are similar, the number of correctly matched segments are sufficiently large. The transformation parameters are almost the same for these pairs of segments and if each set of parameters is mapped to a point in a four-dimensional parameter space, they create a cluster in this space.

The aim of *clustering* is to identify a set of matched segments with similar transformation parameters. In the parameter space, we are looking for those points which fall inside a disk with a radius less than a certain threshold, R_t. Note that each point represents the transformation parameters of a pair of matched segments, one from the input query and the other from the model.

The following algorithm is used to cluster the points in the parameter space.

- For each point, create a single member cluster. Initialize the centre of each cluster as the location of its member. Consider all clusters as *open*.

- Expand each open cluster as described in the next step. If no new member is added to a cluster change its status to *closed*.

- To expand a cluster, add any point which is closer than R_t to its centre and is not already a member of the cluster. Compute the new centre of the cluster as the centre of mass of its members.

- If there exists any open cluster, go to step 2, otherwise stop.

Finally a matching value is associated to the input query and the model based on each cluster. This value is determined by the number of members and also by the way they are scattered around the centre

Figure 5.17. One view of each object of the database.

of the cluster. The latter can be expressed by the standard deviation parameter,

$$s = \frac{\sqrt{\sum_{i=1}^{N} d_i^2}}{N}$$

where d_i refers to distance of the ith member to the centre of the cluster. Note that lower values of s indicate higher degree of similarity. Apart from s, we also need to consider the number of members in the cluster, N. Since larger values of N indicate better similarity, we divide s by N to obtain the matching value, as $m = s/N$. If there are more than one clusters with N greater than a threshold, the matching value is also computed for them and the final matching value is the smallest one among them.

4.4.   Experiments on Retrieval with Occlusion

Our experiments were carried out on a collection of 15 toy aircrafts. A video sequence was prepared for each object using a 3CCD digital video camera. Color images were then grabbed from the video sequence. One view of each object of our database is shown in Fig. 5.17.

On average 50 images of each object were grabbed from different views. When an object is pictured from a large number of viewpoints, it is likely that some of the resulting images are similar and convey no additional information. We used an automatic view selection algorithm [4] to find and discard similar views and reduce the average number of views per object to 25. These images were then processed to extract the segments of their edge contours. The segment features were then computed and stored in hash tables as explained in previous sections.

As for input queries, we prepared another video sequence from the same objects, using different illuminations and backgrounds. From this video sequence, we randomly grabbed a small number of views for each object and used them as input queries to the system. In response to a query, the system returned the n most similar views of the models where n was determined by the user.

Several examples are presented in Figs 5.18 and 5.19 using the first and second approaches to verification, respectively. In these examples, the top images are the input queries following by the outputs of the system. In all cases the input query is different from the outputs with respect to size and orientation. The backgrounds are also different. In some of the examples such as (b) and (c), the shadows of the objects are visible in the input images. If only the first output of the system is considered, the recognition of the input object is carried out correctly in these examples. This is also the case for the vast majority of input queries. In some examples such as Fig 5.18(f) and Fig. 5.19(f) the first output does not represent the input object. However, at least one of the first three outputs represents another view of the input object. The other outputs are also similar to the input queries.

Based on Figs. 5.18 and 5.19, and other observations, we can conclude that the first method of verification performs better but the second method of verification is substantially more efficient. These trade offs should be taken into account in order to select the most suitable method for user applications.

The speed of the system also depends on the width of the windows used to index the hash tables. As explained earlier, this parameter is controlled by δ. Larger values of δ result in larger number of matched segments which in turn results in a slower process of indexing and verification. The initial number of segments per image is also important. This parameter is controlled by the minimum acceptable size of edge contours and edge segments.

As discussed earlier, the advantage of the new method is that it does not require the full recovery of the object boundary which is a very difficult task in the case of images with arbitrary backgrounds as seen in several examples in Fig. 5.20. Two objects from our database were placed in different complex backgrounds and were pictured using a digital camera. These images were then used as inputs to the system. The results are presented in Fig. 5.20. As seen in this Figure, even with a very complex background such as in Fig. 5.20(f) and (g), good results have been obtained. Note that in these cases, many background edge

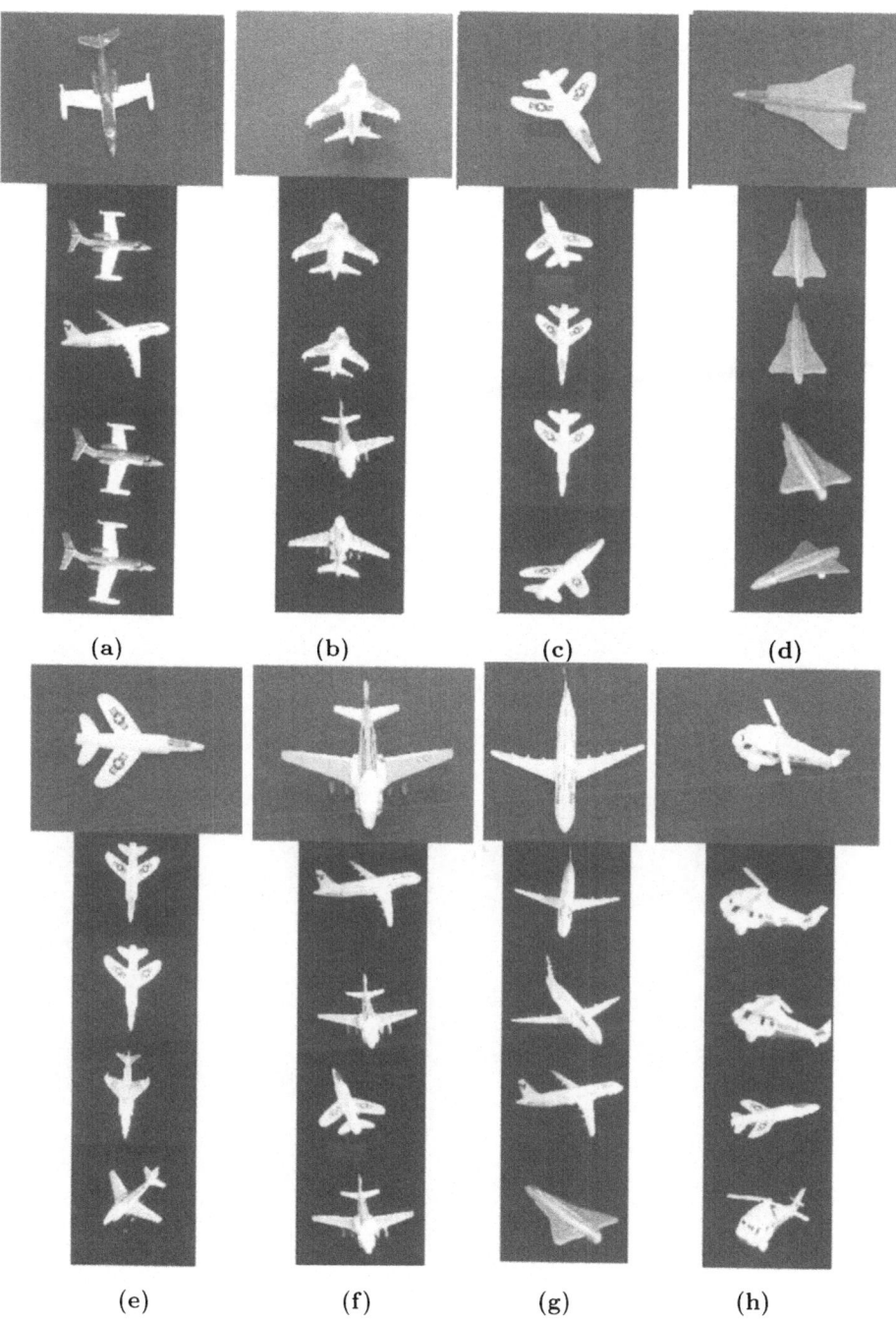

(a)   (b)   (c)   (d)

(e)   (f)   (g)   (h)

Figure 5.18. Query examples, registration method

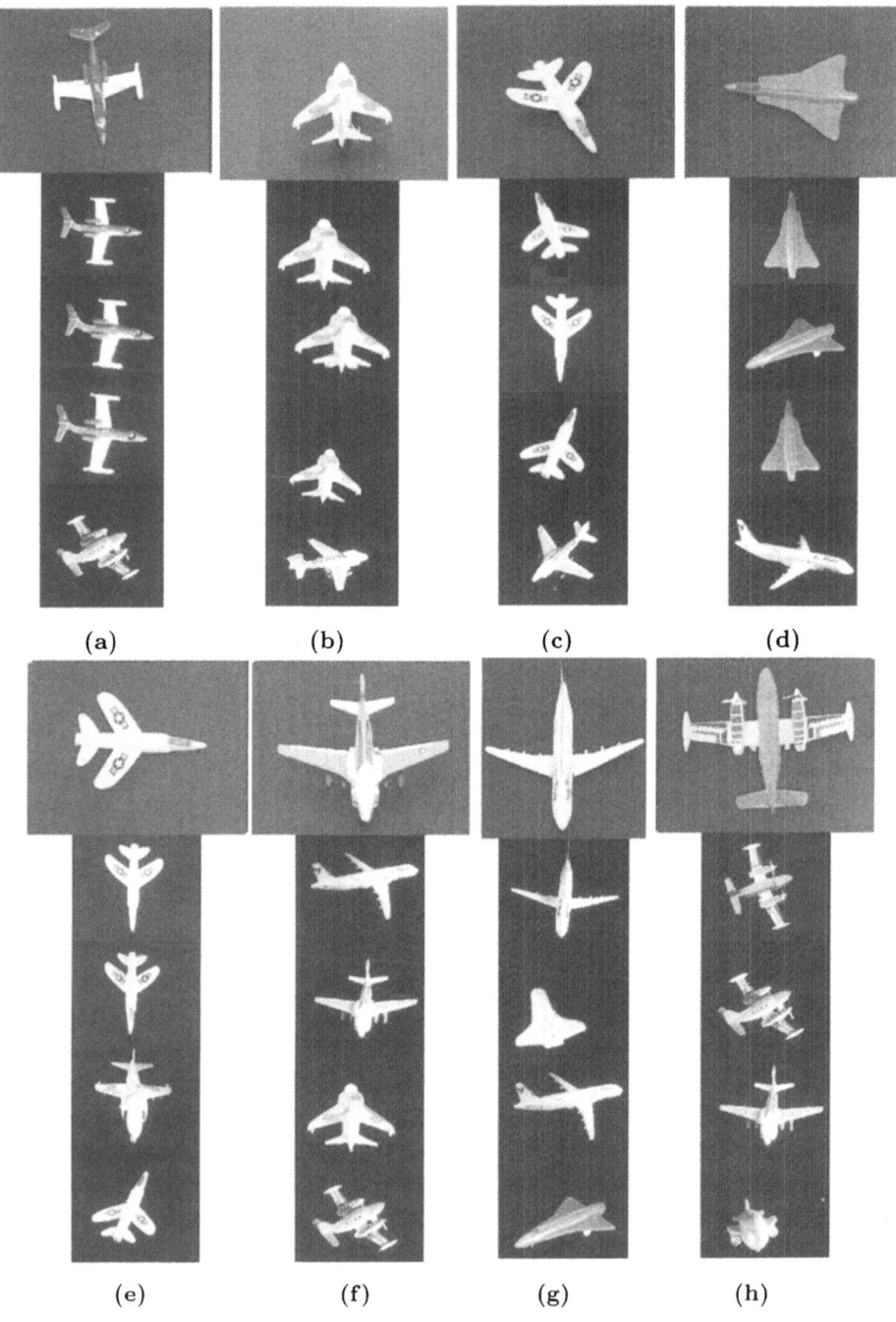

Figure 5.19.   Query examples, clustering method.

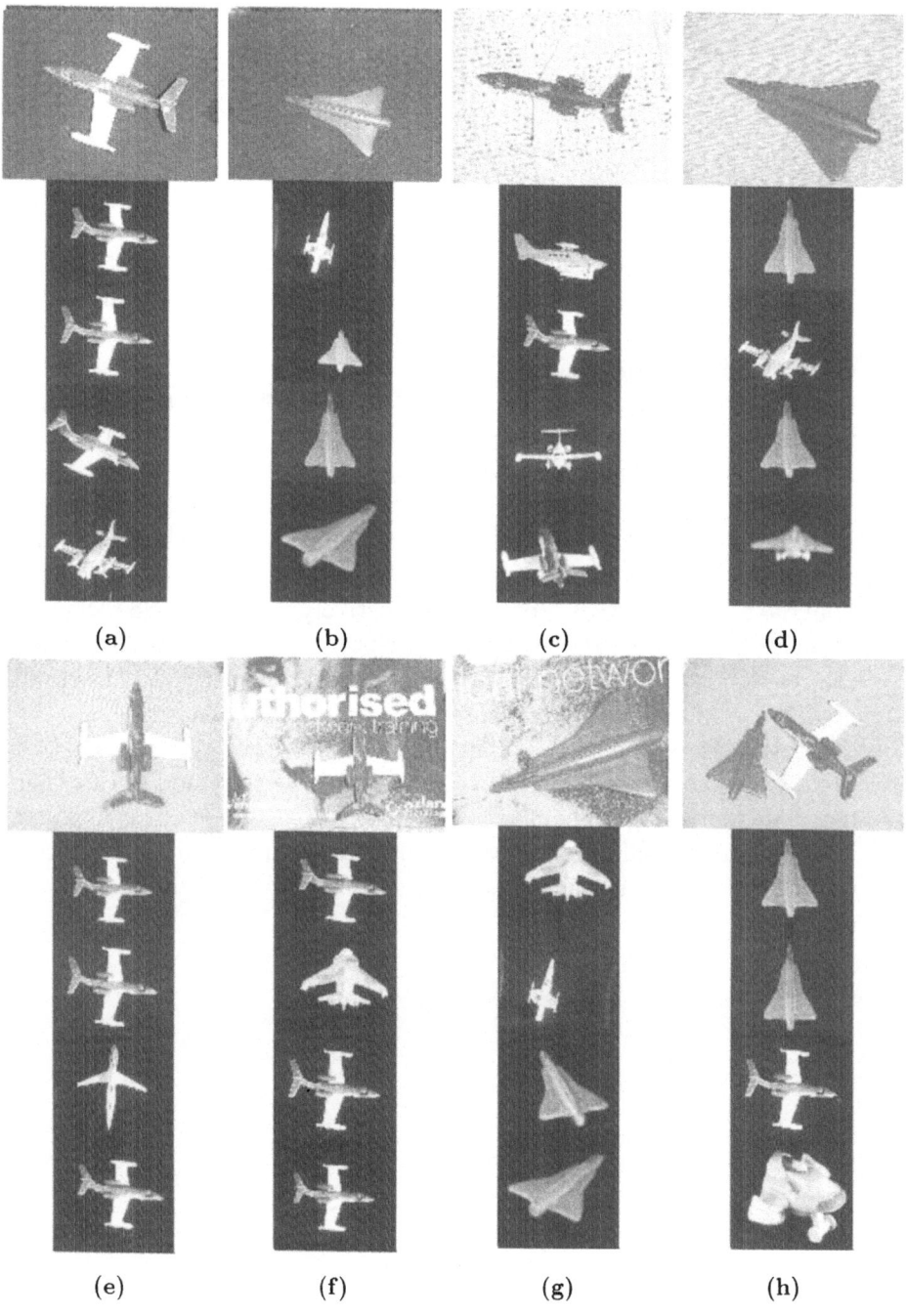

Figure 5.20. Query examples, clustering method.

contours are created. However the pattern of these contours does not match the pattern of any image models. As a result, only object edges create significant clusters in the verification stage. This is also the case when two or more objects are present in the scene, each object's edges create a separate cluster in the parameter space which matches with the corresponding model. Fig. 5.20(h) represents an example.

5.   Conclusions

This chapter addressed the problem of retrieval of free-form 3-D objects using arbitrary views of those objects as queries. It also addressed the issue of automatic selection of the best and the optimum number of views for each object in multi-view 3-D object representation and retrieval. After representing each object by an optimal number of views, an unknown object is recognized by a single image taken from an arbitrary viewpoint using the same shape representation.

The chapter also presented a novel method for fusion of results from multiple shape descriptors. The utilization of this method for multi-view 3-D object representation and retrieval was explored. The object boundary of each view is represented effectively using the CSS technique, Moment Invariants, and Fourier Descriptors. It has been shown that the results obtained from the fusion method are superior to the results obtained from any single technique.

A complete system for free-form 3-D object retrieval using partially occluded or noisy views of those objects was also presented in this chapter. Substantial experiments showed that this system is also robust and reliable for retrieval despite background clutter, noise and partial occlusion of the objects in query images.

Chapter 6

MPEG-7 STANDARDISATION OF THE CURVATURE SCALE SPACE SHAPE DESCRIPTOR

1. Introduction

This chapter is concerned with the standardisation of Curvature Scale Space technology within the MPEG-7 Standard. MPEG-7, also known as the Multimedia Content Description Interface, has been approved by the International Standardisation Organisation in December 2001 and is likely to become the leading Multimedia Description standard. Standardisation of technology is important for technical and commercial reasons. During the standard development stage, different technical solutions to a given problem are proposed, subsequently analysed and subjected to detailed tests and extensive peer-review. This is followed by a collaborative stage, where all participants work jointly on improving the selected technology, possibly by adapting elements from other techniques or simply by introduction of modifications in elements that are considered inferior. Consequently, the solution adopted is likely to offer excellent performance in a broad range of applications and, equally importantly, have the confidence of the industry and users. From the commercial point of view, adoption of a single standardised technology guarantees interoperability of equipment and accelerates or even enables many new applications. Equipment and services costs are reduced due to increased volumes, the users are confident that their new equipment will interoperate and would not become obsolete too quickly, and the manufacturers have a better chance to recoup their development costs and perhaps make some profit.

Previous chapters describe the development of the theory of Curvature Scale Space, and its application to various problems, in particular, to shape description and analysis. This chapter is a logical continuation

- it shows further development and adaptation of the CSS technology for the visual part of the MPEG-7 Standard. This is of interest for several reasons. Firstly, the extensive performance evaluation, including comparison with other state-of-the art techniques, offers many insights into the properties of CSS. Secondly, we present the ISO MPEG-7 contour shape descriptor and discuss in detail various aspects of this technology and its use. Fine points of the technique and its possible implementations, including the usual trade-offs between computational complexity and performance are presented. Many of the details, especially related to non-normative elements of descriptor extraction and matching cannot be found elsewhere. Finally, we show some applications enabled by the use of the contour shape descriptor within the MPEG-7 framework.

The organisation of the remainder of this chapter is as follows. Section 2 presents a short overview of the MPEG-7 standard. The objectives of the standard are discussed and its components briefly introduced, with the emphasis on the visual descriptors, which include shape descriptors. The standardisation process is also briefly outlined. Section 3 introduces MPEG-7 shape descriptors and their application domains. Section 4 describes in detail the MPEG-7 contour shape descriptor, including descriptor syntax, semantics, and the descriptor extraction and matching processes. Section 5 presents the region-based shape descriptor. Details of the performance evaluation methodology and the databases used are described in section 6, while section 7 presents experimental results. Example applications are presented in section 8, and concluding remarks are included in section 9.

2.   MPEG-7 Overview

One of the effects of the increasing use of digital multimedia technologies, systems and services in everyday life is the exponential growth of volume of the data we can access. Sources of digital information include the World Wide Web, audio-visual digital archives, broadcast data streams, and personal and professional databases. The clear trend is that the amount of information, and its quality, will continue to grow. While more users should benefit from these expanding resources, managing content efficiently is becoming increasingly complex, mainly because of its sheer volume. If a piece of information cannot be easily identified and accessed by a user, it is of no benefit. Both professional and private users need advanced tools to support database retrieval and search in digital libraries. There are other areas, for example broadcast channel selection, multimedia editing or multimedia directory services, which have similar requirements. The MPEG-7 Standard was born in response to these demands.

MPEG-7 is an ISO Standard developed by the Moving Picture Experts Group (MPEG) over the period 1999-2001. The MPEG committee is primarily known for the successful development of a series of video compression standards: MPEG-1, MPEG-2, and MPEG-4. The first two standards had profound impact on the electronic industry by enabling new products such as interactive video on CD-ROM and Digital Television. They created a multi-billion dollar industry. MPEG-4, which was only completed in 1999, provides standardised technologies supporting the integration of multimedia content production, distribution and consumption. It supports a wide range of video formats, content manipulation, scalability of the compressed representation and many other useful features.

MPEG-7 differs from its predecessors in that it not concerned with data compression. Its objective is to describe the content of multimedia data, so that it can be efficiently searched, accessed, transformed or adapted to be used by any device. It is also known under the more descriptive name *multimedia Content Description Interface*. MPEG-7 can be viewed as *a few bits about many bits*, and is in fact a specialised short-hand, known as a metadata standard.

The major challenge in its design comes from the combination of two objectives: i) that standard elements should support as broad a range of applications as possible and ii) that it should encompass some degree of interpretation of the *meaning* of the information. A good introduction to the MPEG-7 Standard and its components is presented in [174].

MPEG-7 supports multiple applications, including cases where the multimedia data and descriptions are stored or streamed (e.g. broadcast or 'push-type' applications on the Internet). The descriptors can be extracted in real time or computed off-line and stored for future processing or use. The data and its descriptions do not have to be stored in the same physical location.

There are four basic conceptual components of the Standard: Descriptors (D), Description Schemes (DS), Description Definition Language (DDL) and System Tools (ST). Functions of each component can be summarised as follows:

- Descriptors are representations of some content feature(s), for example, the shape of an object or the pitch of a sound. The Standard defines the syntax and semantics of each component of the representation. When required by interoperability requirements, some elements or even an entire feature extraction procedure may also be defined in the normative part of the Standard. However, for many descriptors the extraction and matching are non-normative. Descriptors are divided into audio and visual Descriptors.

- Description Schemes describe multimedia content using more complex structures, which may use audio-visual descriptors as components. They may also specify content features that are not specific to video or audio, for example recording time or copyright owner. Description Schemes may even include semantics of the relationship between their components, which may be Descriptors or Description Schemes.

- A Description Definition Language (DDL) is defined to support creation of new Description Schemes by the users. Some useful Description Schemes are already created by the Standard; they can be further extended or modified with DDL.

- System Tools are developed to support efficient transmission and manipulation of the MPEG-7 entities. This includes efficient coding in textual and binary formats, multiplexing of the descriptions, synchronisation of the descriptions with content and transmission mechanisms.

Figure 6.1 illustrates relations between the elements of the Standard.

2.1.   Parts of the MPEG-7 Standard

The MPEG-7 Standard consists of eight parts:

1 *MPEG-7 Systems* cover the tools that are needed to prepare MPEG-7 Descriptions for efficient transport and storage, and to allow synchronisation between content and descriptions. It also supports tools related to managing and protecting intellectual property.

2 *MPEG-7 Description Definition Language* defines language to express new Description Schemes and new Descriptors.

3 *MPEG-7 audio* specifies basic Descriptors concerned solely with audio features.

4 *MPEG-7 video* defines basic Descriptors concerned solely with Visual features and more complex Visual Description schemes.

5 *MPEG-7 multimedia Description Schemes* specifies the Descriptors and Description Schemes dealing with Generic features, such as time, location, semantic relations, etc. Based on basic audio and video descriptors, it defines multi-modal multi-media descriptions.

6 *MPEG-7 Reference Software* contains an example software implementation of the relevant parts of the MPEG-7 Standard. It is also known as eXperimental Model or XM for short.

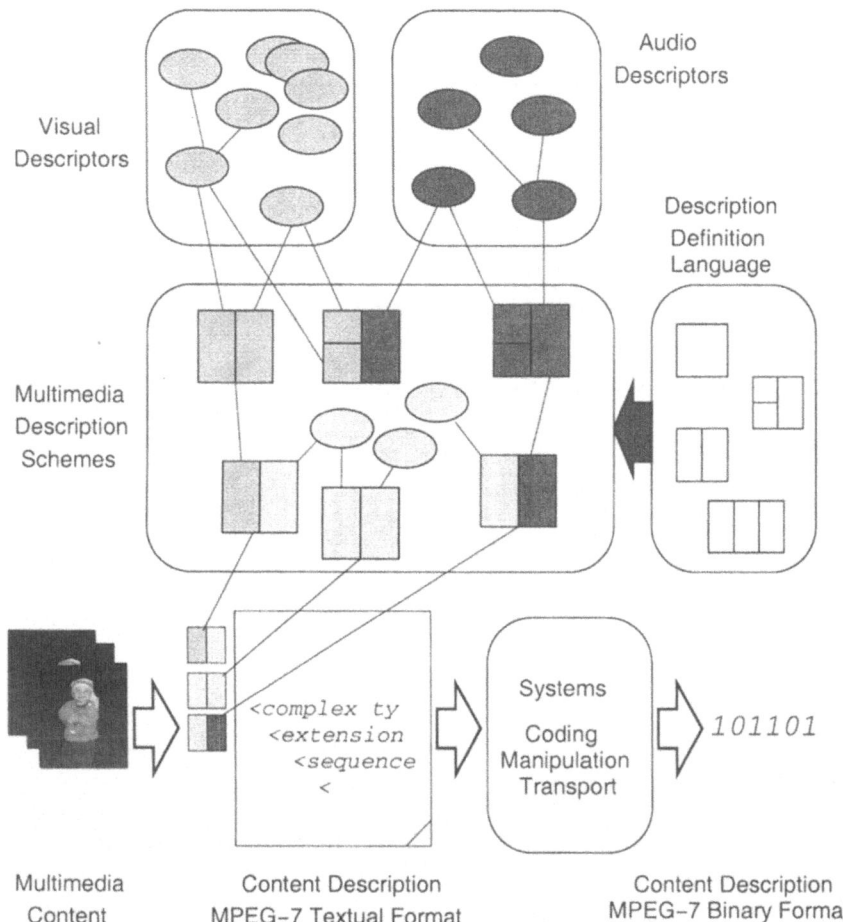

Figure 6.1.   Elements of the MPEG-7 Standard and how they interact in the process of description creation

7 *MPEG-7 Conformance* specifies guidelines and procedures for testing conformance of MPEG-7 implementations.

8 *Extraction and Use of Descriptions* is an informative part, containing non-normative guidelines about the extraction procedures and use of some of the Description Tools.

The Standard was made possible thanks to contributions from many groups, including content creators and managers, broadcasters, publishers, electronic-equipment manufacturers and intellectual property rights managers. The technology underpinning the standard has also been developed by research institutions and academia, based on the state-of-the art technology in computer vision, signal processing, pattern recognition, data coding and many other disciplines.

In the following we will mainly focus on Part-3 of the MPEG-7 Standard, which specifies visual Descriptors. We will review in detail the Contour shape descriptor based on Curvature Scale Space and other shape descriptors.

2.2.   MPEG-7 Visual Part

MPEG-7 Visual Part includes Descriptors concerned with images and video data. It can be divided into the following groups, based on the type of visual features they describe:

- Basic Structures

- Colour

- Texture

- Shape

- Motion

- Localisation

- Others

Each of the above categories usually contains a few descriptors, addressing different functionality or exhibiting a different level of sophistication. We will naturally focus our attention on 2D shape descriptors; the Visual Part of the MPEG-7 Standard defines the *Contour shape descriptor* and the *Regions shape descriptor* [39]. In the following sections we will review the theory underpinning the contour shape and region-shape descriptors and analyse their features and application domains. We will also present experimental results showing the strength and application domains for each of them.

3.   MPEG-7 Shape Descriptors

Object shape provides very powerful features for similarity search and retrieval, and is often strongly linked to object functionality and identity. Shape often carries semantic information, and objects can be recognised solely from their shapes. For example the reader will have no problems with recognising, the objects presented in Figure 6.2, based on their shapes. This would not be possible using solely other visual features, such as object colour, texture or motion. Object recognition, even at a limited level, requires a flexible approach with strong generalisation properties. Consequently, demanding requirements are imposed on the properties and performance of the shape descriptors.

Figure 6.2. Many objects can be easily recognised based solely on their shapes

Figure 6.3 illustrates that there are two notions of similarity: one based on object contour and one based on the distribution of all the pixels within the object [39]. Objects in the first row can be considered similar based on the region-based criterion because they have similar spatial distribution of pixels. The narrow cuts in the object hardly have any effect on the overall distribution of pixels, but they can change object contour dramatically. It is clear that the objects in the first row have different outline contours. When contour-based similarity is concerned, objects can be grouped in columns.

MPEG-7 supports both notions of similarity, with its *region-based* and *contour-based* shape descriptors.

In the following sections we will review the theory underpinning the contour shape and region-shape descriptors and analyse their features and application domains. We will also present experimental results showing the strength and application domains for each of them.

4. Contour-Based Shape Descriptor

The contour shape descriptor selected for MPEG-7 is based on the Curvature Scale Space representation of the contour, which is described in detail in Chapter 1. Before proposing CSS for the standardisation process, a number of approaches to shape representation was considered. The CSS-based technique was chosen due to several unique properties:

- It emulates well the human perception of shape similarity.

- It can cope with partial occlusion.

- It is robust to significant non-rigid deformations.

- It supports matching under perspective transformations.

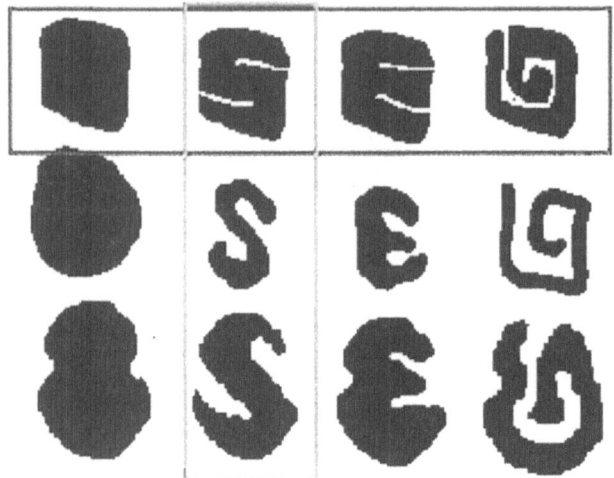

Figure 6.3.   Examples of contour-based and region-based similarity, (c) 2001 IEEE, from [39]

Several important modifications and additions were introduced to the basic CSS concept. Firstly, a different curve evolution approach was applied, based on a simple recursive filter. Secondly, two additional global parameters were added in order to further improve matching performance and speed. Thirdly, the size of the descriptor was reduced two-fold by a sophisticated quantisation approach. Finally, new matching strategies were developed to reduce matching complexity and further improve the retrieval performance.

4.1.   The Contour Shape Descriptor

The Contour shape descriptor consists of five elements:

- Circularity of the original contour

- Eccentricity of the original contour

- Circularity of the prototype contour. The prototype contour is defined as the contour at the final stage of evolution, i.e. when it becomes first convex.

- Eccentricity of the prototype contour

- Set of CSS peak parameters

4.1.1 Global Parameters

As noted above, two additional parameters were introduced to the representation presented in Chapter 3. In addition to eccentricity and circularity of the original contour, we also use eccentricity and circularity extracted from a so-called prototype contour. The prototype contour is defined as the contour at the final stage of evolution, i.e. when it becomes convex. The global parameters of the prototype contour carry important additional information about global properties of the contour, thus improving the overall retrieval performance. Note that unlike the parameters extracted from the "original" curve, the global parameters of the prototype curve are less sensitive to noise or quantisation of the curve.

For completeness, below we provide the definition of the circularity and eccentricity. Circularity is the ratio of the perimeter squared to the area:

$$circularity = \frac{perimeter^2}{area} \qquad (6.1)$$

Eccentricity is a parameter based on object region moments and can be calculated as follows:

$$eccentricity = \sqrt{(\frac{i_{20} + i_{02} + \sqrt{i_{20}^2 + i_{02}^2 - 2i_{20}i_{02} + 4i_{11}^2}}{i_{20} + i_{02} - \sqrt{i_{20}^2 + i_{02}^2 - 2i_{20}i_{02} + 4i_{11}^2}})} \qquad (6.2)$$

where $i_{p,q}$ are the central moments, e.g.

$$i_{02} = \sum_{k=1}^{N}(y_k - y_c)^2 \qquad (6.3)$$

$$i_{11} = \sum_{k=1}^{N}(x_k - x_c)(y_k - y_c) \qquad (6.4)$$

$$i_{20} = \sum_{k=1}^{N}(x_k - x_c)^2 \qquad (6.5)$$

The above sums are computed over all points on or inside the contour shape $\{p_1, ..., p_N\}$ and (x_c, y_c) is the centre of mass of the shape.

4.1.2 CSS Peak Parameters

Humans, when comparing shapes, tend to decompose contours into concave and convex sections. This is followed by finding the best match between different sections of the compared contours. The overall shape similarity is then assessed based on the similarity of the corresponding sections. A good match is achieved when each section on one contour has a corresponding section on the second contour. The corresponding sections should have similar perceptual significance, and their order and relative positions on the contour should be preserved. As has been shown in the previous chapters, the CSS representation exhibits remarkably similar properties. The inflection points, or curvature zero-crossing points, effectively segment the contour into concave and convex sections. This process is repeated at multiple scales (as the contour evolves), providing information about the perceptual significance of each section. Sections that disappear at early stages in the evolution, corresponding to shallow and short concavities, are less perceptually significant than sections surviving longer.

In order to extract positions of the peaks in the CSS image, the contour is progressively smoothed until it becomes convex. The smoothing process is controlled by a parameter σ, which may reflect the parameters of the filter or, as in this case, the number of passes of a digital filter. Evolution of a contour during filtering is shown in the left column in Figure 6.4. The right column shows the curvature functions associated with contours at various stages of evolution. The curvature of a contour at any given point is defined as the derivative of the tangent angle to the contour. For a closed, planar contour, which is parametrised by the normalised arc length parameter, the curvature can be calculated as:

$$k(u) \quad = \quad \dot{x}(u)\ddot{y}(u) - \ddot{x}(u)\dot{y}(u), \tag{6.6}$$

where $\dot{x}(u)$ and $\ddot{x}(u)$ are the first and second order derivatives of $x(u)$ with respect to the parameter u. The CSS image is defined by the zero-crossings of the curvature functions computed for a number of contours at various stages of evolution. Figure 6.4 shows three contours after 3, 29 and 100 iterations and the corresponding curvature functions with zero crossings marked as black dots. As expected, the number of zero crossings is reduced by filtering: there are 6 zero crossings on the top contour (marked as A,B,C,D,E,F), 4 zero crossings on the middle contour and no zero crossing on the lower contour, which is convex. A two dimensional vector (s, l) can be associated with each zero crossing. The first coordinate s corresponds to the amount of smoothing applied and the second coordinate l corresponds to the position of the zero crossing on the contour. This position can be, for example, the distance along the contour from an arbitrarily chosen point on the contour to the inflection point, when circulating the contour clockwise. The CSS image is constructed by marking "black" all zero crossing points on a 2-dimensional plane, where the vertical axis corresponds to the amount of filtering and the horizontal axis corresponds to the position on the contour. Each contour defines one horizontal line in the CSS image, and the number of rows in the image is equal to the number of contours (each at different resolutions) used in this construction. The CSS image generated by the evolution of the fish is shown at the bottom of figure 6.4.

Now we describe how to compute peak parameters according to the MPEG-7 recommendations. Since only the the position of the peaks are needed to construct the descriptor, the CSS image does not have to be explicitly extracted. We assume that the contour from which we would like to extract the parameters is represented by a set of M contour pixels. In the first step, N equi-distant points are selected on the contour starting from an arbitrary point (called reference point). The contour is followed clockwise. Next, the x-coordinates and y-coordinates of the selected N points are grouped into two series X and Y. The evolution of the contour is performed by repetitive application of a low-pass filter with the kernel $(0.25, 0.5, 0.25)$ to x and y coordinates of the selected N contour points. During contour evolution, its concave parts gradually flatten out. As the evolution of the contour progresses, the curvature function and its zero-crossings are computed. Curvature estimation on the evolved contour is based on difference techniques. The positions of the zero crossings are tracked across scales, and the peaks are detected when two neighbouring zero-crossings "disappear" or merge. The position of the peak is recorded as (i_k, j_k), where i_k is the normalised arc-length distance on the contour between the reference point and the zero-crossing, moving clockwise on the contour, and j_k is the number of applications of the low-pass filter. The filtering process terminates when the contour becomes convex, which can be detected by the lack of zero-crossings in the corresponding curvature function. Note that the procedure outlined above is not optimised for speed, and that faster approaches exist.

The peak location values are suitably normalised so that they are invariant to the number of sampling points and the size of the contours. The peaks are ordered based on decreasing values of y_CSS, transformed using a non-linear transformation and quantised. The details of the quantisation are explained in Section 4.2.

4.2.   Efficient Representation of Descriptor Parameters

Compactness of the descriptor is a very important feature. We have conducted detailed studies of the number of bits needed to represent the coordinates of the peaks. They show that the use of 6 and 7 bits for x

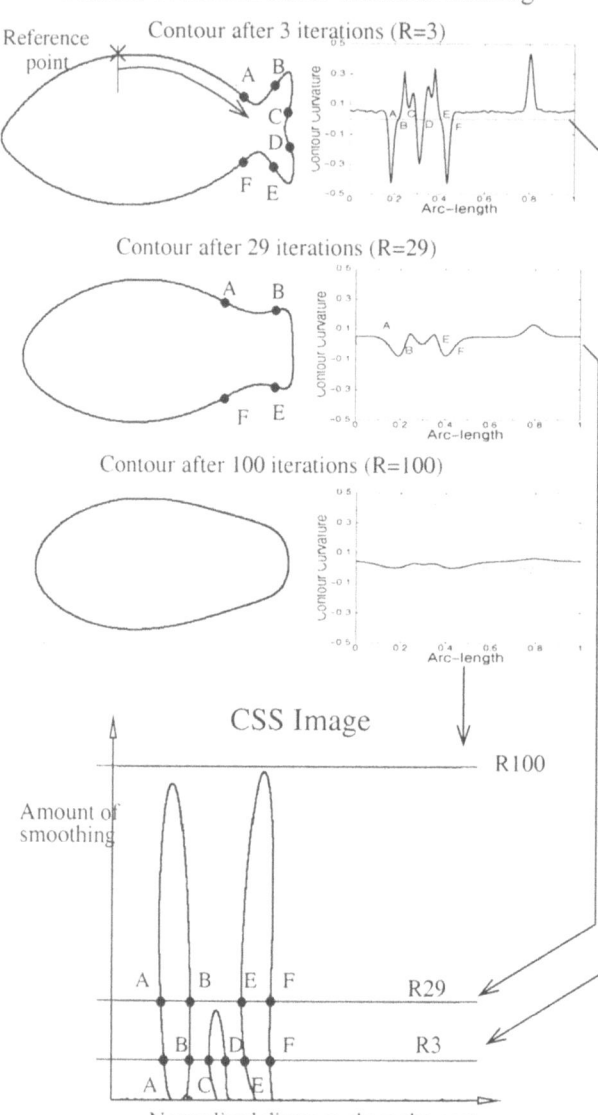

Figure 6.4. The contour evolution process for a fish, showing contours during evolution, their curvature functions and the CSS image, copyright (c) 2002 John Wiley and Sons Limited, Reproduced with permission from [174]

and y components respectively, causes only a minimal deterioration in the retrieval performance. Using this quantisation approach requires 13 bits per peak. However, by using two additional modifications, further optimisation can be obtained.

As explained earlier, changes to the reference (starting) point on the contour correspond to a circular shift of the CSS image along the $x-axis$. We can therefore shift the positions of the peaks so that one of the peaks (e.g. the highest one) has $x = 0$. There is no need to store the x position of the highest peak resulting in savings of 6 bits.

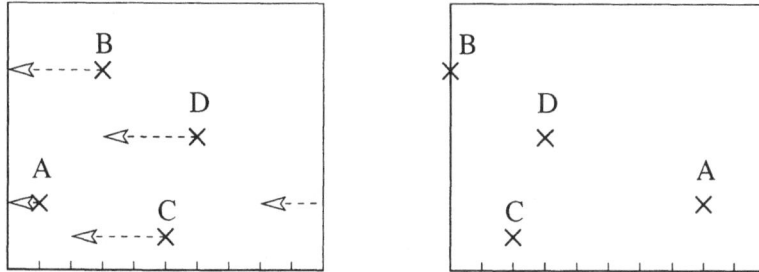

Figure 6.5. The highest peak is shifted so that it has $x = 0$

In order to further improve compactness of the representation, we devised a special relational quantisation scheme for the peak heights parameters (Figure 6.6). Firstly the peaks are ordered based on their height. The first peak is quantised in the range $[0, y_{MAX}]$ using 7 bits. Each remaining peak is quantised in the range determined by the height of the previous, higher peak, using only 3 bits. For example peak D is quantised in the range $[0, y_B]$. As a result of the relational quantisation of the peak height, the number of bits used to code peak parameters was reduced from 13 to 9.

The circularity parameter of the original or prototype curve is uniformly quantised to 6 bits in the range $[12, 110]$. Simple calculations show that from all closed curves, the circle gives a minimum circularity value equal to 4π. It is possible to construct a curve with an arbitrarily large circularity value, however, the circularity of common contours is rarely greater than 50. The maximum value which can be expressed within the standard is 110. If the circularity value calculated is above that limit, the value is clipped to 110.

The eccentricity parameter of the original or prototype curve is uniformly quantised to 6 bits in the range $[1, 10]$. It follows from the def-

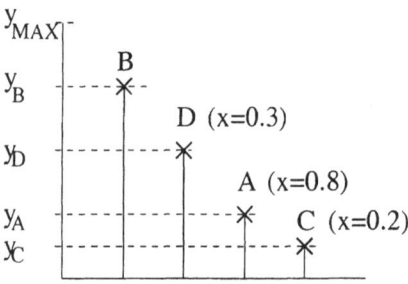

Figure 6.6.   Peak heights are quantised based on peaks ratio

inition of the eccentricity that its value is never less than 1. If the eccentricity value calculated is above 10, the value is clipped to 10.

4.3.   Matching of the Contour Shape Descriptors

A search in a database is performed by evaluating the similarity measure between the descriptor of the query shape and the descriptors associated with the shapes present in the database (sometimes referred to as model shapes).

The definition of the similarity measure is included in Part-8 of the MPEG-7 Standard [125], which is informative only. While a high-performance similarity measure is recommended by the Standard, users are free to select a different approach for their application and the application will still be compatible with the Standard. There could be several reasons for doing so. For example, when a specific class of shapes is concerned, a different similarity measure may better reflect the shape properties within the class and the required notion of similarity. Another example is when an application developer reduces the computational complexity of the descriptor matching stage, perhaps at a cost of some reduction in performance.

Here we describe the matching procedure included in the non-normative part of the Standard and also present optimised approaches which achieve improved performance with reduced matching complexity.

The shape descriptor $\mathcal{D}$ of a shape consists of several components.

$$\mathcal{D} \;=\; \{Ec, Cc, Ep, Cp, yp_0, \{(xp_i, yp_i)\}; i = 1, .., k\} \qquad (6.7)$$

where Ec, Ep are the eccentricities of the original and prototype contours, Cc, Cp are the circularities of the original and prototype contours,

yp_0 is the height of the most prominent peak (if contour is concave) and $\{(xp_i, yp_i)\}; i = 1, .., k\}$ is a set of remaining peaks (possibly empty).

The matching procedure is a two-stage process. Firstly, the global parameters of the query and model shapes are compared, and if they differ significantly, no further comparison is performed and shapes are considered dissimilar. This approach reduces dramatically the amount of calculations required, by rejecting contours which differ significantly. In the second stage, a more detailed comparison of shape is performed, based on the relative positions of CSS peaks in the query and model descriptors.

4.3.1 First Stage of Matching

In the first stage of matching, the similarity between the global parameters of the query and model is computed. As there are four global parameters (eccentricity and circularity of the original and prototype shapes), four similarity coefficients $S_{Ec}, S_{Cc}, S_{Ep}, S_{Cp}$ can be computed, as defined by equations below,

$$S_{Ec} = \frac{\mid Ec^q - Ec^m \mid}{max(Ec^q, Ec^m)} \tag{6.8}$$

$$S_{Cc} = \frac{\mid Cc^q - Cc^m \mid}{max(Cc^q, Cc^m)} \tag{6.9}$$

$$S_{Ep} = \frac{\mid Ep^q - Ep^m \mid}{max(Ep^q, Ep^m)} \tag{6.10}$$

$$S_{Cp} = \frac{\mid Cp^q - Cp^m \mid}{max(Cp^q, Cp^m)} \tag{6.11}$$

where superscripts m and q indicate the model and query parameters respectively.

According to the procedure recommended in [125] two shapes pass the first-stage matching if they satisfy the following conditions:

$$S_{Ec} < T_{Ec} \tag{6.12}$$
$$S_{Cc} < T_{Cc} \tag{6.13}$$

where the two thresholds are: $T_{Ec} = 0.6$ and $T_{Cc} = 1.0$. Table 6.1 summarises results obtained with this and other strategies applied in the first stage of matching. It shows the Overall performance, the retrieval performance in the Similarity test (described in detail in section 6) and the execution time per match. The details of the performance evaluation methodology and the data test-sets used are presented in Section 6.1. The execution time quoted is the average time per individual match,

Pre-filtering parameters	Overall performance	Similarity	Matching time [μS]
S_{Ec}, S_{Cc}	89.69%	79.15%	178
$S_{Ec}, S_{Cc}, S_{Ep}, S_{Ec}$	90.85%	81.12%	137
$S_{Ec}, S_{Cc}, S_{Ep}, S_{Ec}, S_{P0y}, S_{P1y}$	90.62%	80.54%	57

Table 6.1. Comparison of the retrieval performance and speed for several pre-filtering strategies

based on the XM software implementation (version 5.5) runing on the Pentium PC with a 2GHz CPU clock. This time is an indication of the matching complexity, but more efficient implementations than the XM exist, so it should not be taken as an absolute measure of the complexity. The first row in the table presents results when two global parameters S_{Ec}, S_{Cc} are used for thresholding in stage one, as described in [125].

The best overall performance with a reduced matching and search time can be achieved by implementing an alternative strategy where the similarity of the global parameters Ep, Cp of the prototype contour are also used in the filtering stage:

$$\begin{aligned} S_{Ep} &< T_{Ep} \\ S_{Cp} &< T_{Cp} \end{aligned} \tag{6.14}$$

Good threshold values for the global parameters of the prototype contour are: $T_{Ep} = 0.5$ and $T_{Cp} = 0.25$. As can be seen in Table 6.1 (second row), the pre-filtering of shapes based on all four global parameters gives the best performance in each part, and consequently the best overall performance. It also reduces the computational complexity of the matching procedure by approximately 25%.

The matching speed can be reduced even further by additional pre-filtering, based on the heights of the two most-prominent peaks.

$$\begin{aligned} S_{P0y} &= \mid yp_0^m - yp_0^q \mid < T_{P0y} \\ S_{P1y} &= \mid yp_1^m - yp_1^q \mid < T_{P1y} \end{aligned} \tag{6.15}$$

where yp_0^m and yp_0^q are the heights of the highest peaks in the model and query respectively. The yp_1^m and yp_1^q are the heights of the second-highest peaks associated with the the model and query shapes. The threshold values used are: $T_{P0y} = T_{P1y} = 0.2$. If one or more of the peaks in equations 6.15 is not present, it is assumed that the condition is

Coefficient	Value
α_1	0.6
α_2	1.0
α_3	0.6
α_4	1.6

Table 6.2. Weighting coefficients used for the computation of overall similarity (second stage)

fulfilled. The results achieved for this approach are shown in the bottom row of Table 6.1. In this case, there is a significant, three-fold, increase in the matching speed with a minimal drop in performance, which is still better than that of the reference approach. All three strategies outlined used the same procedure and parameter values in the second stage matching, which is described in the following subsection. Further improvements in the processing speed can be achieved by modifying parameters in the second stage of matching.

4.3.2 Second Stage of Matching

In the second stage, matching of the CSS peaks is performed. The overall similarity measure S is computed as as a weighted sum of the similarity measures between the global contour parameters $(S_{Ec}, S_{Cc}, S_{Ep}, S_{Cp})$ and the similarity measure component S_{CSS} computed from the CSS peaks associated with the model and query shapes

$$S \;=\; \alpha_1 * S_{Ec} + \alpha_2 * S_{Cc} + \alpha_3 * S_{Ep} + \alpha_4 * S_{Cp} + S_{CSS} \quad (6.16)$$

Good values for the weighting coefficients α_i are given in Table 6.2.

The similarity value S_{CSS} is essentially an L_2 distance measure between the matching peaks with an additional penalty for each unmatched peak equivalent to the unmatched peak's height. The peak positions are, of course, de-quantised before the calculation of the similarity value takes place. The matching between the model and query peaks is one to one, that is each model peak may have only one query peak assigned to it and vice-versa. However, it is allowed for some of the model or query peaks to not have any peaks assigned to them. For two peaks I and J to be assigned (or matched), the distance between their x coordinates has to be less then a threshold equal 0.1. The similarity value S_{CSS} is computed from the model-query peak assignment which minimises the overall matching cost.

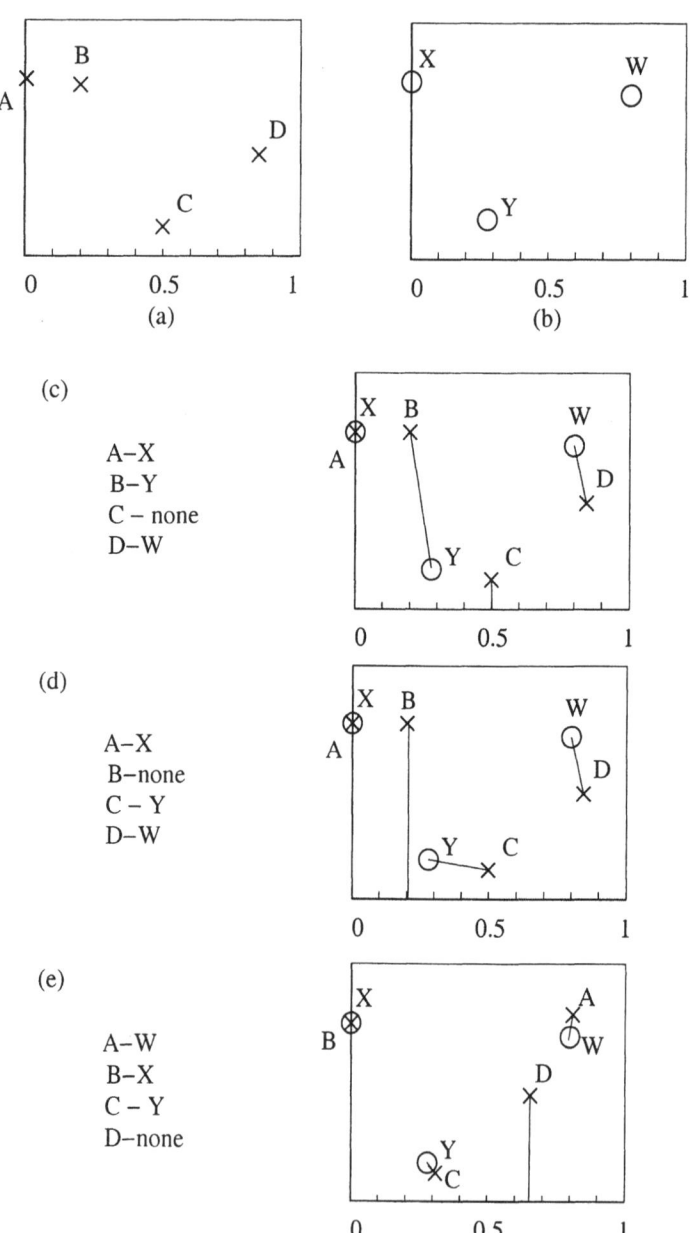

Figure 6.7. Matching between CSS peaks associated with the query and model descriptors

Figure 6.7 shows an example of matching of two sets of CSS peaks. Sub-figure (a) shows four peaks A, B, C, D associated with the query shape and sub-figure (b) shows three peaks X, Y, W associated with the model shape. Sub-figures (c), (d), (e), show example matching between peaks. In sub-figure (c) peak A is assigned to peak X, B is assigned to Y and D is assigned to W. In this case

$$S_{CSS} = d(A, X) + d(B, Y) + d(D, W) + P(C) \qquad (6.17)$$

where $d(I, J)$ is the $L2$ distance between the peaks I, J and $P(C)$ is the penalty for the un-assigned peak C, equivalent to its height y_C. Another possible peak assignment is shown in sub-figure (d), however the pair $C - Y$ does not fulfill the distance condition and therefore it is rejected. The overall assignment between peaks which minimises the error measure S_{CSS} is shown in sub-figure (e). Here, the highest peak X in the model shape is assigned to the second highest peak in the query shape B.

It is clear from the examples shown that there can be many assignments between peaks and a good strategy to find the assignment which minimises the total distance is needed. An exhaustive search for all possible peak assignments is not practical nor necessary. Very good results are obtained by aligning the highest peaks in the query and model shapes. Peaks used for such alignment are called the "reference peaks". In the process of alignment, the query and model reference peaks are shifted vertically to the position $x = 0$ and all other peaks in the query and model descriptors are also shifted by the corresponding vectors (e.g. all peaks in the query descriptor are shifted by the same vector as the query reference peak and all peaks in the model descriptor are shifted by the same vector as the model reference peak). Please note that this is a circular shift, as illustrated in Figure 6.9.

The rationale behind such an approach is that highest peaks correspond to the deepest concavities and are therefore stable and least affected by noise or other shape distortions. However, there can be other peaks with heights comparable to the highest one and they should also be considered when deciding on peak alignment. Consequently, the strategy adopted is based on three principles:

- All peaks in model descriptor that are of similar height as the highest model peak should be considered as possible "reference peaks"

- All peaks in query descriptor that are of similar height as the highest query peak should be considered as possible "reference peaks"

- A peak from the query and a peak from the model create a reference pair if their heights are similar

Thus any model peak P_i^m that fulfills the following condition

$$\frac{yp_i^m}{yp_0^m} > T_{PEAK} \qquad (6.18)$$

is considered as possible reference model peak. The threshold T_{PEAK} is 0.5. A similar condition is imposed on the query peaks to determine which of them can be considered as query reference peaks.

$$\frac{yp_i^q}{yp_0^q} > T_{PEAK} \qquad (6.19)$$

In the next step, all pairs consisting of one query and one model reference peak (that fulfill the conditions 6.18 and 6.19 respectively) are considered. Two peaks yp_i^m, yp_j^q in a pair are considered similar if

$$\frac{|\, yp_i^m - yp_j^q \,|}{\max(yp_i^m, yp_j^q)} > T'_{PEAK} \qquad (6.20)$$

where the threshold T'_{PEAK} is taken as 0.5.

Figure 6.8 shows the reference peak selection strategy for the query and model peaks from Figure 6.7. Subfigure (a) shows the process of selection of the candidate model and query reference peaks. All peaks located in the shaded area (defined by the equations 6.18 and 6.19 and the value of T_{PEAK}) are considered as candidates, in this case peaks A, B, D are the candidate query reference peaks and the peaks X, W are the candidate model reference peaks. Subfigure (b) illustrates what happens when the value of T_{PEAK} is changed to 0.7 - in this case, peak D is excluded from consideration. However, assuming that peak D is included, there are six different reference pairs: $(A, X), (A, W), (B, X), (B, W), (D, X), (D, W)$. Now the condition 6.20 is checked for each pair. Subfigure (c) shows as shaded area the region from which the model peaks fulfil the condition expressed by equation 6.20 with respect to the query peak D. It can be seen that the pair (D, X) does not fulfil the condition and therefore is not considered as a reference pair.

4.4.   Properties of the Contour Shape Descriptor

The contour shape descriptor exhibits many useful properties. Some of the properties were already analysed in the previous chapters, and they will be only briefly mentioned here. It was already shown that the CSS representation is invariant to contour orientation and scale, and to the selection of the reference point and the number of representative

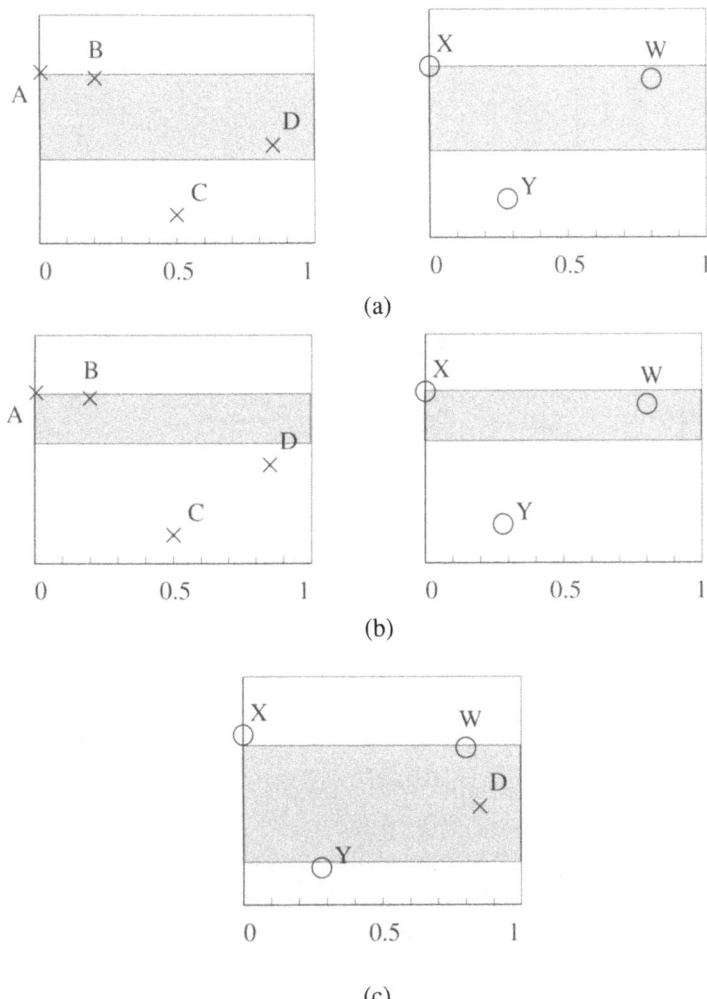

Figure 6.8. Reference peaks selection strategy for the query and model descriptors

points. Mirror shapes can be retrieved by using a similarity measure which takes into account "mirrored peaks". It has also been shown that the representation is robust to noise-like perturbations in the contour. The descriptor inherits all these useful properties.

Another useful feature is that the representation retains the local information about the contour. This is due to the fact that every concavity and convexity on the contour has its own corresponding contour in the CSS image. This feature can be used at the retrieval stage, for example one can determine the number of concave and convex elements on the

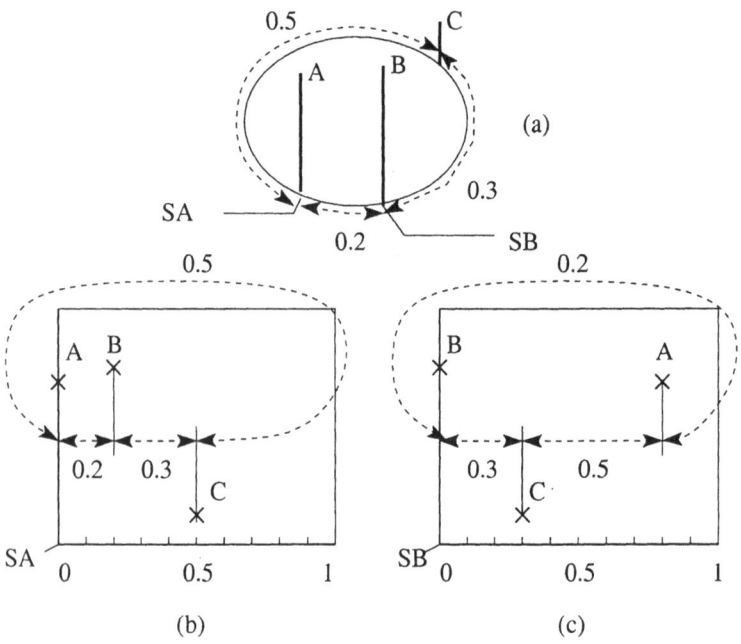

Figure 6.9. Circular rotation of the peak positions is equivalent to selection of different reference points

contour just by checking the number of the peaks. Such queries can not be solved using the region-based descriptor.

As mentioned earlier, shape is one of few visual features which can support - albeit to a limited extent - retrieval of semantically similar objects. As the MPEG-7 testing has shown, the contour shape descriptor outperforms, in that respect, all other techniques tested. It supports search for shapes that are semantically similar for humans, even when significant intra-class variability exists.

The descriptor is also very efficient in applications where high variability in the contour shape is expected. This is often the case in video sequences when a non-rigid object deforms due to motion. Note that even the shapes of rigid objects may undergo significant changes due to perspective transformation, for example as a result of camera motion.

The above properties are clearly visible in the MPEG-7 experiments, described in Section 6.

5.   Region-Based Shape Descriptor

The region-based shape descriptor defined by the MPEG-7 standard belongs to the broad class of shape analysis techniques based on moments [305]. It is based on an orthogonal unitary transform referred to as ART

(Angular Radial Transform). The basis functions of this transform, V_{nm} are defined on a unit disk by a complete set of orthonormal sinusoidal functions expressed in polar coordinates. They are separable along the angular and radial directions:

$$V_{nm}(\rho, \theta) = A_m(\theta) R_n(\rho) \tag{6.21}$$

where

$$A_m(\theta) = \frac{1}{2\pi} \exp(jm\theta) \tag{6.22}$$

$$R_n(\rho) = \begin{cases} 1 & \text{if } n = 0 \\ 2\cos(\pi n \rho) & \text{otherwise} \end{cases} \tag{6.23}$$

The ART coefficients F_{nm} are defined as:

$$F_{nm} = \int_0^{2\pi} \int_0^1 V_{nm}(\rho, \theta) f(\rho, \theta) \rho \, d\rho \, d\theta \tag{6.24}$$

where F_{nm} is an ART coefficient of order n and m, $f(\rho, \theta)$ is an image function in polar coordinates, and V_{nm} is the ART basis function of order n, m.

Real and imaginary parts of ART basis functions are shown in Figure 6.10.

5.1.   The ART Descriptor

The ART descriptor is defined as a set of normalized magnitudes of 35 ART coefficients. Twelve angular and three radial functions are used, thus defining 36 basis functions and resulting in 36 ART coefficients. The coefficient of order $n = 0$, $m = 0$ is excluded from the descriptor because it is used for scale normalisation. All remaining 35 ART coefficients are divided by the magnitude of ART coefficient of order $n = 0, m = 0$. To keep the descriptor size to a minimum, non-linear quantisation is applied to each coefficient using four bits per coefficient. The details of the quantisation are provided in [126].

5.2.   Similarity Measure

The distance between two shapes with ART descriptors is calculated by summing of the absolute differences between all 35 descriptor elements:

$$dissimilarity = \sum_{i=1}^{35} \| M_d[i] - M_q[i] \| \tag{6.25}$$

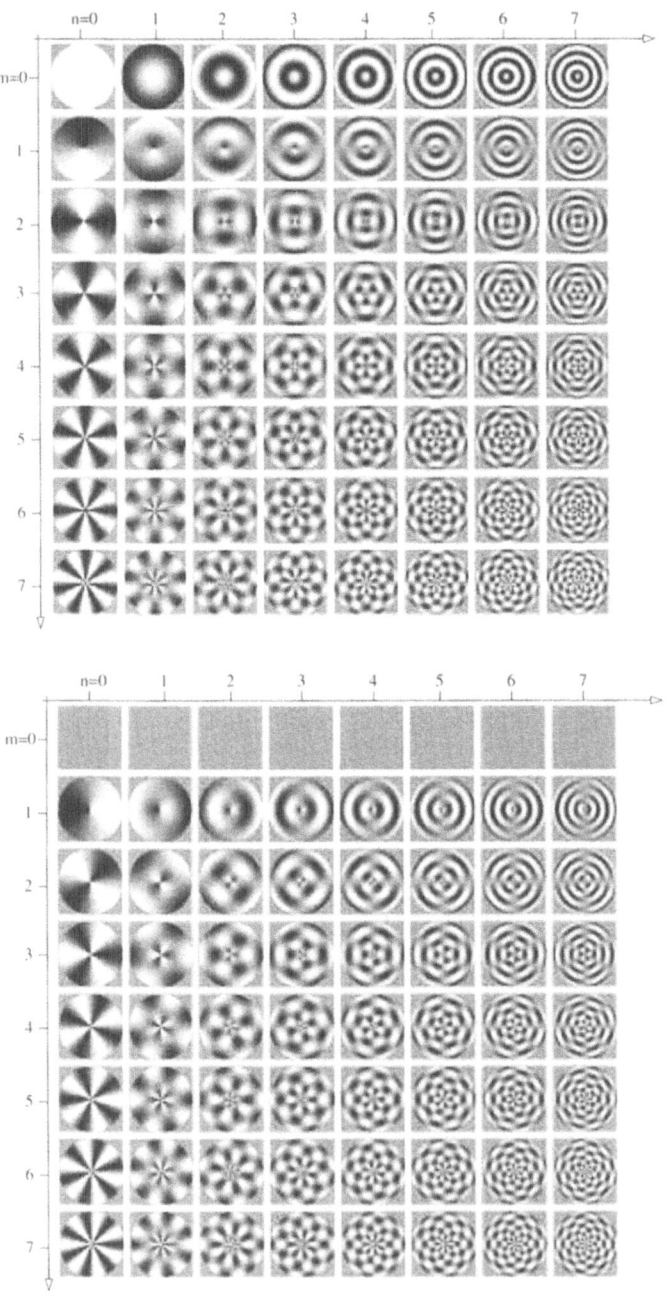

Figure 6.10.   Real and Imaginary parts of ART basis functions

where the subscripts d and q represent image in the database and query image, respectively. M is the array of 35 de-quantised ART descriptor values.

6. MPEG-7 Performance Testing Methodology and Test-Sets

A very detailed evaluation of the performance of the proposed shape description methods was conducted before the final selection was made. In this section we describe in detail the testing methodology and the shape databases used for testing. The experimental results are presented in Section 7.

A good descriptor captures silent perceptual aspects of a shape and stores them in a compact representation, which renders itself easily to fast search and browsing. Representations that mimic the human notion of shape similarity are desired. Two shapes that are considered similar by a viewer should have representations topologically close in the representation space. Since it is not required that the original shape can be reconstructed from the descriptor, significant compactness of the representation can be achieved. The representation should be invariant to scaling, translation and rotation. It should also be robust to distortions commonly encountered in digital images and video. For example scaling of objects to a relatively small size (as compared to the sampling grid) often results in significant shape distortions. Other examples include contour distortions due to segmentation errors, occlusion or perspective deformations. It is critical that the shape representation selected is robust to such effects. Many real-life objects are non-rigid and therefore may be subject to deformations. A good shape representation should be able to cope with such non-rigid deformations. In the next section we describe the components of the MPEG-7 shape dataset, which was designed to support the testing of various elements of the descriptor performance. We then present the evaluation results for a number of descriptors tested, which should help the reader to understand their properties.

6.1. The MPEG-7 Test Database

Performance evaluation of the contour-based descriptors consists of four tests designed to evaluate similarity-based retrieval, but also robustness to rotation, robustness to scaling and robustness to non-rigid deformations.

6.1.1   Similarity-Based Retrieval

The performance in similarity-based retrieval is perhaps the most important of all tests performed. This is because many different shape representations offer good scale and rotational invariance, but very few score highly in the similarity-based test. Moreover, since the notion of similarity is subjective, the assessment of performance is rather difficult. A new way of evaluating the performance was developed in MPEG. The key idea is to start with a database constructed in such a way that similar shapes create clusters. For example, a cluster may include instances of shapes of horses, all depicted from the same view. In such a database, the similarity between shapes belonging to the same cluster is greater that the similarity between shapes belonging to different clusters. In the

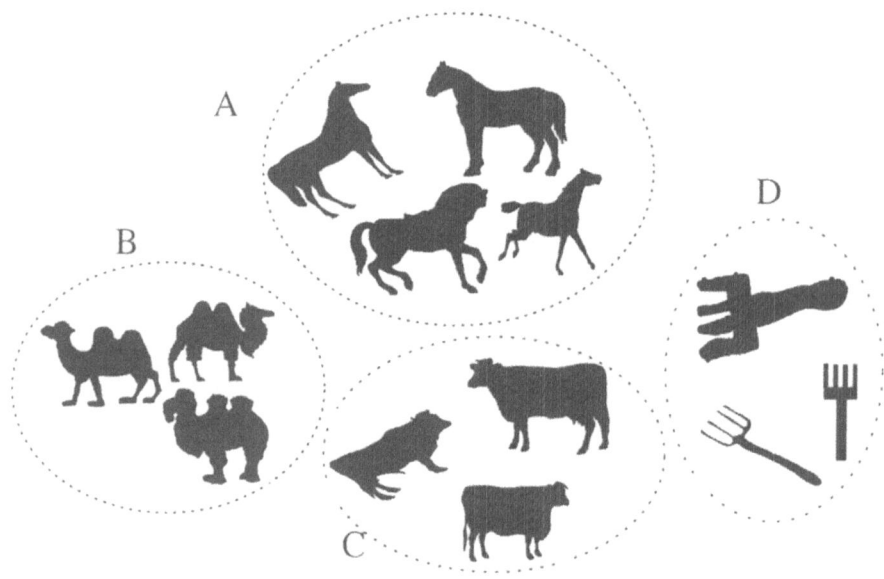

Figure 6.11. Similarity-based retrieval test-set contains groups of perceptually similar shapes, copyright (c) 2002 John Wiley and Sons Limited, Reproduced with permission from [174]

similarity-based retrieval test-set there are 70 classes, four of which are presented in Figure 6.11. In the figure, cluster A contains shapes depicting horses (side view, including articulated motion), cluster B contains shapes of camels (different species), cluster C includes cattle, and cluster D shows man-made objects - forks. Clearly, despite the fact that a significant shape variability exists in each cluster, for example in the position of horse legs or tails, the overall degree of similarity between objects within one class is greater then the similarity between objects belonging to two different classes. An assessment of the retrieval performance is straightforward - a query posed with a shape from a class X should result in all remaining shapes from the same cluster X retrieved with top rankings. The database is rich and it includes examples of natural objects with significant shape variability, such as butterflies or beetles (Figure 6.12, first and second rows); man-made objects (third and fourth rows), objects extracted from cartoons with lesser amounts of shape variability (fifth row) and manually drawn objects with high variation of the contour but simillar according to region-based similarity (last two rows). Overall, there are 1400 images grouped in 70 classes. The testing involves posing a query with each shape in the database and calculating the number of correctly retrieved shapes (i.e. shapes belonging to the same class as the query shape) in the top 40 shapes retrieved.

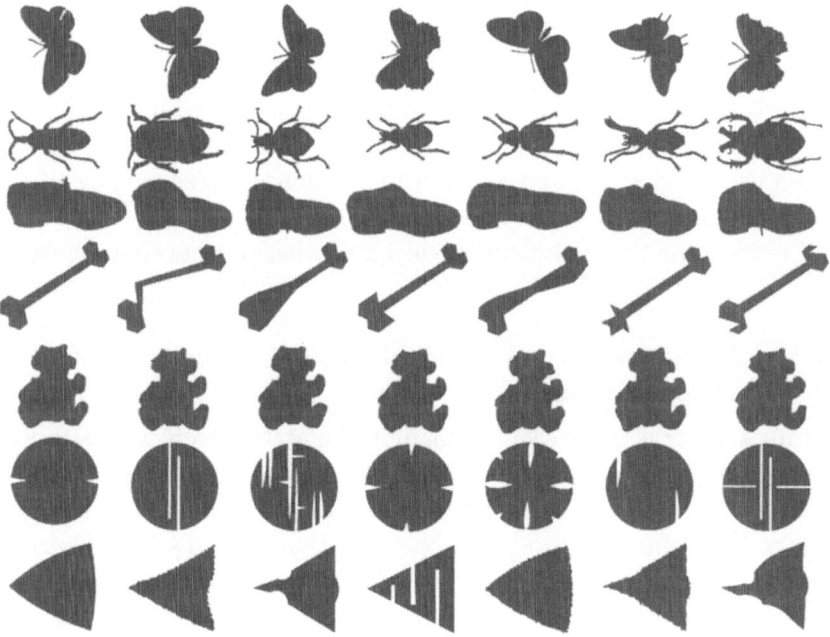

Figure 6.12. Shape clusters represent various notions of similarity

6.1.2 Rotational and Scaling Invariance

While all the techniques tested in the MPEG evaluation theoretically supported rotational and scaling invariance, experimental testing was performed to investigate their behaviour in the presence of shape distortions due to image re-sampling and quantisation of the descriptor representation. Two separate databases were constructed to evaluate scale-invariance and rotational-invariance. In the rotational-invariance test-set, one image from each of the 70 classes was used to obtain zoomed-in and zoomed-out versions. The image is reduced by factors: 0.3, 0.25, 0.2 and 0.1 and enlarged by a factor of 2. The test-set used for rotation-testing contains versions of the reference images rotated digitally anti-clockwise by angles: 9, 36, 45, 90 and 150 degrees. Again, one shape from each class was used as a base for obtaing all the rotated versions. The scale-invariant database and the rotation-invariant database contain 420 shapes each. In both experiments, the query is posed using every image in the respective database, and the number of correct matches in the top 6 ranked shapes is calculated for each query.

The scale-invariance test is very challenging because many objects are severely distorted when reduced by factors of 0.2 or 0.1. However, as the test set also included shapes with less severe scaling, it gave good indication how much degradation the tested technique could cope with. Figure 6.13 shows several examples of shapes where scaling by a factor of 0.1 caused severe distortions. For example, the outline of the running person was divided into two disconnected regions after scaling by a factor 0.1. One may note that some shapes are distorted to the extend that even readers may have difficulties deciding what objects they represent.

6.1.3   Robustness to Non-Rigid Motion as well as Other Deformations

Human observers can easily recognise or match objects which undergo non-rigid motion, which shows that the characteristic shape features are usually retained under such motion. Non-rigid motion is common in video content and therefore it is important that a good shape descriptor can cope with such deformations. For this experiment, a test-set consists of 200 frames depicting a swimming Bream fish combined with 1100 shapes of marine creatures from the SQUID database. The SQUID database contains many examples of fish similar to the Bream fish. The outline of the Bream fish undergoes significant non-rigid deformations due to swimming. Moreover, at some stage the fish turns causing more dramatic changes to its shape. In the test, the first image of the Bream fish is used to pose the query and the number of instances of other Bream fish in the top 200 retrievals is computed.

Figure 6.14 shows the query shape bream-000, several shapes from the bream fish sequence showing non-rigid deformations (top row) and

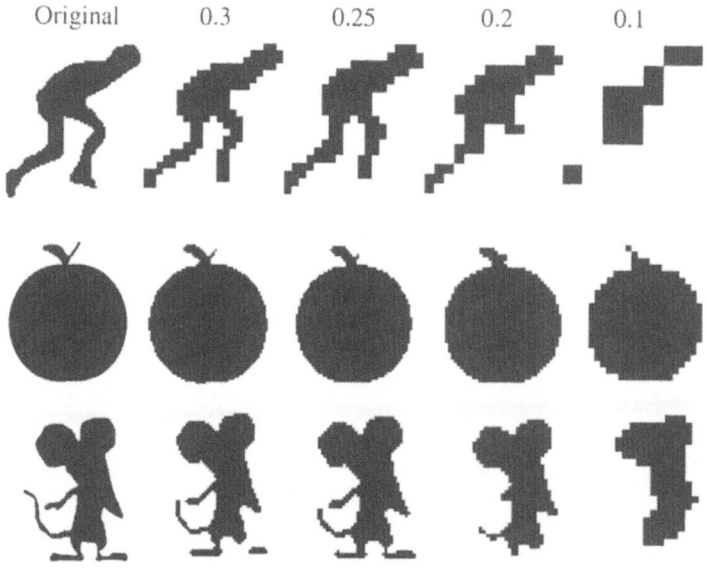

Figure 6.13. Shapes of running person, apple and rat and their scaled-down and re-sampled versions

examples of detractor shapes present in the SQUID database (bottom row).

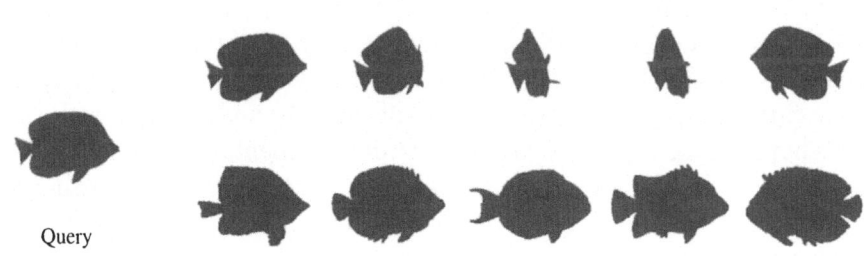

Figure 6.14. Examples of the shape variations in the Bream sequence and the detractor shapes in the SQUID data-set

7.   Experimental Performance Analysis and MPEG-7 Selection Process

This section presents a brief overview of all the shape representation techniques tested in the official MPEG-7 core experiments, followed by an analysis of the results obtained from those techniques.

7.1.   Techniques Participating in MPEG-7 Testing

Several shape representation techniques were tested in the official MPEG core experiments. Besides the CSS-based and the ART-transform-based descriptors, the following approaches were also tested:

- Wavelet-based representation

- Polygon-based representation

- Fourier-based representation

- Multilayer eigenvector representation

- Zernike Moment-based representation

7.1.1   Wavelet-Based Shape Descriptor

To compute the wavelet-based shape descriptor, the contour is first transformed into polar coordinates. Next, a one-dimensional wavelet transform is applied to the magnitude values ρ only. Biorthogonal Daubechie wavelets are used and the first 16 Scale and 48 wavelet coefficients form the descriptor [236]. The descriptor also includes a global parameter called modification ratio.

7.1.2   Polygon-Based Representation

A discrete curve evolution method using polygonal representation was proposed in [157]. They start with an arbitrarily accurate approximation of the curve, (e.g. using a large number of vertices) and in every evolution step they replace a pair of consecutive line segments with a single line segment joining the endpoints. The substitution is done according to a relevance measure. A new similarity measure is defined for contours which have been simplified by the discrete curve evolution. It is based on matching of convex boundary arcs.

7.1.3   Fourier-Based Representation

The contour is described as a polygon, and the segment lengths and turning angles are computed. The DFT transform is then applied to this function, producing a frquency-based representation. Only magnitudes of the Fourier coefficients are considered, which results in rotational invariance. The contour length is normalised in the transformation and thus the descriptor is also invariant with scale [334].

7.1.4 Multilayer Eigenvector Shape Descriptor

In this method the region is repetitively subdivided into four child regions defined by the eigenvectors of the original shape [139]. Four features are computed for each component region, namely the ratio of the eigenvalues, the normal angle, the center of mass and the compactness.

7.1.5 Zernike Moment-Based Representation

This approach belongs to the class of moment-based shape analysis techniques [138], and is similar to the ART approach selected for the region-based descriptor. The main difference lies in the angular component of the basis functions used: the Zernike Moments are derived from Zernike polynomials while ART moments are based on cosine functions.

The Zernike moment of order n with repetition m is defined as:

$$A_{nm} = \frac{n+1}{\pi} \int \int_{x^2+y^2 \leq 1} f(x,y) V_{nm}^*(\rho,\theta) dxdy \qquad (6.26)$$

where $V_{nm}^*(\rho,\theta)$ is the set of Zernike polynomials defined as:

$$V_{nm}^*(\rho,\theta) = R_{nm}(\rho) e^{jm\Theta} \qquad (6.27)$$

and $n \geq 0, m \in Z, |m| \leq n$ and $n - |m|$ is even. The radial component R_{nm} is a polynomial:

$$R_{nm} = \sum_{k=0}^{\frac{n-|m|}{2}} (-1)^k \frac{(n-k)!}{k!(\frac{n+|m|}{2} - k)!(\frac{n-|m|}{2} - k)!} \rho^{n-2k} \qquad (6.28)$$

7.2. Experimental Results

MPEG core experiment results for the evaluated techniques are presented in Table 6.3, top section. The botom part of the table presents other experimental results, which followed the MPEG testing methodology, but were performed outside MPEG. Such results are not cross-verified, and sometimes are not complete, but still give a good indication of the performance of newer techniques. For the MPEG tests, the results in each of the four tests: similarity-based retrieval, robustness to scaling, rotation and non-rigid motion/deformation are presented. The overall

Representation	Similarity	Scaling	Rotation	Non-rigid	Overall
CSS*	81.12%	92.86%	100%	95%	**90.85%**
CSS	79.15%	91.03%	100%	96%	**90.22%**
Zernike Moments	68.61%	97.14%	100%	95%	**87.39%**
Polygon	75.7%	88.05%	99.8%	92.5%	**87.37%**
Wavelet	67.76%	88.04%	97.46%	93%	**86.93%**
ART	68.34%	97.60%	100%	92%	**86.40%**
Multilayer EV	70.32%	92.42%	100%	88%	**84.84%**
Fourier	68.18%	86.35%	100%	85.5%	**82.28%**
Shape Context [30]	76.51%				

Table 6.3.   Comparison of the retrieval performance

Representation	Descriptor size (Bytes)
CSS	14
ART	17.5
Polygon	31.5
Fourier	64
Zernike Moments	140
Wavelet	257
Multilayer EV	804

Table 6.4.   Comparison of the descriptor size

performance for each method is calculated as a weighted average of the performances in each part:

$$overall = ((scaling + rotation)/2 + similarity + non-rigid)/3 \quad (6.29)$$

The results for the CSS and ART descriptor were obtained using the MPEG-7 eXperimental Software version 5.0. The CSS* results are obtained with the Contour shape descriptor using optimised matching strategy based on all four global parameters, which is described in Section 4.3.

Comparison of the descriptor sizes is shown in Table 6.4. Note that some of the descriptors may not have been fully optimised for compactness.

It can be clearly seen from Table 6.3 that the CSS-based descriptor gives the overall best performance. It significantly outperforms all other techniques, including the Polygon-based, Zernike-based and ART-based shape representations, specially in similarity based retrieval.

As previously stated, the technique based on Zernike moments is a region-based approach, exhibiting similar properties to the ART-based method. Overall, the ART descriptor performed worse in core experiments for contour shapes. Nevertheless, the ART-based representation was selected for the region-based descriptor in MPEG-7 because it performed better than Zernike in other experiments with complex shapes (i.e. shapes with multiple regions, holes, etc, such as trade-marks). The region-based techniques are generally less sensitive to changes in contour than contour-based approaches, which gives them a little more robustness to extreme distortions due to scaling, but at a price of significantly worse performance in similarity-based retrieval.

The polygon-based representation belongs to the contour-based class of descriptors and the two approaches exhibit some conceptual similarities, for example they share the concept of contour evolution. These similarities are also reflected in their performance. Nevertheless, the polygon representation is less compact (twice the size of the CSS-based representation) and it is significantly more complex to perform matching. It is interesting to note that the polygon-based approach did not achieve 100% retrieval in the rotation experiment, which was rather easy for the majority of the techniques tested. This points to the fact that the representation may not be stable with respect to the size and orientation of the sampling grid.

We will now examine different properties of the MPEG-7 contour-based and region-based descriptors. Compared to the ART shape descriptor, the CSS descriptor is on average 15% more effective in similarity based retrieval. However, the average performance does not give a full picture of the algorithms' behaviors. This is because the performance is averaged across the test-set, which contains several relatively easy classes where all the algorithms tested obtained high performance close to 100%. In order to understand where the significant differences in performance are, a detailed class-by-class analysis was performed. It was assumed that a descriptor performs considerably better when it retrieves 10% or more correct shapes compared to a competing descriptor. The difference in performance was judged as very significant when it reached 50% for a given class. We found that out of 70 shape classes, the contour shape descriptor outperforms the region shape descriptor in 42 cases while the region shape descriptor outperforms the contour-based descriptor in 8 cases. Here we show example shapes from classes where one of the descriptors outperforms its counterpart significantly. The contour shape descriptor outperforms the region shape descriptor by 50% or more for 17 classes, as shown in figure 6.15. The region shape descriptor outperformed contour shape descriptor in 4 classes - see figure 6.16.

Let us now understand why the contour shape descriptor exhibits such good performance in shapes depicted in Figure 6.15. The first class includes images of a running person, subjected to significant non-rigid deformations. As mentioned earlier, the contour based technique offers excellent robustness to such deformations. Fork is a good example of a shape where characteristic features are conveyed in the contour. The region-based approach tends to confuse it with other classes containing elongated objects, such as Spoon or Hammer. The Butterfly has some characteristic shape features in its antennae, which are not sufficiently reflected in the region-based approach due to their relatively small area. Finally, several classes contain objects that show naturally large variations in their shape. The classes of Beetles, Horses and Elephants belong to this group and demonstrate that such variability is much better captured by the contour-based approach.

Figure 6.16 shows the four classes where the region based approach significantly outperforms contour based approach. Three classes contain *simple convex contours* such as Circle, Triangle and Pentagon that were altered by cutting out a variable number of narrow openings of different shapes. Shapes in each class have similar global distribution of pixels, beacuse the narrow cuts have little impact on it. However, they affect object contours dramatically, especially that the number and shape of the openings varries within each class. The fourth *simple shape* is a square with rounded edges. Since it is a convex shape there are no peaks in the CSS image and the contour conveys little characteristic information. The contour-based technique tends to confuse it with other convex objects, such as the Circle.

8.   Example Applications of the CSS Shape Descriptor

In this section we show two example applications of the MPEG-7 shape descriptors: a cartoon search engine and a smart sensor object-recognition system.

8.1.   Cartoon Search Engine

Figure 6.17 shows a GUI to a Web-based cartoon clip search application developed by Mitsubishi Electric [40]. A collection of short cartoon stories is stored in MPEG-4 format on a web server. Each cartoon is described using MPEG-7 visual descriptors. First, key-frames are ex-

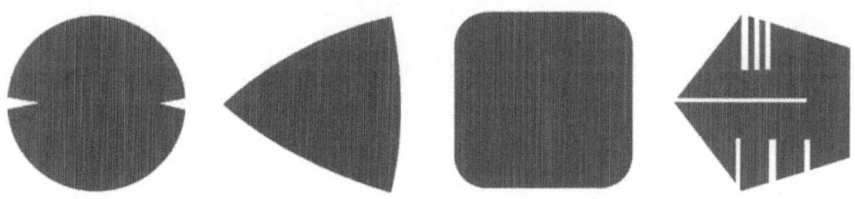

Figure 6.15. Shapes where contour based descriptor outperforms region based descriptor by 50% or more

Figure 6.16. Classes where the region shape descriptor significantly outperforms the contour shape descriptor, Copyright (c) 2002 John Wiley and Sons Limited, Reproduced with permission from [174]

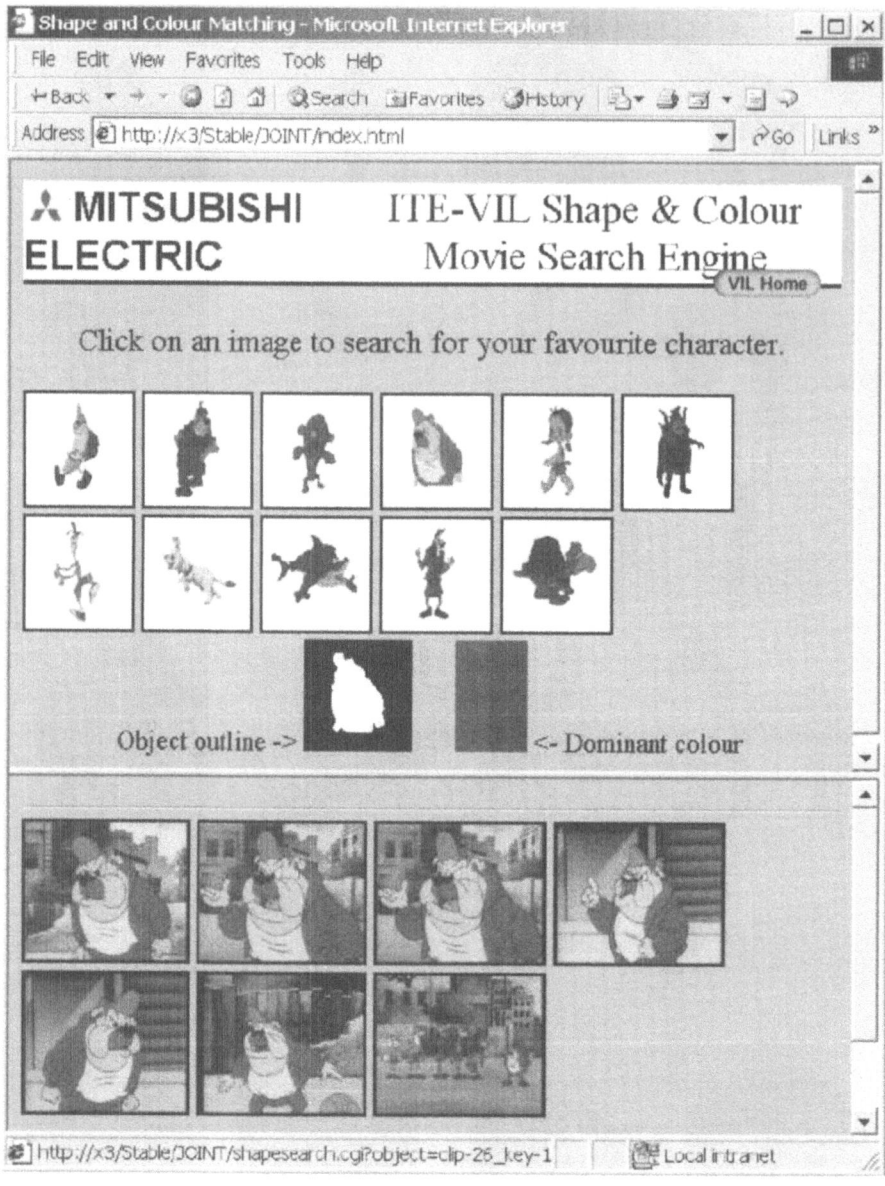

Figure 6.17.   Cartoon search engine, Copyright (c) 2002 Mitsubishi Electric ITE B.V. and cartoon images (c) D'Ocon Film Productions, permitted reproduction

tracted from each video and cartoon characters and objects present in each key-frame are segmented. The contour shape and dominant colour descriptors are then extracted from each region of interest and stored as the MPEG-7 descriptors.

When users wish to find a video or a key frame depicting their favorite characters, they either select a required cartoon character from the group of images in the upper part of the display, or draw the required shape and select corresponding colours with a mouse. A search is then performed on the descriptions stored on the servers. The key frames from the video clips containing similar characters are displayed to the user in a lower row of images. The user can then select the required clip, which is transmitted over the network and played back. Interestingly, a good performance is achieved with a combination of just two descriptors. The combination of the dominant colour and the contour shape descriptor gives robust performance, even in the case of partial occlusion.

8.2.   Object Recognition System

3-D object recognition has many important applications and consequently it has been intensively studied in the last three decades. Despite this significant effort, and many interesting results that have been demonstrated, a system that performs object recognition reliably in an unconstrained environment is yet to be developed.

The performance of the recognition system is determined to a significant extent by the power and flexibility of the underlying object representation employed by the system. Consequently this demanding application is a suitable way to test the properties of the shape representations.

Here we show a 3-D toy-world object recognition system named "Smart Sensor". The system belongs to the class of appearance-based methods. It should be stressed that the main reason this system was built was to investigate the performance and the properties of the MPEG-7 descriptors, and in particular the Contour shape descriptor (CSD). In addition to the Contour shape descriptor, the system also uses global colour features expressed by the Dominant colour Descriptor (DCD). Figure 6.18 shows how a Multiple-View Descriptor is used to build a model of a 3D object by combining descriptors associated with various views. Eight arbitrary views were used for each object and the database contains about 40 objects.

Figure 6.19 shows the block diagram of the system. The images from the camera are analysed by the motion Detection and Segmentation module, which detects and segments new objects that are introduced into the scene. Once a new object is detected, the DCD and CSD descriptors are computed by the Feature Extractors and compared against the entries in the database by the search engine. The search engine finds

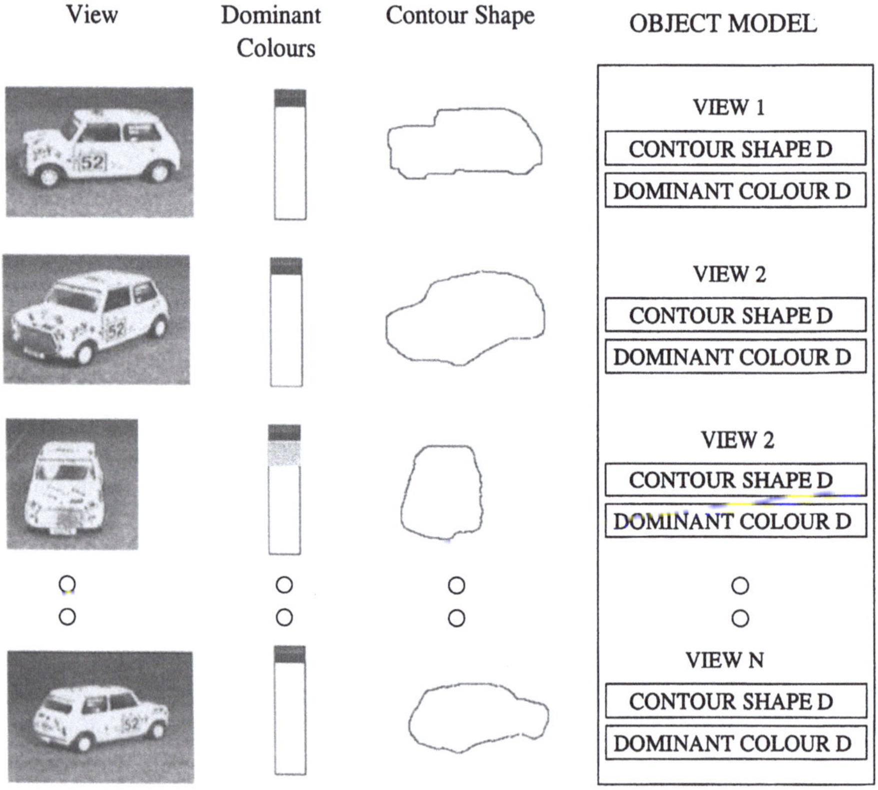

View	Dominant Colours	Contour Shape	OBJECT MODEL

Figure 6.18 content:

			VIEW 1
			CONTOUR SHAPE D
			DOMINANT COLOUR D
			VIEW 2
			CONTOUR SHAPE D
			DOMINANT COLOUR D
			VIEW 2
			CONTOUR SHAPE D
			DOMINANT COLOUR D
			VIEW N
			CONTOUR SHAPE D
			DOMINANT COLOUR D

Figure 6.18. Smart Sensor Object model includes Contour Shape and Dominant colour Descriptors

the object with an associated view that is most similar to the camera view in terms of the outline shape and dominant colours. The identity of the recognised object and information about the object is displayed to the user. Figure 6.20 shows the Graphical User Interface implemented for the system; all stages of the processing are shown: the result of segmentation process, the dominant colour descriptor and contour shape descriptor extracted from the object, the candidate matches and the final decision of the system. There are two rows of numbers below the icons with candidate matches; for each candidate, the top number indicates contour-shape similarity and the bottom one similarity based on the dominant colour description (small values indicate high similarity).

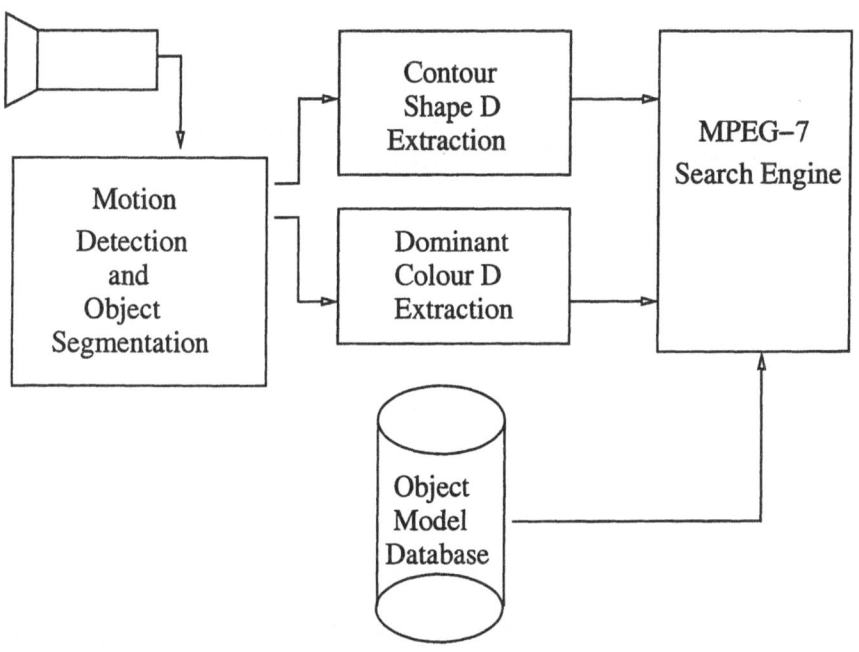

Figure 6.19. Block diagram of the Smart Sensor system

The system is implemented on a standard PC running Windows 2000 OS. The processor speed is not critical, as only a fraction of the processing resources are used. The system can reliably recognise 40 complex three-dimensional objects. The training stage took less than one hour. The entire learning process is automated, except for the placing of the objects in front of the camera. The recognition loop includes object segmentation, extraction of the descriptors, matching and object identity selection, and is completed in less then a second.

Figure 6.21 shows some example objects from the database. Objects in each row have very similar colours, so that this system had to rely solely on the shape properties to distinguish between objects in each row. For example the system could in most cases distinguish between

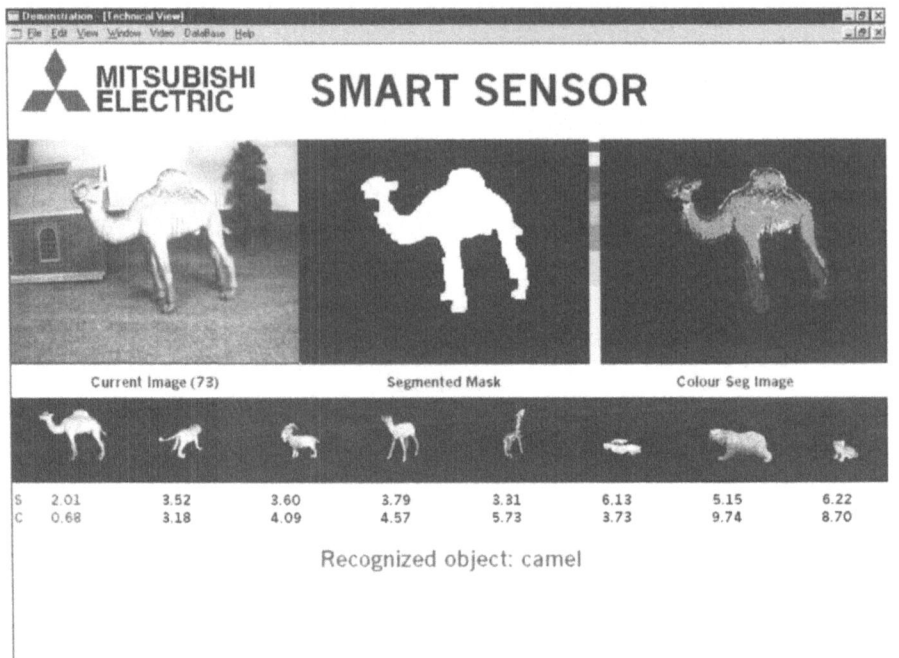

Figure 6.20.   The GUI of the Smart Sensor system, Copyright (c) 2002 Mitsubishi Electric ITE B.V., permitted reproduction

the three red model cars depicted in the top row: the Ford Granada, Toyota Celica and Audi A4 Avant, provided the camera was presented with a side view. There is more variability in the shapes of the animals shown in the second row (all are similar shade of brown), and the system demonstrated reliable recognition from any view. Finally, in the third row we have a TNT van and truck, with identical colours. Again the system could distinguish between these two vechicles from a side view and often from a perspective view. The contour shape descriptor showed robustness to contour distortions due to segmentation errors and also a very good performance under perspective transformation.

9.   Conclusions

This chapter described the state-of-the art in shape analysis, in the context of a new Multimedia Interface Standard called MPEG-7. Two 2-D shape descriptors, contour-based and region-based, are included in the MPEG-7 Visual specification. The contour shape descriptor is based on the Curvature Scale Space theory. New components and features, such as additional global parameters and the quantisation scheme were devel-

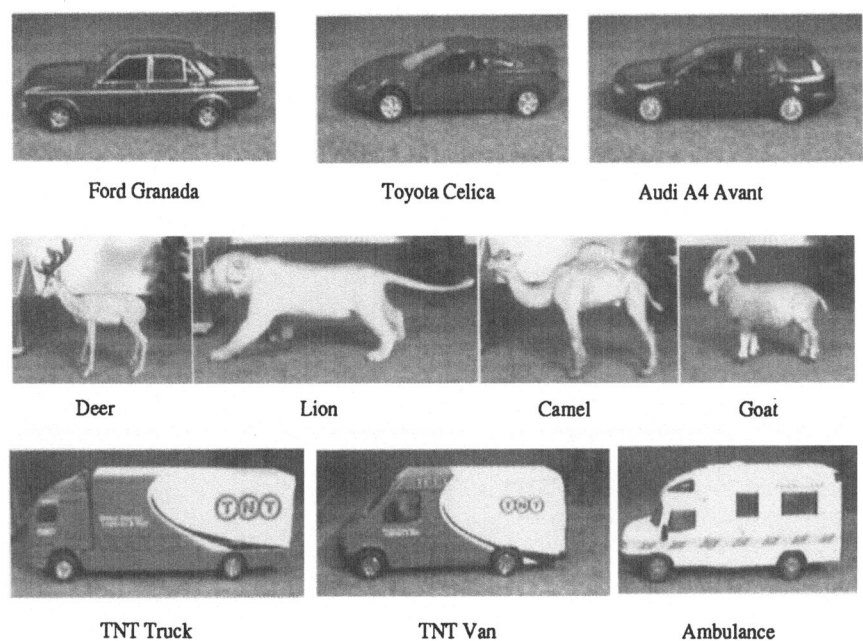

Ford Granada   Toyota Celica   Audi A4 Avant

Deer   Lion   Camel   Goat

TNT Truck   TNT Van   Ambulance

Figure 6.21. Example objects with similar colours from the database, Copyright (c) 2002 Mitsubishi Electric ITE B.V., permitted reproduction

oped to adapt the CSS technology to the requirements of the Standard and to enhance its performance. We described in detail the MPEG-7 contour descriptor, its extraction and use, and presented several novel matching strategies with improved performance and reduced complexity. The performance evaluation behind the extensive comparative testing performed within MPEG-7 was also described, and the test results which clearly show that the contour shape descriptor offers the best performance from the state-of-the art techniques tested. In particular, the CSS-based descriptor shows excellent generalisation properties, similar to that of the human visual system. Two examples of real-time applications of the contour shape descriptor were also presented, indicating that it is well suited for emerging applications. We expect to see a wide range of novel applications, services and devices which will include the MPEG-7 the shape description technology and other visual descriptors in the near future.

Chapter 7

ROBUST IMAGE CORNER DETECTION THROUGH CURVATURE SCALE SPACE

This chapter describes a novel method for image corner detection based on the CSS representation. The first step is to extract edges from the original image using a Canny detector. The corner points of an image are defined as points where image edges have their maxima of absolute curvature. The corner points are detected at a high scale of the CSS and tracked through multiple lower scales to improve localization. This method is very robust to noise and we believe that it performs better than existing corner detectors. An improvement to Canny edge detector's response to 45° and 135° edges is also proposed. Furthermore, the CSS detector can provide additional point features (curvature zero-crossings of image edge contours) in addition to the traditional corners.

This chapter also presents a later enhancement to the CSS corner detector and proposes two new criteria for the performance evaluation of corner detection algorithms.

1. Introduction

Corner detection is an important task in various computer vision and image understanding systems. Applications include motion tracking [184, 185], object recognition, and stereo matching. corner detection should satisfy a number of important criteria:

- All the true corners should be detected.

- No false corners should be detected.

- Corner points should be well localized.

- Corner detector should be robust with respect to noise.

■ Corner detector should be efficient.

This chapter describes a new corner detection method [229] based on the CSS technique. The CSS technique is suitable for extraction of curvature features from an input contour at a continuum of scales. This corner detection method requires image edge contours. In the implementation of the CSS detector a Canny edge detector [45] was used. Note, however, that the Canny edge detector is not a crucial part of the technique: it can be replaced with another edge-detection algorithm. Nevertheless, with Canny's good edge detection, we believe our corner detector performs better than existing ones.

Much work has been carried out on corner detection and Section 2 gives an overview. Section 3 briefly describes the Canny detector and the improvement made to its response on edges at $45°$ or $135°$ angles. Section 4 describes the original CSS corner detector and section 5 presents the experimental results for that detector. Section 6 discusses an enhancement to the CSS corner detector, and section 7 presents the experimental results for the enhanced detector. Section 8 contains a discussion of performance evaluation of corner detection algorithms. It then proposes two new criteria for that purpose, and presents evaluation results for five corner detectors. Section 9 presents the concluding remarks.

2.   Literature Survey

Considerable research has been carried out on corner detection in recent years. This section briefly reviews a number of proposed algorithms. Moravec [235] observed that the difference between the adjacent pixels of an edge or a uniform part of the image is small but at the corner the difference is significantly high in all directions. Harris [104] implemented a technique referred to as Plessey algorithm. The technique was an improvement of the Moravec algorithm. Beaudet [28] proposed a determinant (DET) operator which has significant values only near corners. Dreschler and Nagel [69] used Beaudet's concepts in their detector. Kitchen and Rosenfeld [143] presented a few corner detection methods. The work included methods based on gradient magnitude of gradient direction, change of direction along edge, angle between most similar neighbours and turning of the fitted surface. Lai and Wu [151] considered edge-corner detection for defective images. Tsai [309] proposed a method for boundary-based corner detection using neural networks. Ji and Haralick [128] presented a technique for corner detection with covariance propagation. Lee and Bien [159] applied fuzzy logic to corner detection.

Fang and Huang [79] proposed a method which was an improvement on the gradient magnitude of the gradient angle method by Kitchen and

Rosenfeld. Chen and Rockett utilized Bayesian labelling of corners using a grey-level corner image model in [52]. Wu and Rosenfeld [324] proposed a technique which examines the slope discontinuities of the x and y projections of an image to find the possible corner candidates. Paler et al. [247] proposed a technique based on features extracted from the local distribution of grey level values. Rangarajan et al. [262] proposed a detector which tries to find an analytical expression for an optimal function whose convolution with the windows of an image has significant values at corner points. [21] introduced corner detection by local histograms of contour chain code. [284] worked on ridge's corner detection and correspondence. [240] considered corner location measurement. [290] proposed a mean field annealing approach to corner detection.

Zhang and Zhao [335] considered a parallel algorithm for detecting dominant points on multiple digital curves. Kohlmann [147] applied the 2-D Hilbert transform to corner detection. Mehrotra et al. [180] proposed two algorithms for edge and corner detection. The first is based on the first directional derivative of the Gaussian and the second is based on the second directional derivative of the Gaussian. Davies [65] applied the generalised Hough transform to corner detection. Zuniga and Haralick [339] utilized the facet model for corner detection. Smith and Brady [289] used a circular mask for corner detection. No derivatives were used. Orange and Groen [246] proposed a model-based corner detector. Other corner detectors have been proposed in [308, 245, 316]. Our survey suggested that the Plessey corner detector, the Kitchen and Rosenfeld detector, and the SUSAN detector [289] have demonstrated good performance. These detectors were therefore chosen as our test detectors.

3. Canny Edge Detector

The CSS based image corner detector uses the Canny [45] edge detector. During the implementation of the CSS corner detector it was found that Canny edge detector produced a thick edge when edge orientation was 45° or 135°.

Canny edge detector uses a Gaussian function to compute the first derivatives from an image. The process produces two similar gradient values at either side of an edge if the areas at each side of the edge have a constant brightness level. Non-maximum suppression is meant to ensure that the edge line is thinned and is only one pixel wide. Canny's non-maximum suppression uses the direction of the gradient at an edge point to look at neighbouring pixels. If the chosen neighbouring pixels have larger gradient values than the examined point, the point is removed from the edge map. When there is a 45° or 135° edge with uniform

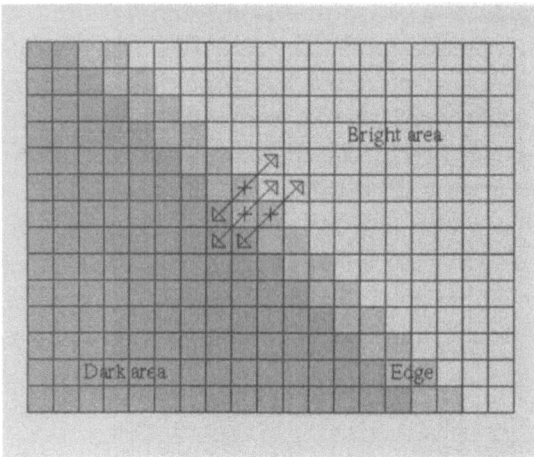

Figure 7.1. Canny's non-maximum suppression with 135° edges

areas on either side, the non-maximum suppression produces a thick edge. This problem is caused by the fact that the gradient direction at the edge point points to non-edge pixels. This can be seen in figure 7.1. The edge points which are examined during the non-maximum suppression do not see their neighbouring edge pixels due to the 45° or 135° orientation.

This problem is solved with a small modification to the Canny edge detector algorithm. The final stage should be to compare each edge pixel which has an edge orientation of 45° or 135° to one of its horizontal or vertical neighbours. If the neighbour has the same orientation, the other point can be removed.

4. Original CSS Corner Detection Method

This section contains a description of the original CSS corner detection technique [231]. It starts with an overview of that technique followed by an outline of the algorithm. Each stage of the method is then described in more detail.

4.1. Overview

The corners are defined as the local maxima of the absolute value of curvature. At a very fine scale there exists many such maxima due to noise on the digital contour. As the scale is increased, the noise is smoothed away and only the maxima corresponding to the real corners

remain. The CSS corner detection method finds the corners at these local maxima.

As the contour evolves, the actual locations of the corners change. If the detection is achieved at a large scale the localization of the corners may be poor. To overcome this problem tracking is introduced in the detection. The corners are located at a high scale σ_{high}, assuring that the corner detection is not affected by noise. σ is then reduced and the same corner points are examined at lower scales. As a result, location of corners may be updated. This is continued until the scale is very low and the operation is very local. This improves localization and the computational cost is low, as curvature values at scales lower than σ_{high} do not need to be computed at every contour point but only in a small neighbourhood of the detected corners.

There are local maxima on the evolved contours due to rounded corners or noise. These can be removed by introducing a threshold value t. The curvature of a sharp corner is higher than that of a rounded corner. There is one final addition to the corner candidate declaration. Each local maximum of the curvature is compared to its two neighbouring local minima. The curvature of a corner point should be double the curvature of a neighboring extremum. This is necessary since if the contour is continuous and round, the curvature values can be well above the threshold value t and false corners may be declared.

4.2. Outline

The process of CSS image corner detection is as following:

- Utilize the Canny edge detector to extract edges from the original image.

- Extract the edge contours from the edge image:

  - Fill the gaps in the edge contours.
  - Find the T-junctions and mark them as T-corners.

- Compute the curvature at highest scale σ_{high} and determine the corner candidates by comparing the maxima of curvature to the threshold t and the neighboring minima.

- Track the corners to the lowest scale to improve localization.

- Compare the T-corners to the corners found using the curvature procedure and remove corners which are very close.

The following is an explanation of each stage of the CSS corner detector.

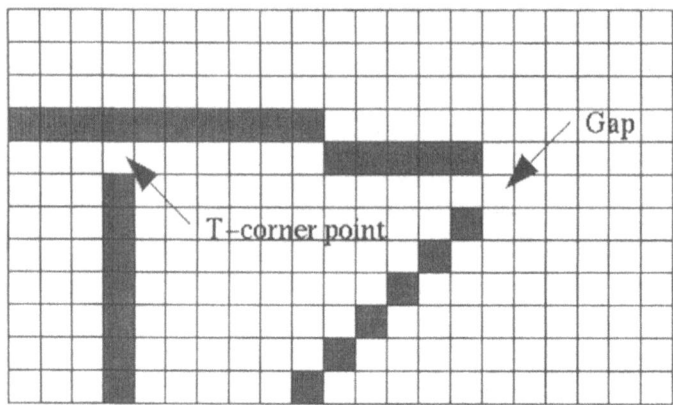

Figure 7.2.   The two cases of gaps in the edge contours

4.3.   Canny Edge Detection

The first stage of the CSS corner detection method is edge detection. A Canny edge detector was chosen for this implementation due to its good performance. A small σ was used for Canny to obtain good edge localization.

4.4.   Filling the Gaps and T-junctions

Canny detector can cause gaps at T-junctions and the corners might not be found with the CSS method. Canny can also cause gaps in otherwise continuous edges. When the edge extraction method arrives at the endpoint of a contour it performs two checks:

- If the endpoint is nearly connected to another endpoint, fill the gap and continue the extraction.

- If the endpoint is nearly connected to an edge contour, but not to another endpoint, mark this point as a T-junction corner.

In figure 7.2 both cases of gaps are shown. The T-junction gap is marked as a corner and the gap between two contour ends is filled.

4.5.   Initial Corner Points

The edge contours are extracted from the edge image and the absolute value of curvature is computed at the initial scale σ_{high}. The local maxima of absolute curvature are the possible candidates for corner points.

A local maximum is either a corner, the top value of a rounded corner or a peak due to noise. The latter two should not be detected as corners. The curvature of a real corner point has a higher value than that of a rounded corner or noise. The corner points are also compared to the two neighbouring local minima. The curvature of a corner should be twice that of one of the neighboring local minima. This is because when the shape of the contour is very round, contour curvature can be above the threshold t. The threshold t depends on σ_{high} used and it is set according to it.

4.6. Tracking

After the initial corner points are located, tracking is introduced to the detection. As the corners were detected at scale σ_{high} the corner localization might not be optimal. We compute curvature at a lower scale and examine the corner candidates in a small neighborhood of the previous corners. Corner locations are updated, if needed, in this neighborhood. Tracking continues until scale is very low. This process results in very good localization. No thresholding is needed for the tracking. The number of corners is determined at the initial σ_{high} and tracking only affects the localization, not the number of corners. In fact, tracking improves the localization of the corners. Corners do not move significantly during tracking and only a few other curvature values need to be computed.

4.7. Removing False Corners

As described before, corners are declared using two methods and in some cases the two methods mark the same corner. In figure 7.3, the case where one corner is marked twice is shown. The edge extraction algorithm examines a small neighborhood when it arrives at the end of a contour. The corner in figure 7.3 is a Y junction and is marked twice. The CSS method finds a corner on the continuous contour and the edge extraction algorithm marks a T-corner at the end of the other contour as it is nearly connected to a continuous edge contour. The final part of the algorithm is to examine the points marked by the edge extraction algorithm. These T-junction corners are compared to the corner points found with the CSS method and if they are very close to each other, the T-junction corners are removed. In the implementation, a 5x5 neighborhood was used.

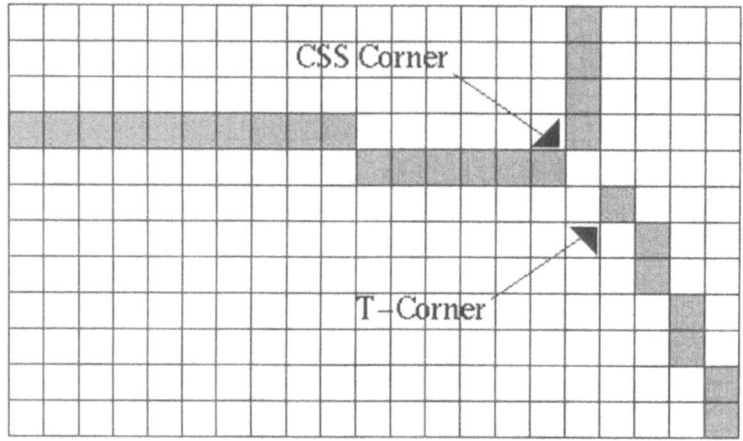

Figure 7.3.   Case where one corner is marked twice

5.   Original CSS Experimental Results and Discussion

The CSS corner detector was tested using four different images and the results are compared with the output of three other corner detectors: Kitchen and Rosenfeld, SUSAN and Plessey corner detectors. Note that we attempted to obtain the best possible results for each corner detector tested by searching for parameter values that appeared to yield the best results. The first test image is an artificially created one with significant Gaussian noise added. The second test image is a real image of blocks. Much texture and noise is present in the image. The third test image is an image of a house. This image has a lot of small details and texture in the brick wall. Finally an image of a lab is used.

The results showed that the CSS corner detector gave the best results in each of the four cases, and that it was robust to image noise.

The CSS detector performed very well on the noisy artificial image, but the other three detectors did not perform well, as seen in figure 7.4. The real blocks image corner detection was a more difficult task for the detectors. Again the CSS corner detector gave the best results amongst the four. The results are seen in figure 7.5.

House image was a difficult task for all the detectors as the details are very varied. Overall, the CSS detector still performed better. Figure 7.6 shows the results. Finally the results on the lab image are shown in figure 7.7. The CSS detector performed very well with the image,

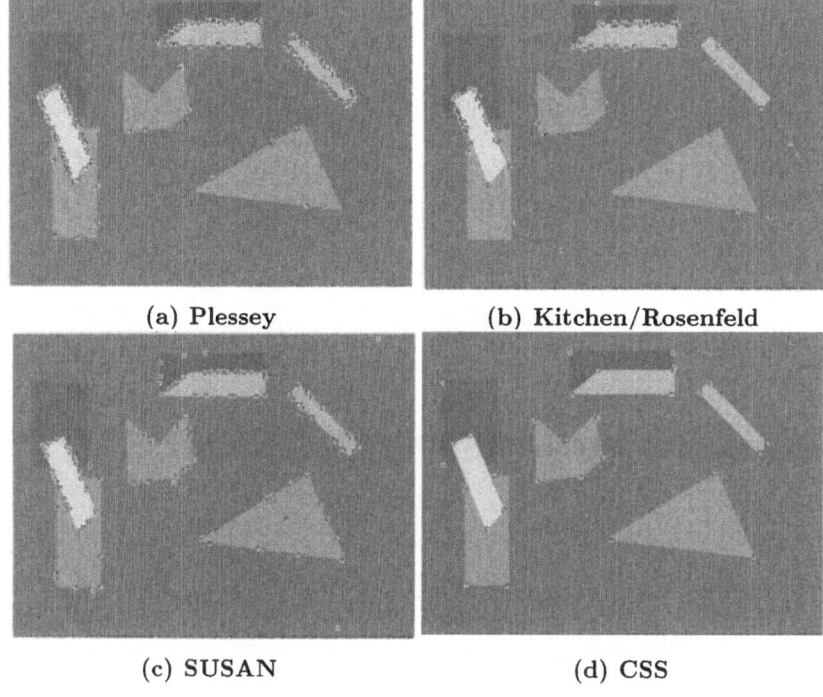

(a) **Plessey**

(b) **Kitchen/Rosenfeld**

(c) **SUSAN**

(d) **CSS**

Figure 7.4. Artificial test image with noise

but the three other detectors had serious problems with very obvious corners.

The speed of the corner detectors was measured on a Sun SPARC-station 5. The Kitchen and Rosenfeld detector was the fastest of these detectors, but the rest of the detectors had quite similar speeds. All the detectors are implemented in C++. Over 80% of the time used by the CSS detector is spent in edge detection.

The CSS corner detector uses only two important parameters. Experiments showed that $\sigma_{high} = 4$ gave good results with almost all images. The threshold t depends on the value of σ_{high} and with $\sigma_{high} = 4$ the threshold can be set to 0.03. Other values of σ_{high} are also possible and for a very noisy image $\sigma_{high} = 8$ and threshold $t = 0.02$ can be used. Starting with $\sigma_{high} = 4$, tracking can be accomplished at $\sigma = 2$, $\sigma = 1$ and $\sigma_{final} = 0.7$. The final scale σ_{final} should be as local as possible to ensure good localization. It was found that the results were not sensitive to the exact values of the parameters, and that the same values worked well for the different test images used except for one that was very noisy by intention. Note however that the *detection* of corners can be carried

(a) **Plessey**   (b) **Kitchen/Rosenfeld**

(c) **SUSAN**   (d) **CSS**

Figure 7.5.   Blocks image

out at multiple scales. As a result, by adjusting the scale, the number of corner points recovered can increase or decrease, depending on the requirements of later processes. For example, in a motion tracking system, object detail is not needed when tracking in a non-cluttered scene, and a small number of corners will be sufficient. However, when part of the object becomes occluded, a larger number of corners will be required.

It has been argued that corner detectors that perform directly on images may be preferrable since they do not depend on the results of an earlier stage (such as edge detection). It should be pointed out that most corner detectors carry out some form of edge detection either implicitly or explicitly. As a result, even when they appear to be directly applicable to the input image, the results are affected by the implicit edge detection. The CSS detector simply makes the process explicit.

(a) Plessey   (b) Kitchen/Rosenfeld

(c) SUSAN   (d) CSS

Figure 7.6. House image

The CSS detector makes both image edges and image corners available for later processes. It can also provide additional point features as well as the traditional corners [232]. The new features are the curvature zero-crossings or inflection points of the image edge contours recovered in a similar way as the corners. They can complement the traditional corners when used by later processes. For example, they can be utilized by motion tracking systems in an area of the image where there is a lack of corner features.

6. Enhanced CSS Corner Detection Method

In the original CSS corner detection algorithm, σ_{high} must be large enough to remove the noise but retain the real corners. In fact, the value of this parameter was held constant for all contours of the image. However, we noticed that using a large σ_{high} results in the removal of some

(a) Plessey   (b) Kitchen/Rosenfeld

(c) SUSAN   (d) CSS

Figure 7.7. Lab image

of the real corners on short contours. At the same time, lower values of σ_{high}, produce a number of false corners on long contours. The same problem exists in comparing corner candidates with their neighboring minima. While applying this criterion may result in removal of false corners from long contours, some real corners on short contours may also be removed. The remedy is to choose different values of σ_{high} for contours with different lengths as described in this section.

Another problem of the original method is that its performance depends on the selection of the threshold value, t. The proper value of t may change from one image to another. It is also subject to change for a

particular image which transforms under rotation or scaling. Therefore, methods which use thresholds are not always robust and their performance depends on careful selection of the threshold values.

The New-CSS corner detector mainly addresses these problems [223]. The outline of New-CSS corner detector is as following:

- Extract edges from the original image.

- Extract image edge contours, filling the gaps and finding T-junctions.

- Use different scales of the CSS for contours with different lengths.

- Compute absolute curvature on the smoothed contours.

- Smooth the absolute curvature function for long contours.

- Detect initial local maxima of the absolute curvature for short contours.

- Detect initial local maxima of the smoothed absolute curvature functions for long contours.

- Consider those local maxima as initial corners whose absolute curvatures are more than twice as much as one of the neighbouring local minima.

- Track the corners down to the lowest scale for each contour to improve localization.

- Compare the T-junction corners to the corners found using the curvature procedure to unify close corners.

The new steps have been described in detail in the remainder of this section.

6.1.   Using Different Scales of CSS

After extracting image edge contours, the number of contours and the number of points on each contour are known.

Based on the number of points on each contour, our algorithm categorizes all the image edge contours into three categories: long, medium and short contours. We set σ_{high} at 4, 3 and 2 for long, medium and short contours respectively. As a result, short contours are not smoothed excessively which could remove their corners, but long contours are smoothed sufficiently. In Figure 7.8 the edge contours of a test image with two marked contours, short C1 and long C2 have been illustrated. The effect of selecting different scales in computation of absolute curvature for long and short contours can be seen in Figure 7.9.

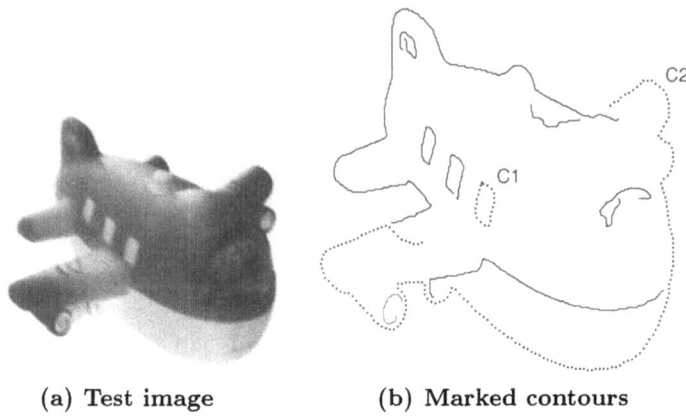

(a) Test image   (b) Marked contours

Figure 7.8.   Test image and two marked contours in edge contours of that image, C1: short contour, and C2: long contour

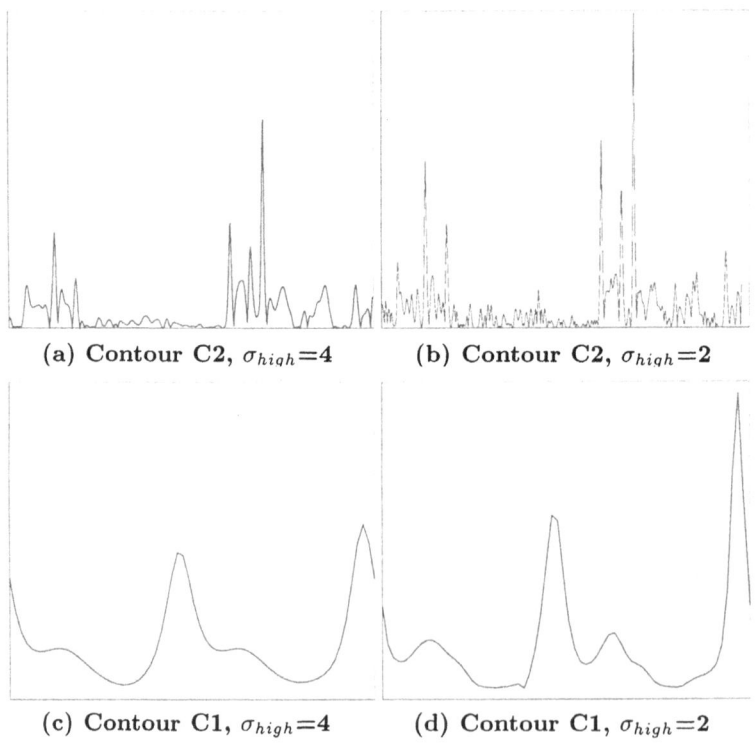

(a) Contour C2, $\sigma_{high}=4$ (b) Contour C2, $\sigma_{high}=2$

(c) Contour C1, $\sigma_{high}=4$ (d) Contour C1, $\sigma_{high}=2$

Figure 7.9.   Computation of absolute curvature

In Figure 7.9(a), the absolute curvature of contour C2 with $\sigma_{high}=4$ has fewer false maxima due to noise in comparison to Figure 7.9(b) which shows the computation of absolute curvature of contour C2 with $\sigma_{high}=2$. Obviously for long contours, computation of absolute curvature should be done using $\sigma_{high}=4$. Also Figure 7.9(c) illustrates the computation of absolute curvature of contour C1 with $\sigma_{high}=4$. If we use the local maxima of this absolute curvature function for detecting corners, only two corners are detected. Therefore using high scale for smoothing short contours removes some local maxima of absolute curvature of these contours that correspond to real corners. In other words, on short contours, computation of absolute curvature should be done using $\sigma_{high}=2$. Figure 7.9(d) shows the absolute curvature of short contour C1 with $\sigma_{high}=2$. It can be seen that four corners are available from this figure.

The results of this section are as following:

- The problem of false local maxima on long contours is more critical than short contours.

- Computation of absolute curvature function on long contours should be carried out at high scale.

- Computation of absolute curvature function on short contours should be carried out at low scale.

6.2. Smoothing the Absolute Curvature Function of Long Contours

In this stage, after smoothing edge contours for computation of absolute curvature, some false maxima due to noise can still be seen (see Figure 7.9(a)). The simplest solution is to compute the absolute curvature of C2 with higher σ such as 8. But as mentioned earlier, if higher σ is chosen not only false corners but also many real corners are removed as well. Therefore our solution in this stage is to smooth the absolute curvature function of long contours using $\sigma=4$. This has been illustrated in Figure 7.10(a). Note that smoothing of the absolute curvature function is different from smoothing of contours. In Figure 7.10(a) in comparison to Figure 7.9(a), after smoothing, many false maxima of absolute curvature are removed.

In this step, if curvature function of short contours becomes smoothed, as seen in Figure 7.10(b), a number of real corners are lost. As we can see for contour C1, in Figure 7.10(b), only two local maxima remain that indicate two corners, whereas contour C1 is the window of airplane with four corners. Final criterion for removing false corners, after initializing

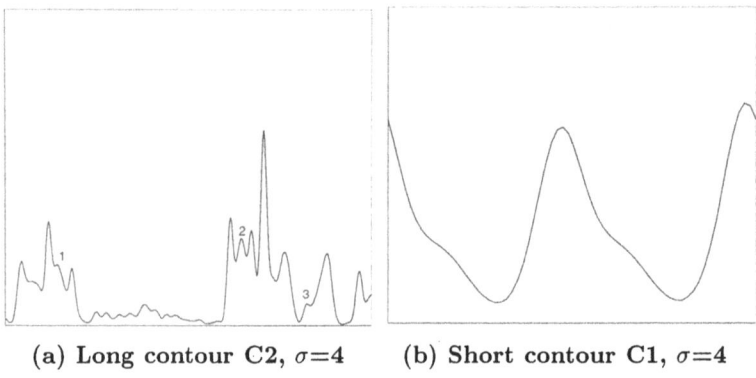

(a) Long contour C2, $\sigma=4$ (b) Short contour C1, $\sigma=4$

Figure 7.10.   Smoothing of absolute curvature function

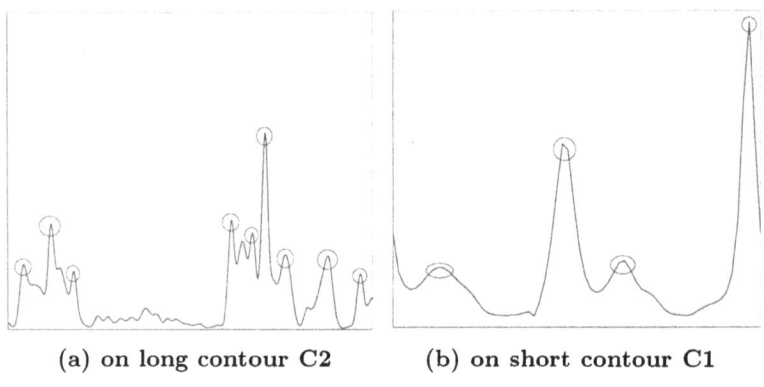

(a) on long contour C2   (b) on short contour C1

Figure 7.11.   Comparing the initial local maxima with two neighbouring local minima

local maxima points is to compare the initial local maxima with two neighbouring local minima. The curvature of a corner should be more than twice as much as one of the neighbouring local minima.

Using this criterion, false corners such as 1,2 and 3 (see Figure 7.10(a)) are removed after comparison to their neighbouring local minima of absolute curvature in this figure. The positions of initial corners of figures 7.10(a) and 7.9(d) after taking this criterion into account have been illustrated in Figure 7.11(a) and 7.11(b) respectively. The method finds four corners on contour C1 and no false corners on contour C2. Contour C2 should have nine corners that can be seen in Figure 7.11(a). Remember that for short contours first our method computes absolute curvature with $\sigma_{high}=2$, then uses the final criterion discussed above. The positions of corners of short contour C1 are marked in Figure 7.9(d) and have been illustrated in Figure 7.11(b).

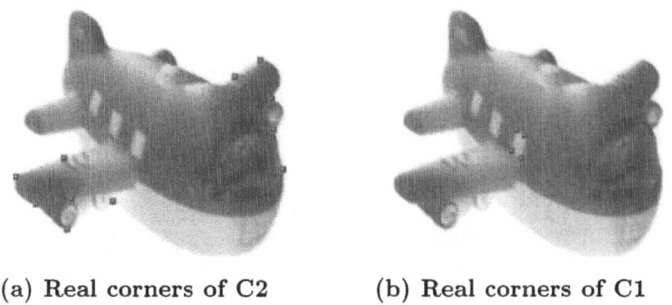

(a) **Real corners of C2**   (b) **Real corners of C1**

Figure 7.12.   Tracking corners through low scales for good localisation

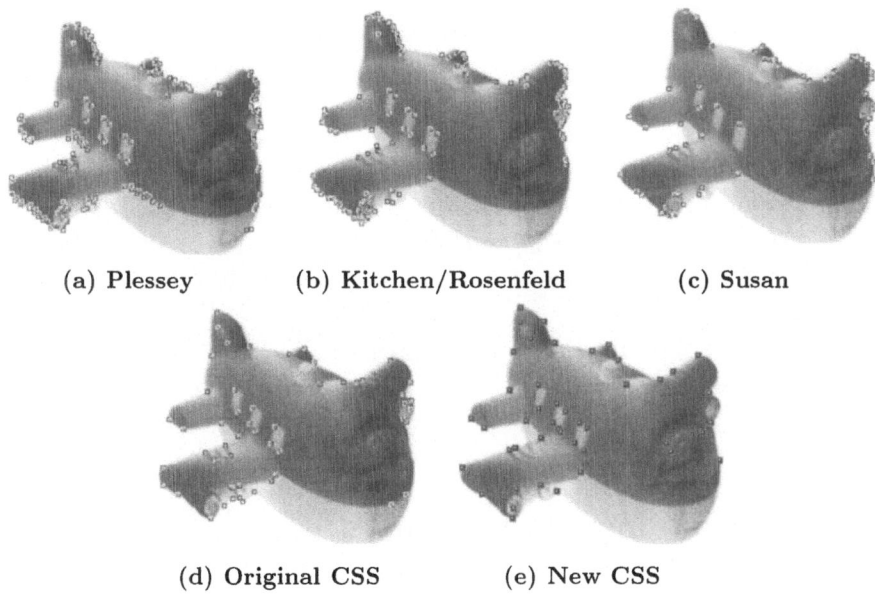

(a) **Plessey**   (b) **Kitchen/Rosenfeld**   (c) **Susan**

(d) **Original CSS**   (e) **New CSS**

Figure 7.13.   Airplane image

6.3.   Tracking

After the initial corner points are located, tracking is applied to the detected corners. As the corners were detected at scale σ_{high}, corner localization might not be optimal. We compute curvature at a lower scale and examine the corner candidates in a small neighbourhood of the previous corners.

Corner locations are updated, if needed, in this neighbourhood. Note that if initial corners on one contour are recovered at $\sigma_{high}=4$, tracking

(a) Plessey   (b) Kitchen/Rosenfeld   (c) Susan

(d) Original CSS   (e) New CSS

Figure 7.14.   Another view of airplane image

for this contour can be accomplished at $\sigma=3$, $\sigma=2$ and $\sigma_{final}=1$. If initial corners are extracted at $\sigma_{high}=2$, tracking can be accomplished at $\sigma_{final}=1$. In other words, tracking continues down to a very low scale. The localization of corners for contour C2 and C1 after tracking has been shown in Figure 7.12.

This process results in excellent localization. The number of corners is determined at the initial σ_{high} and tracking only changes the localization, not the number of corners. Since corners do not move significantly during tracking, only a few other curvature values need to be computed.

6.4.   Unifying Close Corners

As described earlier, corners are detected using the enhanced CSS technique taking T-junctions into consideration. In some cases the two methods mark the same corner. The final part of New-CSS is to examine T-junctions and the corners that result from tracking. If they are very close to each other, the T-junction corners are removed.

7.   New-CSS Experimental Results and Discussion

The CSS corner detector was tested on several different images. Only four of them have been illustrated here. The results are compared with the outputs of four other corner detectors: original CSS [230], Kitchen and Rosenfeld [143], Susan [289] and Plessey [104]. Note that we attempted to obtain the best possible results for each corner detector by searching for parameter values that appeared to yield the best results.

The first test image, figure 7.13 is a real image of an airplane. The second test image, figure 7.14 is another view of the first image in order to demonstrate the robustness of New-CSS. There are many blunt corners in these two images. However, original-CSS, the best one among these four detectors, has difficulty finding their correct positions. Another problem of original-CSS that can be seen in these two images is detection of false corners. New-CSS performs better in comparison to these four corner detectors.

The third test image is the house image. This image has many small details and texture in the brick wall and was a difficult task for all the detectors as the details are very varied. Again the New-CSS corner detector gave the best results amongst the four. The results are shown in Figure 7.15. Finally the blocks image (with many sharp corners) was used. Overall, the New-CSS detector still performs better. Figure 7.16 shows the results. These examples show that New-CSS corner detector, especially for blunt corners, performs better than the other methods and that it is robust to image noise, whereas for sharp corners it performs as well as original-CSS and much better than the others.

The New-CSS corner detector uses only one important parameter: σ_{high}. It can also provide additional point features as well as the traditional corners such as the curvature zero-crossings or inflection points of the image edge contours recovered in a similar way as the corners.

8.   Performance Evaluation of Corner Detection Algorithms under Similarity and Affine Transforms

This section presents an evaluation of the performance of New-CSS corner detector and four other corner detectors under similarity and affine transforms [182]. The majority of authors of published corner detectors have not used theoretical criteria to measure the stability and accuracy of their algorithms. They usually only illustrate their results on different test images and compare them to the results of other test corner detectors. A few of them use only one criterion. This criterion is the number of matched corners between original and transformed images, divided by the number of corners in the original image. This criterion is flawed since it favours algorithms which find more false corners in input images.

We propose two new criteria to evaluate the performance of corner detectors. Our proposed criteria are *consistency of corner numbers* and *accuracy*. These criteria were measured using many images and experiments such as rotation, uniform scaling, non-uniform scaling and affine transforms. To measure accuracy, we created ground truth based on

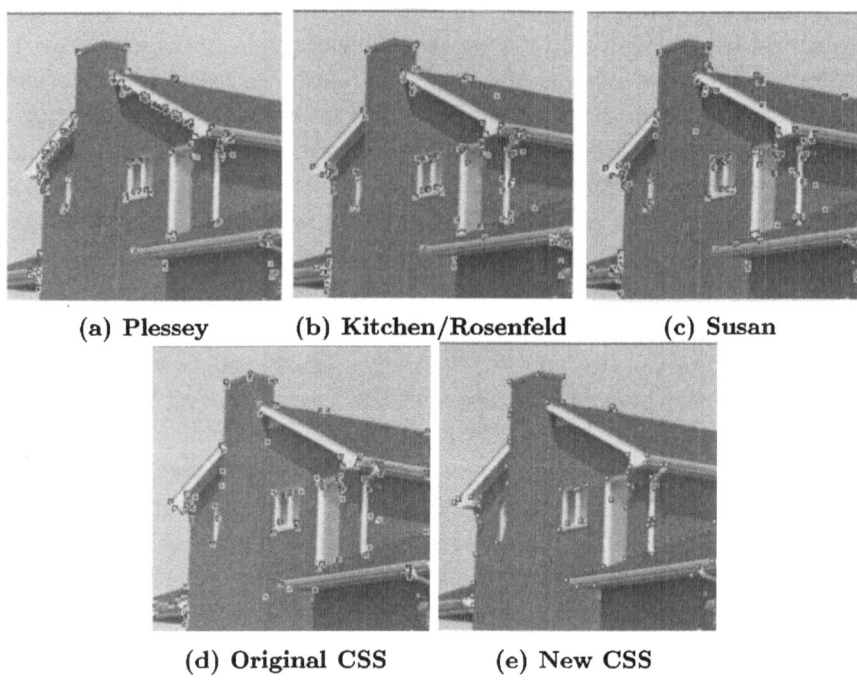

(a) Plessey   (b) Kitchen/Rosenfeld   (c) Susan

(d) Original CSS   (e) New CSS

Figure 7.15.   House image

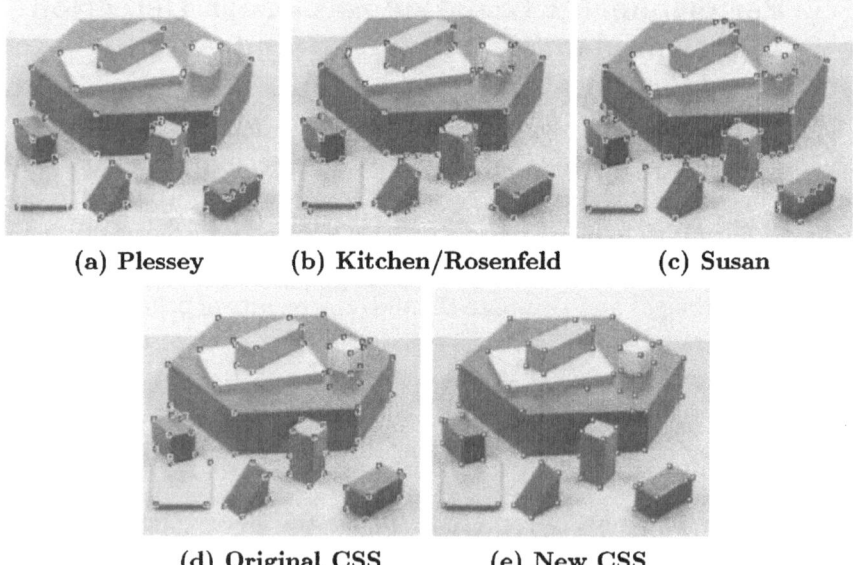

(a) Plessey   (b) Kitchen/Rosenfeld   (c) Susan

(d) Original CSS   (e) New CSS

Figure 7.16.   Blocks image

majority human judgement. The results show that the New-CSS corner detector performs better under similarity and affine transforms.

8.1. Previous Criteria for Performance Measurement

The majority of published corner detectors have not used properly defined criteria for measuring the stability and accuracy of their corner detectors. They have only demonstrated their results on different images in comparison to other test corner detectors. Some published results on corner detection include studies on the effects of noise and parameter variation on the results of their corner detectors. These parameters include Gaussian scale σ [337], [289], [275], [261], σ white noise [337], thresholds [308], [294], signal-to-noise ratio [49], cross-correlation matching [308], cost-function [325] and the width of the gray level transitions in original image [265] but no definition of stability and its results. A few of them have used only one criterion to measure the stability of their corner detectors as follows:

Trajkovic and Hedley [308] used a measure of $k = \frac{N_m}{N_c}$, where N_m and N_c denoted the number of strong matches and number of corners in the original image respectively. In terms of stability, a corner detector was better if k is higher. Schmid and Mohr [275], applied the criterion of the ratio of total matches to the number of points extracted. This ratio varies depending on the image as well as on the type of transformation between the images. A problem with both criteria is that if we have an algorithm which marked all of the pixels in one image as corners then k would become 100%. In other words, algorithms with more false corners tend to have a larger number of matched corners. Therefore this criterion is flawed for measuring the stability of corner detectors. Our criteria are consistency of corner numbers and accuracy. Only with consideration of these criteria together, is it possible to judge correctly on the best corner detectors for tracking and matching tasks.

8.2. Definition of New Criteria

In this section our criteria for measuring the stability and accuracy of corner detectors are defined theoretically. In the following, let N_o be the number of corners in original image (note that $N_o \neq 0$), N_m number of matched corners in each of transformed images when compared to original image corners and N_t number of corners in each of the transformed images.

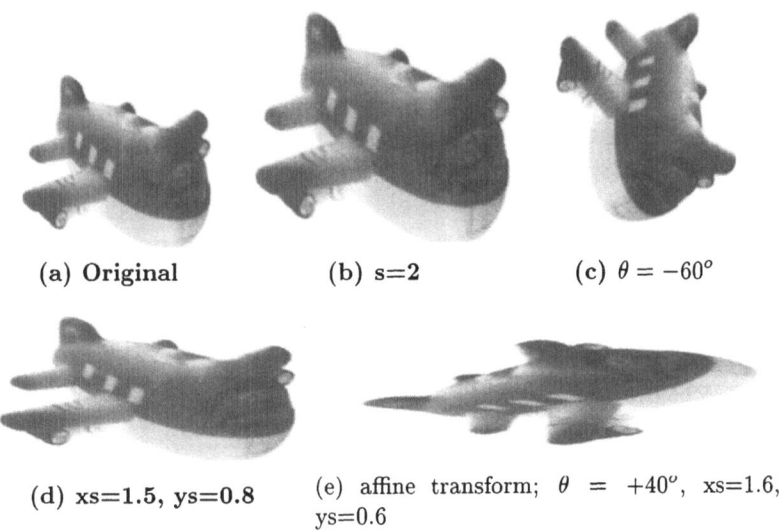

(a) Original   (b) s=2   (c) $\theta = -60°$

(d) xs=1.5, ys=0.8   (e) affine transform; $\theta = +40°$, xs=1.6, ys=0.6

Figure 7.17. Airplane image under similarity and affine transforms. In this figure s, xs, ys and θ stand for uniform scaling, x-scale and y-scale in non-uniform scaling and rotation parameters respectively.

8.2.1   Consistency

Consistency means corner numbers should be insensitive to the combination of noise, rotation, uniform or non-uniform scaling and affine transform. More importantly, corner numbers should not change when multiple images are acquired of the same scene. Previous criterion of consistency has been defined as follows:

$$consistency = \frac{N_m}{N_o} \qquad (7.1)$$

Using this definition, algorithms which find more false corners in input images are favoured since they have higher numbers of matched corners. Therefore we replace this criterion by two new criteria, *consistency of corner numbers* and *accuracy*. We define the criterion of *consistency of corner numbers* as follows:

$$CCN = 100 \times 1.1^{-|N_t - N_o|} \qquad (7.2)$$

where CCN stands for "consistency of corner numbers". Since stable corner detectors do not change the corner numbers from original image to transformed images then in terms of consistency, the value of CCN for stable corner detectors should be close to 100%. This criterion for corner detectors with more false corners becomes closer to zero.

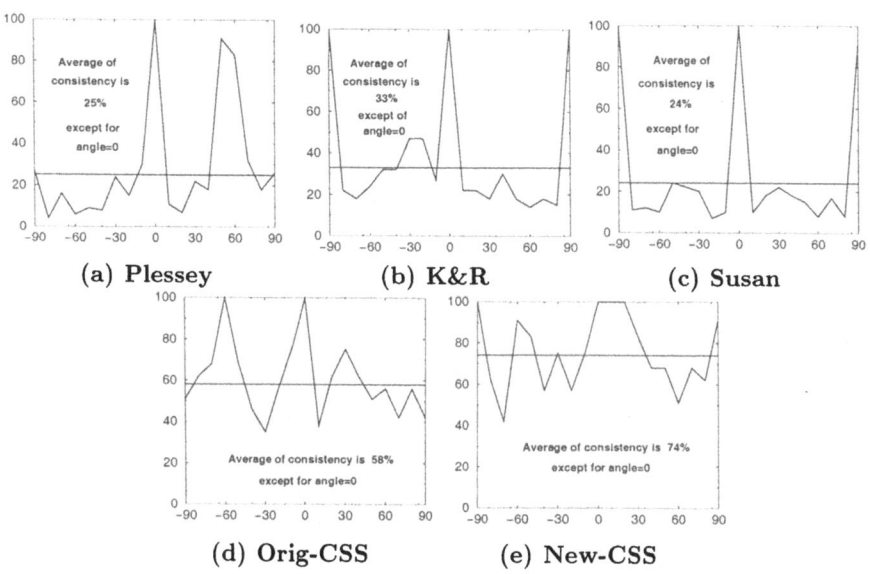

Figure 7.18. Consistency of corner numbers for rotation

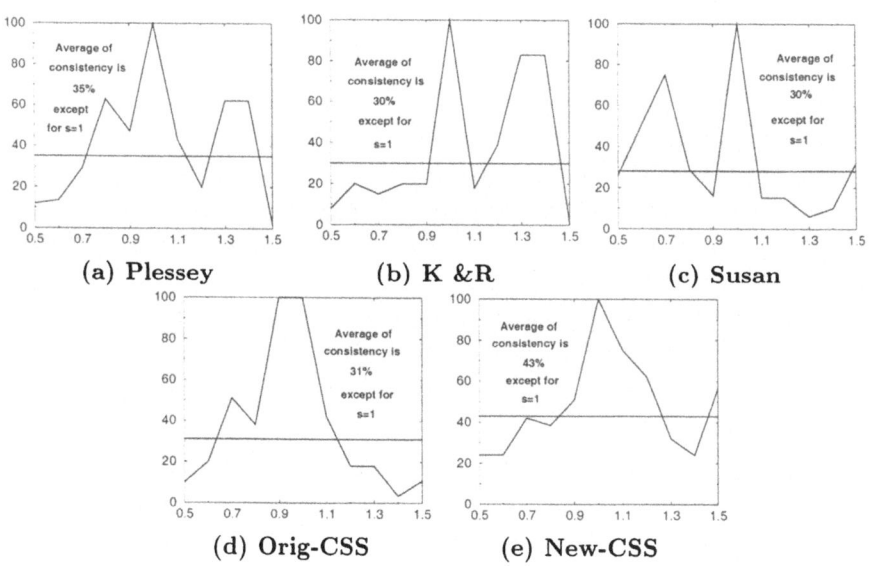

Figure 7.19. Consistency of corner numbers for uniform scaling

8.2.2 Accuracy

Accuracy requires that corners should be detected as close as possible to their correct positions. In a given image, the corner positions and numbers can be different according to different people. Furthermore, as there is no standard procedure to measure accuracy of corner detectors, we adopted a new approach for creating ground-truth. This approach is based on majority human judgement. To create ground-truth, ten persons who were familiar with the task of corner detection were chosen. None of them were familiar with the algorithm used by our corner detector. We asked them individually to mark the corners of an image. The corners marked by at least 70% of individuals were selected as the ground-truth for that image. The position of a corner in the ground-truth was defined as the average of the positions of this corner in individual images marked by those ten persons. We repeated the same for other images. Then by comparing the detected corners using each of five corner detectors to the list of corners in ground-truth, the accuracy was computed as follows:

Let N_o be the number of corners in original image (note that $N_o \neq 0$), N_a the number of matched corners in original image when compared to ground-truth corners and N_g the number of corners in the ground-truth. The criterion of accuracy is defined as:

$$ACU = 100 \times \frac{\frac{N_a}{N_o} + \frac{N_a}{N_g}}{2} \tag{7.3}$$

where ACU stands for "accuracy". In terms of accuracy, the value of ACU for accurate corner detectors should be close to 100%. ACU for corner detectors with lower accuracy is closer to zero. The case of $N_o=0$ occurs if test images have no corners or tested corner detectors can not detect any corners. These situations do not arise in practice as only images with many corners are used in experiments and corner detectors under consideration also find many corners in test images.

8.3.   Performance Evaluation, Results and Discussion

We considered the results of our experiments on several images. Examples of image transforms have been illustrated in Figure 7.17. These experiments were performed as follows:

Experiment 1: In the first experiment, the number and positions of corners in the original image were recovered using the test corner detectors. Next, original image was rotated with rotation angle chosen by uniform sampling of the interval $[-90^o$ to $+90^o]$ excluding zero. Distance between consecutive samples was 10°. Then the numbers and positions of corners in all rotated images were extracted using the test detectors.

Experiment 2: In the second experiment, we did the same for original image and uniform scaling of this image with ten scale factors chosen

Average of CCN for					
	Plessey	K & R	Susan	Orig-CSS	New-CSS
Non-uniform scaling	28%	31%	31%	55%	68%
Affine transform	14%	11%	9%	42%	51%

Table 7.1. Average of consistency of corner numbers for tested corner detectors

by uniform sampling of the interval [0.5, 1.5] excluding 1.0. Distance between consecutive samples was 0.1.

Experiment 3: We repeated the same in the third experiment with non-uniform scaling. Values of x-scale and y-scale were chosen by uniform sampling of the intervals [0.8, 1.0] and [0.5, 1.5] respectively. Distance between consecutive samples was 0.1.

Experiment 4: affine transform was our fourth experiment that applied rotation angles of -10° and $+10^o$ combined with x-scale and y-scale chosen by uniform sampling of the intervals [0.8, 1.0] and [0.5, 1.5] respectively. Distance between consecutive samples was 0.1.

After performing our experiments on rotated, uniformly and non-uniformly scaled and affine transformed images, we computed CCN. The results of these computations for rotation and uniform scaling have been illustrated in Figure 7.18 and Figure 7.19. The average of consistency in non-uniform scaling and affine transform have been shown in Table 7.1.

Final test was performed for computation of accuracy. We computed accuracy using our database which included leaf, airplane, fish, lab, and building images. These images have been illustrated in Figure 7.20. Furthermore, in this figure the corner points of their ground truths have also been shown. The comparison of consistency and accuracy in similarity and affine transforms for test detectors have been illustrated in figure 7.21. Overall, the results of these comparisons show that our corner detector has the better accuracy and stability among these five corner detectors.

9. Conclusions

This chapter described a new corner detection method based on the CSS technique. The edges of a real image were extracted using the Canny edge detector. The gaps between two close contours were examined in order to find T-junction corners or to fill the gap to form a continuous contour. Curvature maxima were extracted at a high scale and the corner locations were tracked at multiple lower scales to improve localization. Finally the T-junction corners were compared to the CSS

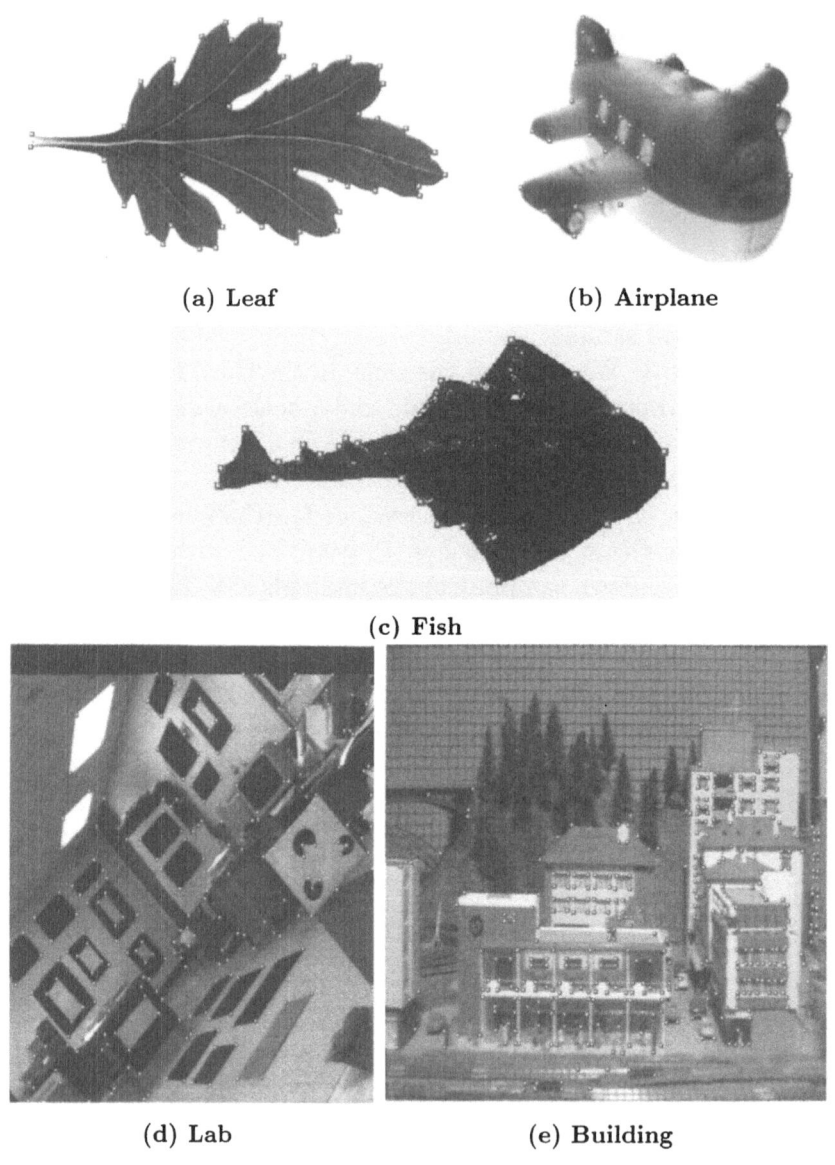

(a) Leaf   (b) Airplane

(c) Fish

(d) Lab   (e) Building

Figure 7.20.   Examples of test images for computation of accuracy

corners in order to remove corners marked twice. The CSS image corner detection method was robust with respect to noise and performed better than the other detectors it was compared to.

A later improvement of the CSS corner detector was also described. The New-CSS corner detector adjusts the degree of smoothing applied

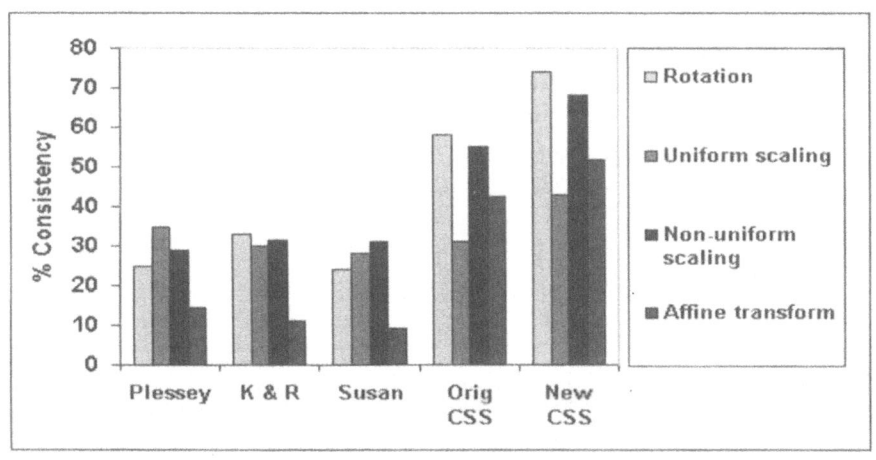

(a) consistency of corner numbers

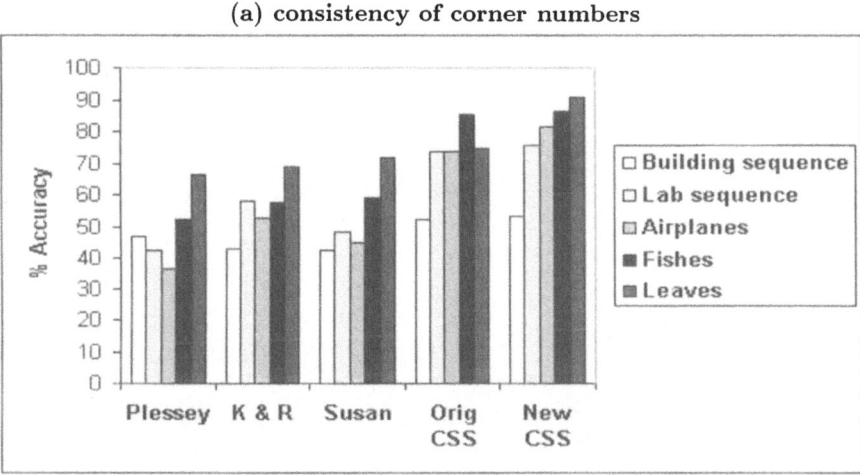

(b) accuracy

Figure 7.21.   Comparison of consistency and accuracy in similarity and affine transforms for tested corner detectors

to image edge contours depending on their lengths. An important advantage of New-CSS is that it does not use any threshold values.

Two new criteria, consistency of corner numbers and accuracy, were proposed for performance evaluation of corner detection algorithms. The New-CSS corner detector, the original CSS corner detector and three other corner detectors were evaluated under similarity and affine transforms using the proposed criteria. A new approach for creation of ground-truth used for computation of accuracy was also proposed. Application of this procedure resulted in the correct number of matched corners due to comparison to ground-truth. As a result, no false corners were taken into account when matching corners between original image and ground-truth. New definition of consistency of corner numbers prevented algorithms which find more false corners in input images from achieving a high score. Overall, the application of these criteria showed that the New-CSS corner detector produced the best results with respect to similarity and affine transforms.

Chapter 8

FAST ACTIVE CONTOUR CONVERGENCE THROUGH CURVATURE SCALE SPACE FILTERING

Contour evolution can be considered an early form of active contours (snakes) since evolving contours are similar in behaviour to snakes without any external constraints. The behaviour of an active contour in energy-minimizing active contour models based on dynamic programming is controlled by its internal and external energies. The second part of internal energy in this model is the smoothness/curvature term of the active contour. When this term is removed, the computation of the total energy of the active contour is simplified. However without the effect of the curvature part in internal energy of the active contour, the final snake loses its smoothness. In this chapter, an accurate and high speed active contour model is proposed based on reformulating internal energy by removing the curvature part and using CSS filtering for smoothing. By applying CSS smoothing, proposed model is more independent of model parameter values and initial snake. One of the advantages of the new model over the existing models is that it has only one parameter that affects the internal energy of the active contour. Images with single and multiple objects are selected to evaluate the speed and performance of the proposed model. The results show that the proposed model converges quickly to the final solution.

1. Introduction

Our interest in active contour models comes from their use as a user interface for image/video database retrieval using shape content. An active contour can be successfully used for these tasks if it has good accuracy and high speed. A number of models have been proposed for active contours. The performance of active contours in these models depends on the proper setting of model parameters and initial snake.

Due to these shortcomings, majority of existing active contour models often fail to converge to the desired solution especially in complex images, see Fig. 8.1. Furthermore they demonstrate low speed and inaccurate output. The first active contour model was proposed by Kass et al. [134]. Its algorithm has a number of shortcomings pointed out in [19] and a solution was presented as a discrete multi-stage decision process using a "time-delayed" discrete dynamic programming algorithm. The behaviour of an active contour is generally controlled by internal and external energies. The internal energy acts to shorten and enforces smoothness of active contour and the external energy moves it towards image features such as image edges. The total energy of the active contour is defined as:

$$\sum_{i=0}^{n-1} E_{int}(i) + E_{ext}(i)$$

where n, E_{int} and E_{ext} are the number of points, internal energy and external energy of active contour respectively. Discretized E_{int} of an active contour represented as $v(s) = (x(s), y(s))$, can be computed by:

$$E_{int}(i) = (\alpha_i |v_i - v_{i-1}|^2 + \beta_i |v_{i+1} - 2v_i + v_{i-1}|^2)/2 \qquad (8.1)$$

where v_i refers to the ith point of the active contour and similarly for v_{i-1} and v_{i+1}. In [18, 19], the minimization of the total energy of the active contour was viewed as a discrete multi-stage decision process. For each stage of this process, the total energy of active contour was computed to be:

$$
\begin{aligned}
E_t(i, j, k) &= \min_{0 \leq m \leq N} \{ E_t(i - 1, k, m) + E_{ext}(v_{i-1} \oplus k) \\
&+ \frac{1}{2}(\alpha |v_{i-1} \oplus k - v_{i-2} \oplus m|^2 \\
&+ \beta |v_i \oplus j - 2v_{i-1} \oplus k + v_{i-2} \oplus m|^2) \}
\end{aligned}
\qquad (8.2)
$$

Note the following points about this formula:

- $E_t(i, j, k)$ denotes the total energy of active contour at iteration t. Energy values are stored in a 3-D matrix. In other words, for any point of the active contour, we have a 2-D matrix with j rows and k columns. i, j and k refer to each point of active contour and its two neighbours respectively. As each point of the active contour can remain at its previous location or move to one of its eight neighbours, the possible range for both k and j is [0-8].

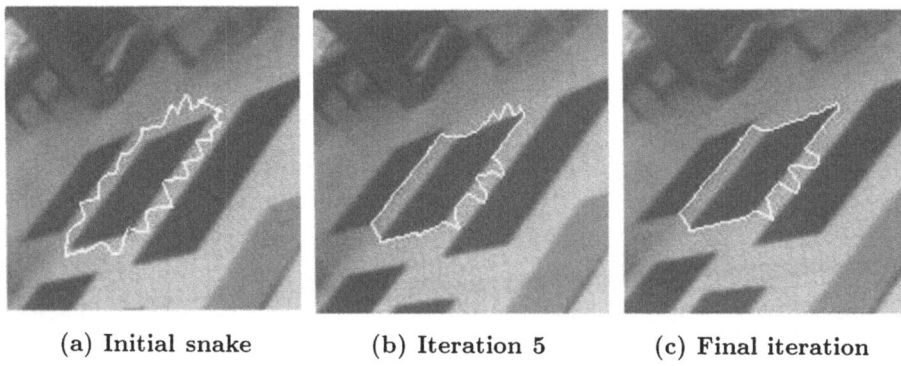

| (a) Initial snake | (b) Iteration 5 | (c) Final iteration |

Figure 8.1.   An example of an active contour failing to converge to the interest object.

- N presents the number of possible directions for each point of active contour, therefore $N=9$.

- v_i and v_{i-1} denote the ith point and its previous point on the active contour. For example the point before the starting point is the endpoint due to the active contour being closed.

- $v_{i-1} \oplus k$ represents the kth neighbour of point i -1 on the active contour. The zeroth neighbour refers to the initial location of a point, not its neighbours.

- $|v_{i-1} \oplus k - v_{i-2} \oplus m|^2$ indicates the distance between the kth neighbour of point i-1 and the mth neighbour of its previous point.

- $E_{ext}(v_{i-1} \oplus k)$ is the external energy of the kth neighbour of point i-1 on the active contour. This is defined as the distance between the kth neighbour of point i-1 on the active contour and the nearest edge of the underlying image.

- $E_t(i-1, k, m)$ represents an element of the energy matrix in iteration t at row k and column m. Actually it is the total energy of active contour at point i-1 in row k and column m.

- $|v_i \oplus j - 2v_{i-1} \oplus k + v_{i-2} \oplus m|$ is the curvature of active contour at point i estimated using the three points; i-2, i-1 and i.

Internal energy of the active contour in these formulas is composed of two parts; the first part helps to reduce the length of the active contour during its movement until it locks on to the interest object. The second term, is the curvature of the active contour and ensures its smoothness through these stages. If we can remove the second term in Eq. 8.2, this formula summarizes to Eq. 8.3 as follows:

$$E_t(i,k) = \min_{0 \le j \le N} \{E_t(i-1,j) + \alpha |v_i \oplus k - v_{(i-1+n)} \oplus j|^2 + E_{ext}(v_i \oplus k)\} \quad (8.3)$$

For an explanation of this formula, see section 3. The total energy of the active contour in this formula can be computed through a 2D matrix instead of computing a 3-D matrix for each point of the active contour. Consequently the entire multi-stage process becomes less time consuming and initial snake converges very quickly to a final solution. But removing the curvature part from the internal energy causes the final active contour to lose smoothness. As a result, the existence of this part in total energy of active contour is necessary. However, an alternative idea is to replace it with another step that ensures smoothness but converges faster.

In this chapter, we present an accurate and high speed algorithm for minimizing the energy of active contour models based on Eq. 8.3 combined with CSS filtering for smoothing [224, 225]. By applying CSS filtering, we smooth the output of each iteration in the new algorithm using Gaussian filtering with a small scale (the standard deviation of Gaussian, σ, is referred to as scale) until the snake reaches underlying image edges [183]. Following that, the remainder of iterations continue without smoothing. The process halts when there is no change in the minimum energy of active contour between two successive iterations. Images with single and multiple objects are selected to evaluate the capability of our proposed method. The results show that our method converges very quickly to the final solution without losing smoothness in the final active contour shape. An alternative to this method is improved curvature estimation for accurate localization of active contours [181].

The following is the organisation of the remainder of this chapter. Section 2 presents an overview of active contour models and their applications. The theory underlying our proposed algorithm is explained in section 3. In section 4 the results of our active contour model on different images are illustrated. The conclusions are presented in section 5.

2. Literature Survey

Considerable research has been carried out on active contour models in recent years. We review a number of proposed methods and some active contour applications in this section. The underlying idea of the energy minimizing active contour model was proposed in [134]. The problems of Kass et al.'s algorithm were discussed in [19] which proposed an algorithm for active contours based on dynamic programming. Their time consuming method needs careful setting of model parameters and does not lock on to objects of interest very accurately. Williams and Shah [320] presented a fast active contour model based on a greedy algorithm with three parameters; α, β and γ and two thresholds. Therefore their method is more dependent on model parameter and threshold values. Brownian Strings algorithm [100] controlled the evolution of the active contour by a simulated annealing process which causes the contour to settle into the global minimum of a non-parametric and image-derived energy function. In [46], a geodesic formulation for active contours was presented. Bayesian wavelet snake [328] was developed for identifying a closed-contour object with a fuzzy and low-contrast boundary. The velocity snake [253] was proposed based on applying velocity control to the class of elasto-dynamic contour models. Hui et al. [168] proposed a robust snake model using the reformulated internal energy and the combination of both region and edge information to enlarge the capture range. Giraldi et al. [93] addressed the limitations of dynamic programming (DP) by reducing the region of interest for image segmentation through the use of Dual-T-snake approach. Kim et al. [140] presented a new contraction energy of active contours independent of the object's form for segmentation. Ngoi and Jia [242] presented a new active contour model for contour extraction in natural scenes. A new shape constraint for active contours based on DP has been proposed by Kang [132]. It allows fast and stable tracking of object boundaries in magnetic resonance imaging (MRI) sequences with an equality constraint. In [329], a new contour detection method based on the snake model was developed. Affine-invariant snake model was presented in [124]. Xu [326] proposed a new external force for active contours called gradient vector flow. In [318], a high performance snake based on spline representation and multiple stage energy minimisation process was proposed.

3. Smoothed Active Contour (SAC)

The proposed active contour model, SAC, is based on reformulating the internal energy of active contour by removing the curvature part of the internal energy and using CSS filtering with a small scale (σ) for smoothing. By applying the CSS technique for smoothing, we smooth the output of each iteration of the discrete multi-stage process for minimizing the energy of active contour until locking on to the underlying image edges occurs. After that the remainder of these iterations continue without smoothing. The process halts when there is no change in the minimum energy of active contour between two successive iterations. The outline of the SAC model has been illustrated in Fig. 8.2. The reformulated internal energy of the SAC model can be viewed in

Eq. 8.3. To better understand this equation, note that the differences in comparison to Eq. 8.2 are as follows:

- $E_t(i, k)$ denotes the total energy of active contour at iteration t. Energy values are stored in a 2D matrix. i and k refer to each point of active contour and its previous point respectively. As in Eq. 8.2, the possible range for k is [0-8]. Each row of this matrix belongs to one point of the active contour.

- v_{i-1+n} designates the previous point of point i on the active contour.

- $|v_i \oplus k - v_{i-1+n} \oplus j|^2$ indicates the distance between the kth neighbour of point i and the jth neighbour of its previous point.

- $E_{ext}(v_i \oplus k)$ indicates the distance between the kth neighbour of point i on the active contour and the nearest image edge. Actually it is the external energy of the kth neighbour of point i on active contour.

- $E_t(i-1, j)$ represents an element of energy matrix in iteration t at row *i-1* and column j. Therefore it is the total energy of active contour at previous point of i in column j.

The outline of the method is as follows: First, the elements of energy matrix are computed using Eq. 8.3. Second, a 2-D location matrix with dimensions $i \times k$ is computed as well. Third, the minimum energy of active contour in first iteration is found. This minimum energy is the minimum entry of the energy matrix in the last row:

$$E_{\min}(t) = \min_k E_t(n - 1, k)$$

. The column of this minimum energy indicates the neighbour that the endpoint on active contour should move towards on the new active contour. Finally the location matrix entry in the end row with the same column as minimum energy indicates the neighbour that point *n-2* on the active contour should move towards on the new active contour. This *backward* process continues until the new position of all points on the new active contour will be determined. The output of this iteration is then smoothed by CSS filtering with a small scale. Finally this output is tested to determine whether the snake has reached the underlying image edges. If the new active contour has not reached underlying image edges, other iterations repeat the same procedure on the new active contour. However, if the new active contour has reached underlying image edges, the CSS filtering step is omitted. The process continues until there is no change in minimum energy of active contour between two successive iterations. Due to using CSS filtering, setting σ to low

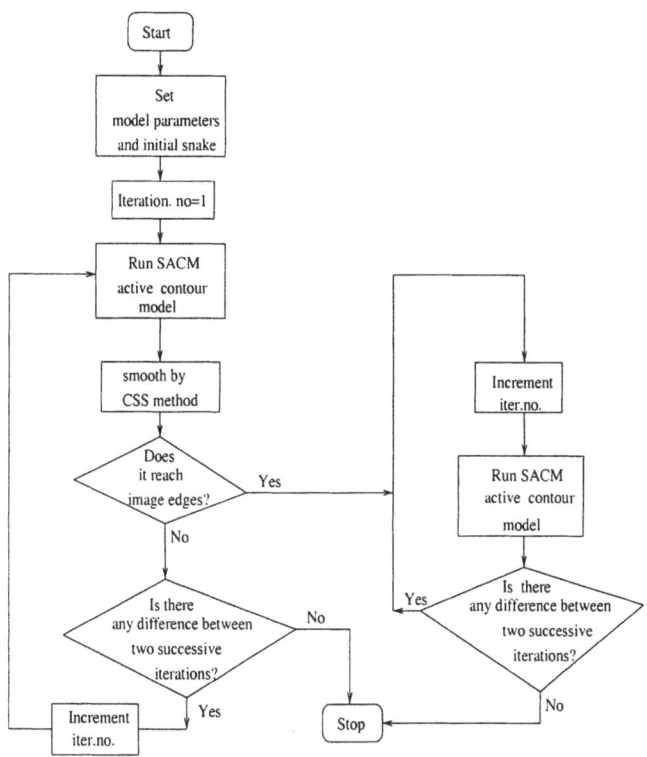

Figure 8.2. The SAC active contour model

value is emphasized. High value of σ may smooth the initial snake so much that it cuts the interest object at some parts of its boundary and in the worst case goes inside the object. Once σ has been set to a low value, it does not need to be set again. Therefore SAC has only one parameter α, that affects the internal energy of the active contour. Setting this parameter to a higher value gives a higher weight to internal energy. In the algorithm proposed in [19], (AMI model) if there are n points on the active contour and m possible directions at each point, the time complexity for each iteration is $O(nm^3)$. However in the SAC model, the time complexity for each iteration is $O(nm^2)$ in both simple and complex images while SAC has just one model parameter. Furthermore the SAC model does not lose smoothness in the shape of final snake and also

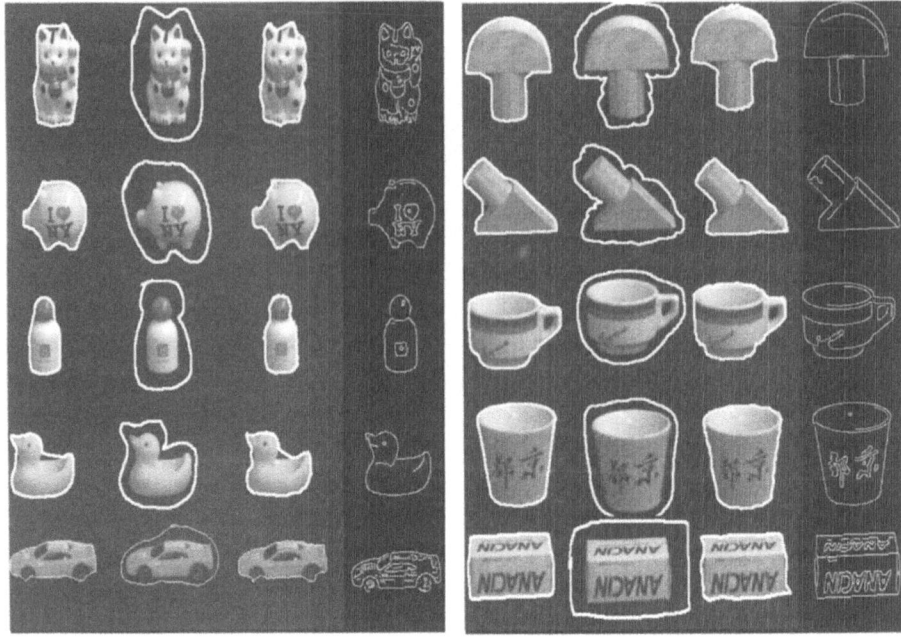

Figure 8.3.   The behaviour of the SAC and the AMI active contour models in terms of the shape of the final snake in simple images. From left to right, first column, second column and third column show the final snakes of the SAC, the initial snakes for both models and the final snakes of the AMI respectively. The fourth column shows the edges of the underlying images. It can be seen from the fourth column, how close the final snake is to the correct solution.

convergence to desired solution is more likely. Overall, the SAC model is a good user interface for image/video database retrieval system.

4.   Experimental Results

We have applied the SAC model to a number of real images. These real images are classified into two categories; simple and complex images. Also in all experiments, initially user specifies the position of active contour and sets model parameters. Our results are demonstrated with consideration of two points; speed and the shape of the final snake. In Fig. 8.3, the behaviour of the SAC and the AMI active contour models in terms of the shape of the final snake in some images of the Coil database [241] have been illustrated. The results of applying both methods to complex images have been shown in Fig. 8.4. Table 8.1 shows the number of iterations and the execution times of the SAC and

Images		Size	SAC		Pars		AMI		Pars	
			Iter.	Time	α	σ	Iter.	Time	α	β
Single object images	Cat	156 × 177	31	5	1	1	25	6.85	1	1
	Piggy	158 × 171	35	7.42	1	1	32	8.80	1	1
	Bottle	162 × 178	22	3.16	1	1	18	4.5	1	1
	Duck	162 × 180	14	1.83	1	1	16	3.56	1	1
	box	187 × 117	22	3.81	1	3	29	6.87	1	1
	Toy1	152 × 175	17	6.47	1	1	18	9.63	1	1
	Toy2	155 × 179	19	4.60	1	1	17	6.63	1	1
	Cup	153 × 177	15	3.17	1	1	12	5.14	1	1
	Glass	150 × 176	15	5.93	1	1	15	6.61	1	1
	Car	154 × 178	19	5.04	1	1	17	7.33	1	1
Multi object images	Lab	175 × 153	9	8.66	8	3	10	9.44	20	5
	Olympic1	176 × 144	11	1.01	4	1	7	1.5	6	1
	Olympic2	176 × 144	5	0.47	1	3	6	1.06	4	1
	Children	176 × 144	9	0.85	1	1	9	2.03	1	1

Table 8.1. Results of execution time (seconds) and number of iterations for the SAC and the AMI on our test images. In this table, Pars, Time and Iter. stand for parameters, execution time and iterations number respectively.

the AMI active contour models using our test images. The experiments have been carried out on an Intel Pentium III. The comparison of speed and number of iterations for these two models have been illustrated in Fig. 8.5 using our test images. All results show better performance of the SAC model in terms of speed and accuracy.

5. Conclusions

This chapter proposed an alternative for the smoothness term of the internal energy of active contours in energy minimizing active contour models to speed up the convergence to the desired solution without increasing the number of model parameters. The external energy at any point of the active contour is computed using the minimum distance of this point to the underlying image edges. The curvature part of the internal energy is removed and replaced by CSS filtering for smoothing. We smooth the output of each iteration in a discrete multi-stage process until locking on to the underlying image edges occurs at least at one point. The remainder of iterations continue without smoothing until minimum energy is reached and the whole process halts. Therefore in the SAC model, initial snake never cuts the boundary of interest object and never goes inside this object. This improves the stability of the SAC model. Application of this new procedure resulted in locking onto the

Figure 8.4. The behaviour of the SAC and the AMI active contour models in terms of the shape of the final snake in complex images. In this figure, the first column shows the final snakes of the SAC model, the second column shows the initial snakes for both models, and the third column shows the final snakes of the AMI model. The edges of underlying images are shown in the fourth column.

interest object quickly and accurately. The SAC model has only one parameter that affects the internal energy of the active contour, therefore it is more independent of model parameters. Also due to using CSS filtering with a small scale for smoothing, the SAC is more independent of initial snake as well. In addition, smoothing helps the initial snake move towards the interest object faster especially in complex images. The SAC was tested using many single and multiple object images. Overall

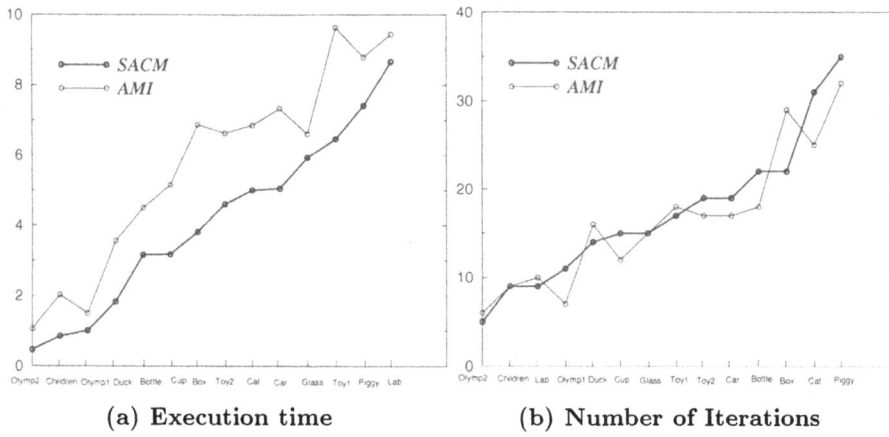

(a) Execution time (b) Number of Iterations

Figure 8.5. Comparison of the execution time and the number of iterations for the SAC and the AMI models

the results show that the SAC model converges to the final solution more quickly without losing smoothness. The proposed active contour model is accurate and efficient which makes it suitable as a user interface for image/video database retrieval using shape content.

Chapter 9

EFFICIENT MULTI-SCALE CONTOUR DATA COMPRESSION AND RECONSTRUCTION USING THE CURVATURE SCALE SPACE IMAGE

The CSS technique has been used in conjunction with Hermite curves for automatic fitting of digitized contours at multiple scales. A parametric representation of the input contour is convolved with Gaussian functions in order to obtain multi-scale descriptions of the contour. Curvature can be computed directly at each point of the smoothed contours. As a result, a set of curvature zero-crossing points can be recovered from each smoothed contour.

Hermite curves were used since each Hermite curve is defined by two endpoints and the tangent vectors at those points. No points external to the input contour are required for Hermite curves. Hermite endpoints are defined as consecutive curvature zero-crossing points extracted at multiple scales using the CSS method. Hermite tangent vectors can also be determined using the CSS technique. The only data stored are the endpoints and the tangent vectors needed by the Hermite curves in order to arrive at an approximate reconstruction of the original contour.

Approximation error and compression ratio are computed at each scale. The graph of compression ratio as a function of approximation error is smoothed to remove noise and small fluctuations. The *bending point* of that function is then defined as the largest maximum of its second derivative. The bending point can be considered as the boundary between the mostly vertical and the mostly horizontal segments of the graph. It can be used for automatic selection of an optimal scale.

1.   Introduction

Existing techniques for contour data reconstruction suffer from a number of shortcomings. Polygons have been used to approximate the shape of free-form contours by several researchers [249, 66, 156]. The vertices

255

of the approximating polygons are then stored for later reconstruction
of the shape. This approach works best when the corners of the input
shapes are detected and chosen as polygon vertices. Robust corner detection is itself a challenging problem that needs to be addressed first.
Another problem with polygon approximation is that polygons are not
suitable for description of free-form contours, and would therefore require
a large number of sides for a good approximation. A spline approach
would yield a better approximation using less data. Fourier Descriptors
[252, 269] are another class of methods which can be utilized for contour
data reconstruction. The first k Fourier coefficients can be computed
and stored for later reconstruction of the contour. Naturally, a larger k
yields a better reconstruction. A shortcoming of Fourier Descriptors is
that a large number of them will be needed to obtain an accurate reconstruction of the input contour. Furthermore, it is not suited to CAD
applications. An algorithm by Schneider [276] fits splines to digital contours but the procedure used for control point selection is quite ad hoc
and will not, in general, yield the best results.

Applications of this technique include efficient contour data compression as well as Computer Aided Design. It is usually assumed in CAD
work that the user will supply all the control points required to generate
the desired shapes. In this case, the user will have to start the design
work from scratch. However, often the user may wish to start from a
known shape which exists in digital form and modify that to obtain the
final desired shape. The proposed technique will enable the user to obtain a spline approximation to that starting digital shape. The control
points can then be adjusted by the user to produce the desired shape.

This chapter presents a new method based on the CSS technique for
automatic fitting of digitized contours [233]. This technique utilizes the
CSS method to recover curvature zero-crossings and tangent vectors at
those zero-crossings which are then used for Hermite curve fitting [311].

Section 2 presents a brief review of spline fitting techniques. Section
3 is on contour data reconstruction through CSS and Hermite curves.
Section 4 discusses the computation of the approximation error and the
compression ratio at multiple scales. Section 5 presents the results and
discussion, and section 6 contains the concluding remarks.

2.   Spline Fitting Techniques

Spline fitting techniques [86, 87] have been used widely in computer
graphics, computer vision and image processing. They are useful since
they can model free-form shapes in a compact way: A relatively small
number of points and/or tangent vectors are sufficient for accurate reconstruction of the original shape.

A number of spline fitting techniques with different properties are available. The most common of these are the family of parametric cubic curves. This family consists of three major types of curves:

Hermite curves Defined by two endpoints and two endpoint tangent vectors.

Bézier curves Defined by two endpoints and two other points (not on the contour) which control the endpoint tangent vectors.

B-splines Defined by four control points (not on the contour).

3.   Contour Data Reconstruction through CSS and Hermite Curves

This section explains how contour data reconstruction can be achieved at multiple scales through a combination of Hermite curves and the CSS technique.

Hermite curves were chosen since they do not require any points external to the input contour. Each Hermite curve segment requires two endpoints and the tangent vectors at those endpoints as input. All of these can be supplied automatically and robustly by the CSS method at multiple scales.

Suppose that $P(x_p, y_p)$ and $Q(x_q, y_q)$ are the endpoints of a Hermite segment. Assume that $\mathbf{u}(x_u, y_u)$ and $\mathbf{v}(x_v, y_v)$ are the tanget vectors at P and Q respectively. The Hermite segment is given by the following equations [87]:

$$
\begin{aligned}
x(w) &= (2w^3 - 3w^2 + 1)x_p + (-2w^3 + 3w^2)x_q \\
&+ (w^3 - 2w^2 + w)x_u + (w^3 - w^2)x_v
\end{aligned}
$$

$$
\begin{aligned}
y(w) &= (2w^3 - 3w^2 + 1)y_p + (-2w^3 + 3w^2)y_q \\
&+ (w^3 - 2w^2 + w)y_u + (w^3 - w^2)y_v
\end{aligned}
$$

where $w \in [0, 1]$.

At a specific scale, the input contour is smoothed, and curvature is computed at each point using the formula given in section 3. This step is followed by the recovery of curvature zero-crossing points. Each pair of adjacent curvature zero-crossings are used as the endpoints of a Hermite curve. The advantage of using curvature zero-crossing points is that they are invariant to many transforms and therefore constitute a natural set of feature points. Furthermore, tangent vector estimation is more robust at inflection points since the contour is locally straight at those points. The directions of the tangent vectors at those endpoints are given by:

$$(\mathcal{X}_u, \mathcal{Y}_u)$$

where $\mathcal{X}_u$ and $\mathcal{Y}_u$ are the first derivatives of the coordinate functions of the contour. Note, however, that in general the lengths of those tangent vectors have to be adjusted in order to obtain the optimal shape for the Hermite curve segment that best fits the input contour segment.

The tangent vector directions must be estimated from a smoothed contour [169] in order to remove the influence of noise on the estimation process. However, their lengths are optimized using the original input data since our intention is to approximate the original contour as best as possible.

This optimization is carried out using an iterative procedure. The initial tangent vectors are multiplied by a real number n. Hermite curve fitting then takes place. The average distance between the Hermite curve segment and the input contour segment is then defined as following:

The distance from each point on the Hermite curve segment to the closest point on the input contour segment is computed. All such distances are added up and divided by the total number of points on the Hermite curve segment to determine the average distance.

The value of n is increased by a step size and Hermite curve fitting is repeated. The average distance is then re-computed. This process continues as long as the average distance continues to decrease. When the process terminates, the optimal length tangents and therefore the optimal Hermite curve has been found.

To enhance efficiency, it is possible to start with a relatively large step size, and to reduce it as the process approaches the optimal value of n. In this approach, if a larger step size causes the average distance to increase, the process backtracks and attempts a smaller step size.

Our experiments indicate that this optimization procedure does converge to a global minimum. We believe the reason is that we can initialize the procedure at a point sufficiently close to the global minimum.

4.   Approximation Error and Compression Ratio

When all Hermite curve segments have been fitted, the total approximation error is defined as the mean of the average distances for all the Hermite curve segments. Furthermore, compression Ratio is defined as the size of the data after compression divided by the size of the original data.

Contour data compression can be carried out at multiple scales. This allows the user to find an appropriate trade-off between approximation error and compression ratio. Clearly, reducing the approximation error would also result in less compression and more accuracy whereas allowing the approximation error to rise would result in more compression and less accuracy.

The graph of compression ratio as a function of approximation error can be smoothed to remove noise and small fluctuations. The *bending point* of that function after smoothing can then be defined as the largest maximum of its second derivative. The bending point can be considered as the boundary between the mostly vertical and the mostly horizontal segments of the graph. It can be used for automatic selection of an optimal or natural scale [267, 268] for contour data reconstruction. As a result, the user will not have to set any parameters in order to use this technique.

5.   Results and Discussion

This section presents results on contour reconstruction through the CSS image. The test data consisted of three contours: Africa, Hokaido, and a design from a Persian carpet. The test contours can be seen in figure 9.1.

Reconstruction by spline fitting was implemented next. Figures 9.2 and 9.3 show the contour reconstruction results for Africa at multiple scales. As the scale increases, the number of curvature zero-crossing points and therefore the number of spline segments decreases, but this is accompanied by an increase in approximation error. Figures 9.4 and 9.5 show the corresponding results for Hokaido and figures 9.6 and 9.7 show the corresponding results for the carpet design.

Finally figures 9.8, 9.9, and 9.10 show the graphs of compression ratio as a function of approximation error for Africa, Hokaido, and carpet design corresponding to spline reconstruction. They demonstrate that as reconstruction accuracy decreases, greater compression of the input data can be achieved. The point marked with a + on the Africa graph indicates the bending point of that graph which corresponds to the optimal scale for reconstruction.

At a specific scale, the complexity of the fitting process is $O(nk)$ where k is the size of the convolution filter and n is the number of points on the input contour.

6.   Conclusions

A novel technique was presented for automatic fitting of digitized contours at multiple scales through the CSS technique used in conjunction with Hermite curves.

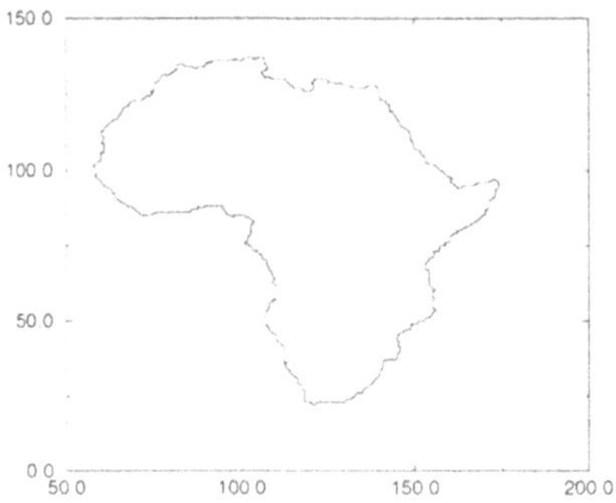

(a) Africa 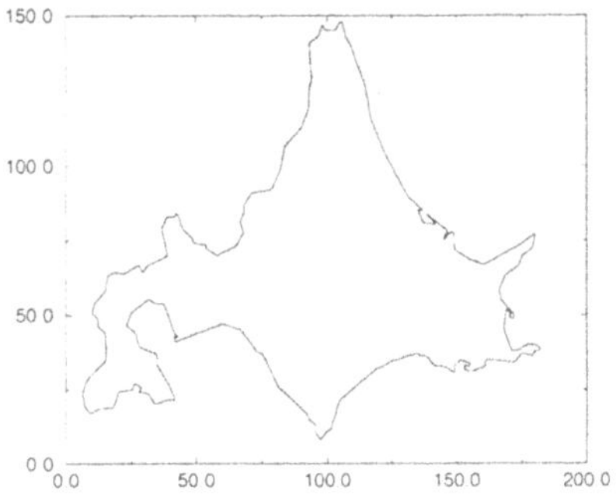   **(b) Hokaido**

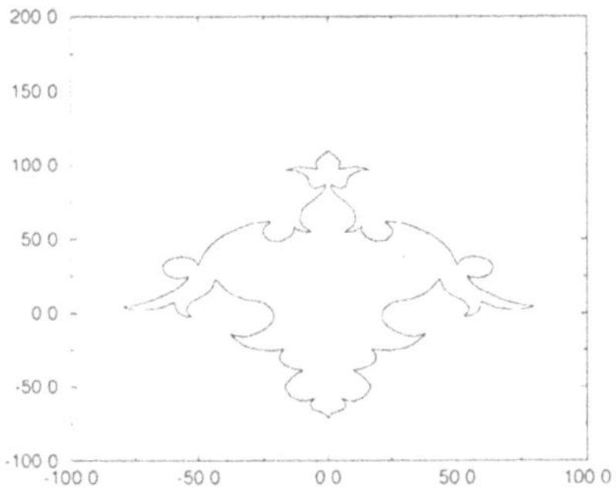

(c) Carpet Design

Figure 9.1.   Test contours

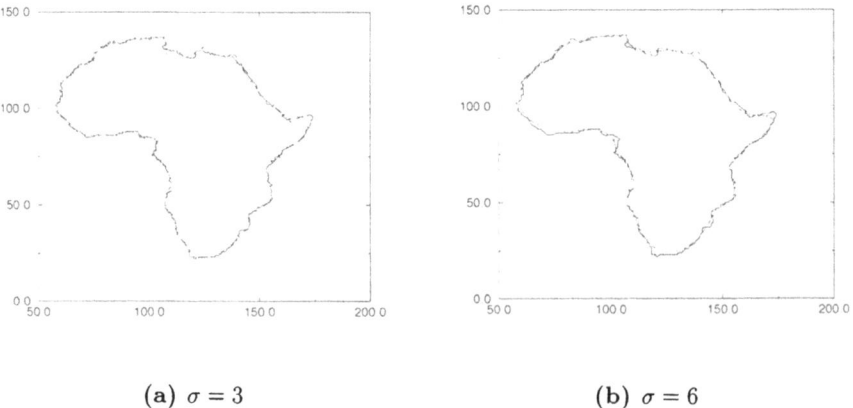

(a) $\sigma = 3$ **(b)** $\sigma = 6$

Figure 9.2.   Reconstruction of Africa

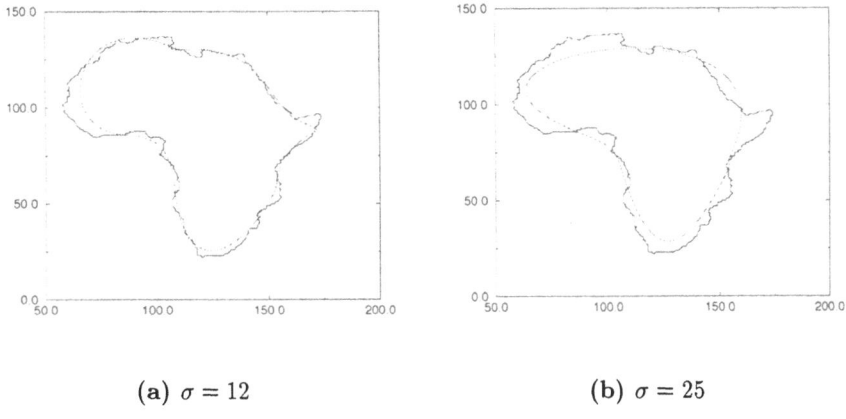

(a) $\sigma = 12$ (b) $\sigma = 25$

Figure 9.3. Reconstruction of Africa (cont.)

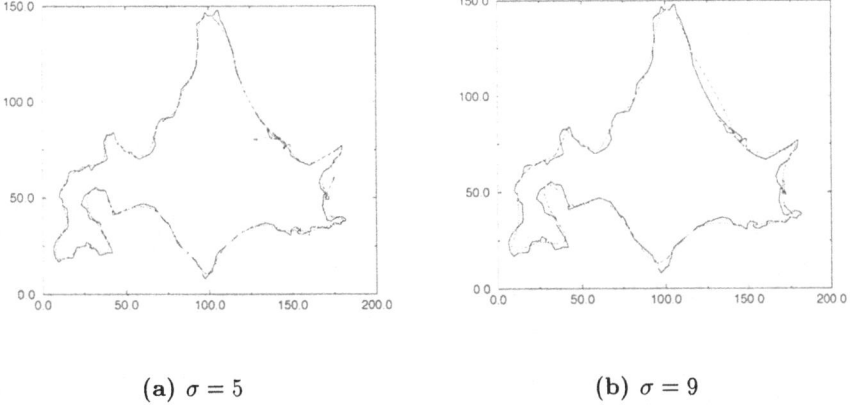

(a) $\sigma = 5$ (b) $\sigma = 9$

Figure 9.4. Reconstruction of Hokaido

Hermite curves were used since each Hermite curve is defined by two endpoints and the tangent vectors at those points. No points external to the input contour are required for Hermite curves. Hermite endpoints were defined as consecutive curvature zero-crossing points recovered at multiple scales using the CSS method. Hermite tangent vectors were also determined using the CSS technique. Contour data compression was achieved by storing only the endpoints and the tangent vectors needed by the Hermite curves in order to arrive at an approximate reconstruction of the original contour.

Approximation error and compression ratio were also computed at each scale. The graph of compression ratio as a function of approxima-

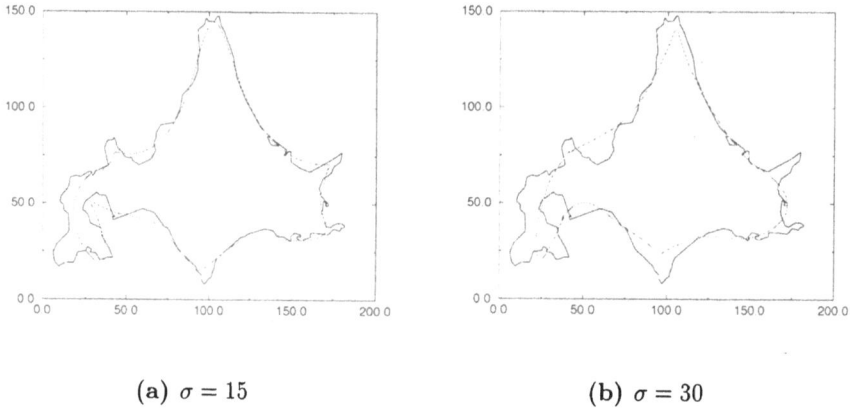

(a) $\sigma = 15$ (b) $\sigma = 30$

Figure 9.5.   Reconstruction of Hokaido (cont.)

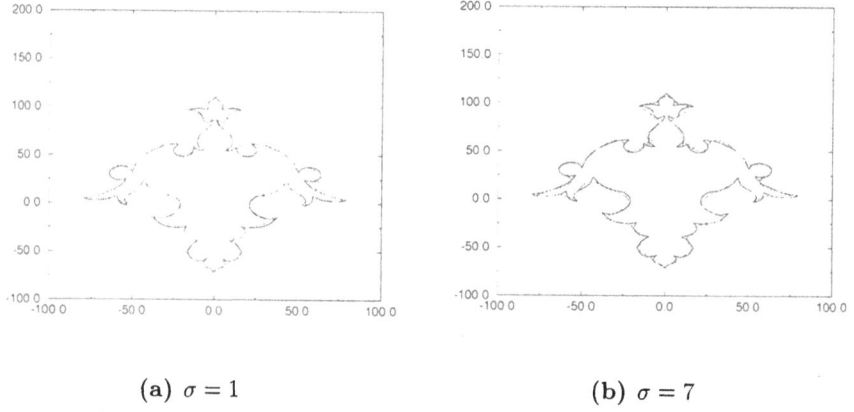

(a) $\sigma = 1$ (b) $\sigma = 7$

Figure 9.6.   Reconstruction of carpet design

tion error was smoothed to remove noise and small fluctuations. The *bending point* of that function was then defined as the largest maximum of its second derivative. The bending point was considered as the boundary between the mostly vertical and the mostly horizontal segments of the graph. It was used for automatic selection of an optimal scale for contour data reconstruction. As a result, the user will not have to set any parameters to use this technique.

Note that this approach can also be extended to the 3-D case since the CSS method has been generalized to 3-D surfaces recently [213, 331, 136, 137].

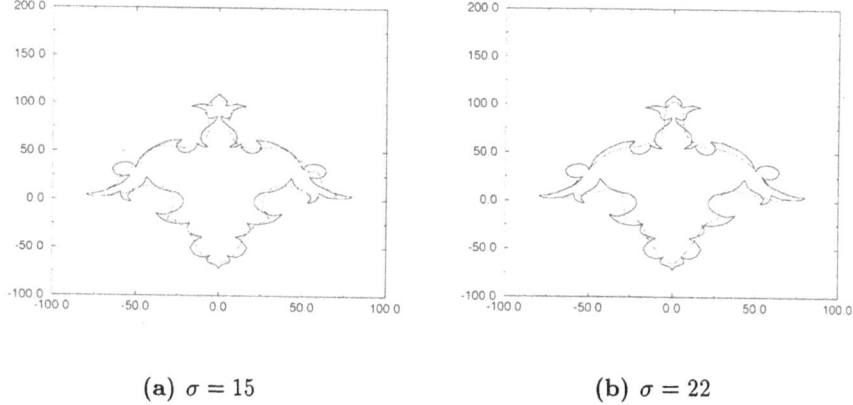

(a) $\sigma = 15$ (b) $\sigma = 22$

Figure 9.7.   Reconstruction of carpet design (cont.)

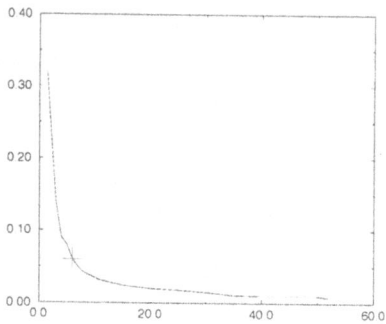

Figure 9.8.   Graph of compression-ratio as a function of approximation-error for spline reconstruction of Africa

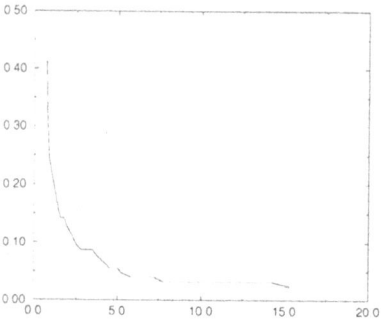

Figure 9.9. Graph of compression-ratio as a function of approximation-error for spline reconstruction of Hokaido

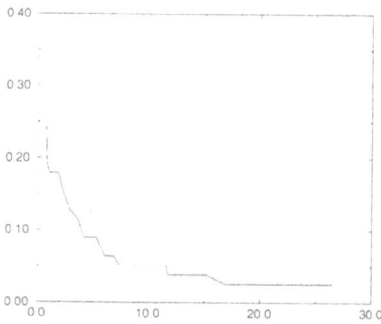

Figure 9.10. Graph of compression-ratio as a function of approximation-error for spline reconstruction of carpet design

Chapter 10

MULTI-SCALE REPRESENTATIONS FOR FREE-FORM SPACE CURVES: THE TORSION SCALE SPACE IMAGE AND ITS PROPERTIES

This chapter introduces a multi-scale shape representation, referred to as the *Torsion Scale Space Image* (TSS), for space curves. It is argued that space curves are useful for representing 3-D surfaces and objects. Experiments show that the representation is robust and suitable for recognition of noisy curves at any scale or orientation.

The method rests on the concept of describing curves at varying levels of detail using invariant geometric features. Three different ways of computing the representation, each with different properties, are described here. A two-phase matching algorithm consisting of TSS matching followed by transformation parameter optimization demonstrates the usefulness of the representations in recognition tasks.

The process of describing curves at increasing levels of abstraction is referred to as the *evolution* of those curves. Several evolution properties of space curves are described in this chapter. Together, these results provide a sound theoretical foundation for the representation methods introduced here.

1.   Introduction

This chapter introduces a theory of multi-scale shape representation based on geometric invariants for space curves [198]. Why study the problem of representing the shape of space curves? Space curves lie on the surfaces of 3-D objects. Representations which have been proposed for 3-D objects are usually based on geometric primitives [250] and are therefore not suitable for many classes of free-form 3-D objects. Systems based on such representations could not be characterized as robust or reliable. Accurate laser range-finders have recently become available at reasonable cost. Recent work [108, 292] has demonstrated the feasibility of fusing multiple range images of a single object in order to automatically build a complete model. Such models can accomodate 3-D objects with arbitrary shapes but they are represented by detailed polygonal meshes with, in general, a very large number of vertices. Handling of many such objects could pose problems in terms of storage and computational requirements. Geometrically invariant contours which lie on the surface of 3-D objects can convey much information about those surfaces and can be utilized as compact, efficient representations for them. Furthermore, space curves admit natural, arc-length parametrizations whereas parametrization of arbitrary 3-D surfaces is far more difficult. The question then is: how can geometrically invariant space curves be defined on the surfaces of general 3-D objects? Note that Brady *et al* [42] and Besl & Jain [35] have demonstrated algorithms based on classical differential geometry for the computation of surface invariants such as curvature features from range images of 3-D surfaces. Observe that:

- Many manufactured objects are composed of locally flat surfaces. Contours formed by surface points where mean curvature has a local maximum, are space curves in general and would correspond to 3-D surface edges on such objects (where large changes in surface normal direction would be observed). These contours would convey a great deal of information about the underlying object.

- Contours of zero Gaussian curvature (parabolic curves) on a 3-D surface are space curves and also rich in information content. These contours are suitable for representation of free-form objects with more complex surfaces. If necessary, these contours can be combined with the previous type in order to generate a richer description of the surfaces under study.

A useful shape representation method in computational vision should make accurate and reliable recognition of an object possible. Therefore such a representation should satisfy a number of necessary criteria. The following is a list of such criteria. Note that two space curves are described as having the *same shape*, if there exists a transformation consisting of uniform scaling, rotation and translation which will cause one of those curves to completely overlap the other. As a result, every point of the first curve will have the same location as a point of the second curve and *vice versa*.

Invariance If two curves have the same shape, they should have the same representation.

Uniqueness If two space curves do not have the same shape, they should have different representations.

Stability If two space curves have a small shape difference, their representations should also have a small difference and if two representations have a small difference, the curves they represent should also have a small shape difference.

The importance of the invariance criterion is that it guarantees that all curves with the same shape will have the same representation. It will therefore be possible to conclude that two curves have different shapes by observing that they have different representations. Without the invariance criterion, two curves with the same shape may have different representations.

The uniqueness criterion is important since it guarantees that two curves with different shapes will have different representations. It will therefore be possible to conclude that two curves have the same shape by observing that they have the same representation. Without the uniqueness criterion, two curves with different shapes may have the same representation.

The significance of the stability criterion is that it guarantees that a small change in the shape of a curve will not cause a large change in its representation and a small difference between two representations does not indicate a large shape difference between the curves they represent. As a result, when two representations are close, the curves they represent are close in shape and when two representations are not close, the curves they represent are not close in shape. When this criterion is satisfied, the representation can be considered to be stable with respect to noise.

It is useful for a shape representation technique to satisfy a number of additional properties in order to become suitable for practical shape recognition tasks. The following is a list of such criteria. Note that similar criteria have been proposed in [220].

Local support Quite often, it is necessary to be able to recognize that the shape of a segment of a curve is the same as the shape of another curve-segment. Only a representation computed using *local* information can provide such an ability.

Efficiency The representation should be efficient to compute and store. This is important since it may be necessary for an object recognition system to perform real-time recognition. By *efficient* we mean that the computational complexity should be a low-order polynomial in time and space (and in the number of processors if a parallel computing architecture is used) with a small constant as a function of the size of the input curve.

Shape Properties It may be useful to have the ability to determine properties of the shape of a curve using its representation. For example, if a curve has a symmetric shape, it may be desirable to be able to determine that fact from its representation (the *symmetry* criterion). Furthermore, if the shape of a space curve or part of a space curve is the same as the shape of part of another space curve, it may be useful to have the ability to determine that relationship using their representations (the *part/whole* criterion).

Implementation If two or more competing representations exist which are very close in other aspects, it is advantageous to choose one of those representations such that the implementation of the computer program which computes that representation requires the least time spent on programming and debugging.

The following is the organization of the remainder of this chapter. Section 2 presents a survey of existing work in this area. The evolution of a space curve and the TSS image are described in section 3. Mackworth & Mokhtarian [170] introduced a modification of the CSS image referred to as the *renormalized CSS image*. The 3-D equivalent of that representation, the renormalized TSS image, is described in section 4. The *resampled TSS image* was introduced in [188]. It is a significant further modification of the renormalized TSS image which is based on the concept of *arc length evolution*. It is shown that the resampled TSS image is more suitable than the renormalized TSS image for recognition of curves with non-uniform noise or when local shape differences exist. The arc length evolution of a space curve and the resampled TSS image are described in detail in section 5. Section 6 contains the evolution and arc length evolution properties of space curves. Almost all these properties are shown to be true about both evolution and arc length evolution. Together, these properties provide a theoretical foundation for the representation methods proposed in this chapter. The proofs of the theorems of section 6 are given in appendix B. Section 7 describes a two-phase space curve matching algorithm which consists of a TSS matching phase to obtain an approximation for the transformation parameters and an optimization phase which iteratively solves for the optimal set of parameters. Section 8 presents a number of experiments carried out to demonstrate the stability of the proposed representations under various conditions of noise and to illustrate the performance of the matching algorithm described in section 7. It also contains a comparison of the regular, renormalized and resampled TSS images and an evaluation of the TSS representations according to the criteria proposed earlier. Section 9 presents the conclusions of this chapter.

2.   Literature Survey

Many shape representation methods for planar curves have been proposed in the computational vision and image processing literature. Some of these methods can be extended to apply to space curves but each fails to satisfy one or more of the criteria outlined above. However, each may be suitable for special purpose shape representation and recognition tasks.

The *Hough transform* has been used to detect arbitrary shapes [25]. Edge elements in the image vote for the parameters of the objects of which they are parts. The votes are accumulated in a parameter space. The peaks of the parameter space then indicate the parameters of the objects searched for. *Chain encoding* [89] techniques approximate a curve using line segments which lie on a grid. *polygonal approximations* [249] of a curve are computed by using various criteria to determine breakpoints that yield the "best" polygon. *Strip trees* [66, 26] are a set of approximating polygons ordered such that each polygon approximates the curve with less approximation error than the previous polygon. *splines* [27] represent a curve using a set of analytic and smooth curves. *Fourier Descriptors* [252] represent a curve by the coefficients of the Fourier expansion of a parametric representation of the curve. The *curvature primal sketch* [23] technique approximates the curve using a library of analytic curves. Then the curvature function of the approximating curve is computed and convolved with a Gaussian of varying standard deviation. The *extended circular image* [116] is the 2-D equivalent of the *extended Gaussian image*. In the extended circular image, one is given the radius of curvature as a function of normal direction. *volumetric diffusion* [146] defines a geometrical object by way of its "characteristic function" $\chi(\mathbf{r})$ which equals unity when the point $\mathbf{r}$ belongs to the object and zero otherwise. The object is then blurred by requiring that its characteristic function satisfy the diffusion equation. The boundary of each blurred object is defined by the equation $\chi(\mathbf{r}) = 0.5$ or by applying the Laplacian operator to the blurred function.

3.   The Torsion Scale Space Image

This section reviews the parametric representation of space curves, and the computation of torsion from that representation. Next, it is shown how to compute torsion on a space curve at varying levels of detail. A torsion-based, multi-scale representation for a space curve is then proposed.

3.1.   The Parametric Representation of a Space Curve

The set of points of a space curve is the values of a continuous, vector-valued function [95]:

$$\Gamma = \Gamma(u) = (x(u), y(u), z(u))$$

where $x(u), y(u)$ and $z(u)$ are the components of $\Gamma(u)$ and u is a function of arc length s of the curve. s is also referred to as the *natural parameter*. The function $\Gamma(u)$ or the triple of functions $(x(u), y(u), z(u))$ is referred to as a *parametric representation* of the curve. Here, when first parametrizing Γ, u is same as s.

3.2.   Computation of Torsion

Intuitively, torsion is a local measure of the non-planarity of a space curve. For a precise definition, see [95]. Since the curve is represented in parametric form, in order to compute torsion at each point on the curve, we need to express it in terms of the derivatives of $x(\ ), y(\ )$ and $z(\ )$. Note that $\dot{f}$ represents the derivative of function f with respect to s or u, $\ddot{f}$ represents its second derivative, and $\dddot{f}$ represents its third derivative.

Torsion of a space curve with arc length parametrization is given by:

$$\tau(s) = \frac{\dot{x}(\ddot{y}\dddot{z} - \ddot{z}\dddot{y}) - \dot{y}(\ddot{x}\dddot{z} - \ddot{z}\dddot{x}) + \dot{z}(\ddot{x}\dddot{y} - \ddot{y}\dddot{x})}{(\ddot{x})^2 + (\ddot{y})^2 + (\ddot{z})^2}.$$

In case of an arbitrary parametrization, torsion is given by:

$$\tau(u) = \frac{\dot{x}(\ddot{y}\dddot{z} - \ddot{z}\dddot{y}) - \dot{y}(\ddot{x}\dddot{z} - \ddot{z}\dddot{x}) + \dot{z}(\ddot{x}\dddot{y} - \ddot{y}\dddot{x})}{(\dot{y}\ddot{z} - \dot{z}\ddot{y})^2 + (\dot{z}\ddot{x} - \dot{x}\ddot{z})^2 + (\dot{x}\ddot{y} - \dot{y}\ddot{x})^2}.$$

3.3.   Computing Torsion at Varying Levels of Detail

In order to compute τ at varying levels of detail of a curve Γ, functions $x(s), y(s)$ and $z(s)$ are convolved with a Gaussian kernel $g(u, \sigma)$ of width σ. The choice of kernel is motivated by its desirable properties as demonstrated by the theorems of section 6. Utilization of the Gaussian kernel results in a TSS image which has the *reconstruction* property (theorem 10.6), the *monotonicity* property (theorem 10.7), and the *convergence* property (theorem 10.12). An alternative to smoothing by convolution would be fitting of 3-D splines to the contour data at multiple scales. Torsion could then be computed from those splines. The drawbacks of this approach would be high computational cost and the variability of results due to arbitrary choice of knot points. The convolved functions together define the *evolved* curve Γ_σ. The process of generating evolved curves Γ_σ as σ varies from 0 to ∞, is referred to as the *evolution* of Γ. Let

$$\mathcal{F}(u,\sigma) = f(u) \otimes g(u,\sigma)$$

where $\otimes$ is the convolution operator. It is known that

$$\dot{\mathcal{F}}(u,\sigma) = \frac{\partial \mathcal{F}(u,\sigma)}{\partial u} = f(u) \otimes \frac{\partial g(u,\sigma)}{\partial u}$$

$$\ddot{\mathcal{F}}(u,\sigma) = \frac{\partial^2 \mathcal{F}(u,\sigma)}{\partial u^2} = f(u) \otimes \frac{\partial^2 g(u,\sigma)}{\partial u^2}$$

and

$$\breve{\mathcal{F}}(u,\sigma) = \frac{\partial^3 \mathcal{F}(u,\sigma)}{\partial u^3} = f(u) \otimes \frac{\partial^3 g(u,\sigma)}{\partial u^3}.$$

These properties of convolution can be used to compute torsion on evolved versions of a space curve. Let

$$\mathcal{X}(u,\sigma) = x(u) \otimes g(u,\sigma)$$

$$\mathcal{Y}(u,\sigma) = y(u) \otimes g(u,\sigma)$$

and

$$\mathcal{Z}(u,\sigma) = z(u) \otimes g(u,\sigma).$$

torsion on evolved curve Γ_σ is given by:

$$\tau(u,\sigma) = \frac{\dot{\mathcal{X}}(\ddot{\mathcal{Y}}\breve{\mathcal{Z}} - \ddot{\mathcal{Z}}\breve{\mathcal{Y}}) - \dot{\mathcal{Y}}(\ddot{\mathcal{X}}\breve{\mathcal{Z}} - \ddot{\mathcal{Z}}\breve{\mathcal{X}}) + \dot{\mathcal{Z}}(\ddot{\mathcal{X}}\breve{\mathcal{Y}} - \ddot{\mathcal{Y}}\breve{\mathcal{X}})}{(\dot{\mathcal{Y}}\ddot{\mathcal{Z}} - \dot{\mathcal{Z}}\ddot{\mathcal{Y}})^2 + (\dot{\mathcal{Z}}\ddot{\mathcal{X}} - \dot{\mathcal{X}}\ddot{\mathcal{Z}})^2 + (\dot{\mathcal{X}}\ddot{\mathcal{Y}} - \dot{\mathcal{Y}}\ddot{\mathcal{X}})^2}.$$

3.4. A Multi-Scale Representation for Space Curves

The curvature and torsion functions of a space curve specify that curve uniquely up to rotation and translation [95]. The *torsion scale space* (TSS) *image* is therefore proposed as a torsion-based multi-scale representation for space curves. To compute the TSS image of a space curve $\Gamma = (x(u), y(u), z(u))$, evolved curves Γ_σ are computed as σ varies from a small to a large value. The torsion zero-crossing points of each Γ_σ are extracted and recorded in the $u - \sigma$ space. The smallest value of σ used is slightly larger than zero. The value of σ used gradually increases until the number of torsion zero-crossings on the evolved curve stabilizes at a small value.

The solution of the equation

$$\tau(u,\sigma) = 0$$

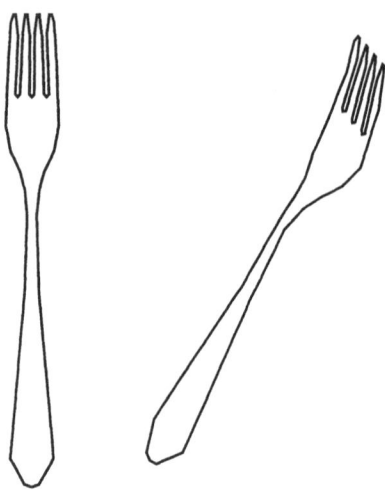

Figure 10.1. Two views of a space curve depicting a fork

forms the torsion scale space image of Γ.

The CSS image of a space curve can be constructed in a similar fashion. The only difference is that level-crossings rather than zero-crossings should be searched for. This is because the curvature of a space curve has only magnitude and no sign. Note, however, that it was decided not to compute the CSS images of the space curve shown in this chapter since those images would be based on curvature level-crossings and therefore not invariant with respect to scale changes of space curves. Furthermore, experiments indicate that the TSS image is a robust tool for representation and recognition of space curves.

Figure 10.1 shows two views of a space curve depicting a fork. Figure 10.2 shows the fork during evolution and figure 10.3 shows the TSS image of the fork. Figure 10.4 shows two views of a space curve depicting an armchair. Figure 10.5 shows the armchair during evolution and figure 10.6 shows the TSS image of the armchair. Figures 10.2 and 10.5 show that as a space curve evolves, more and more high-level descriptions of its shape are obtained. Thus the evolution process results in a continuous fine-to-coarse description of the shape of a space curve. Increasing values of σ remove noise as well as small features from the original contour leaving only the major shape components. Eventually all shape structures disappear at a sufficiently large value of σ and the evolved contour becomes simple.

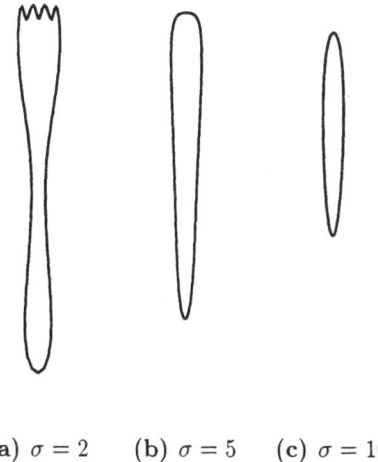

(a) $\sigma = 2$ (b) $\sigma = 5$ (c) $\sigma = 10$

Figure 10.2.   The fork during evolution

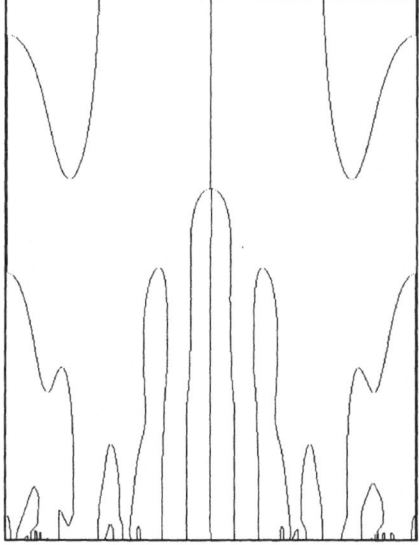

Figure 10.3.   The torsion Scale Space image of the fork

4.   The Renormalized Torsion Scale Space Image

It was observed in [170] that in general, the arc length parameter on a curve shrinks in a non-linear fashion during evolution. This can lead to poor matches when local shape differences exist between two curves.

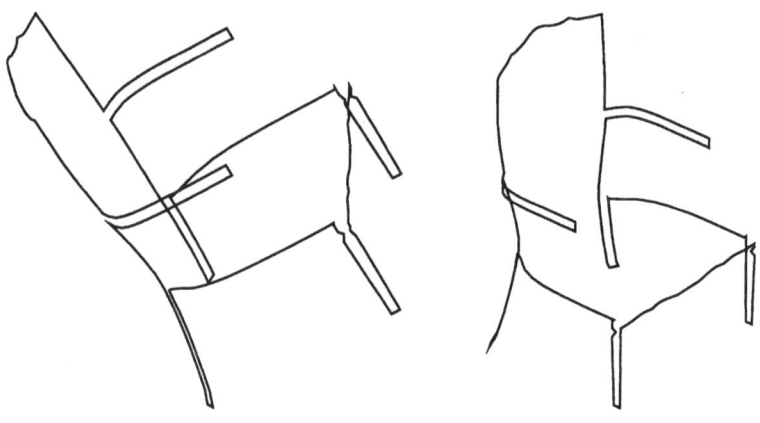

Figure 10.4.   Two views of a space curve depicting an armchair

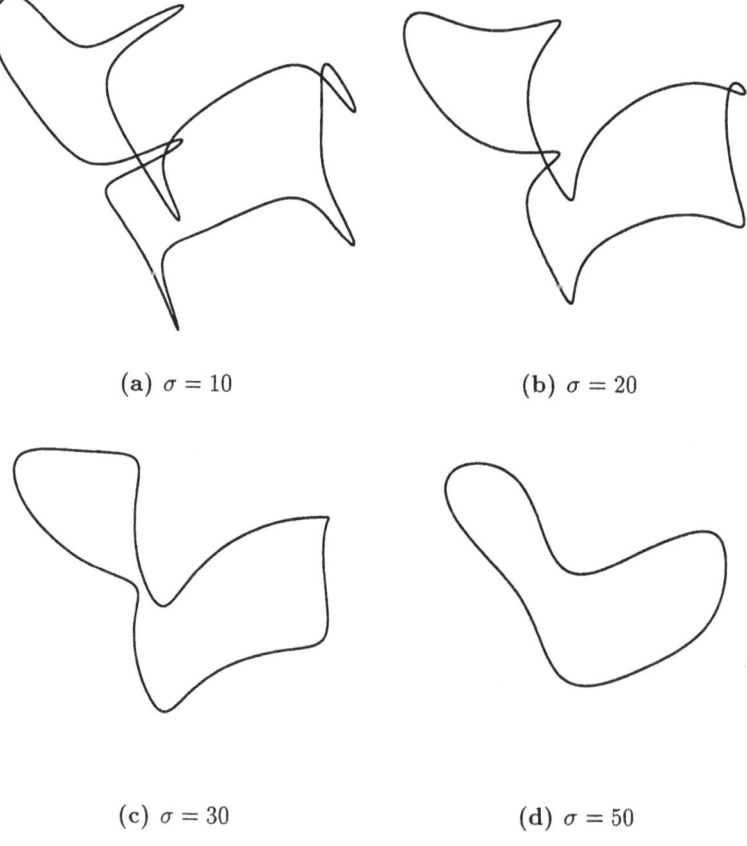

(a) $\sigma = 10$ **(b)** $\sigma = 20$

(c) $\sigma = 30$ **(d)** $\sigma = 50$

Figure 10.5.   The armchair during evolution

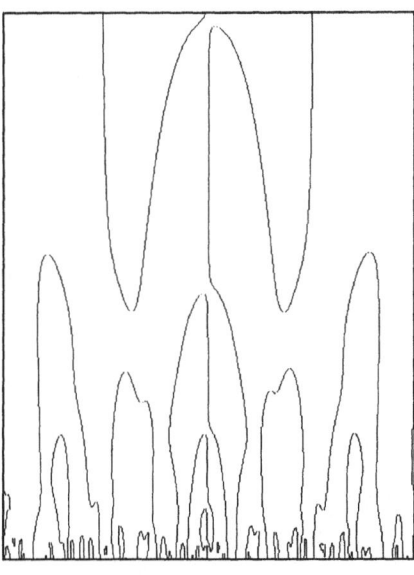

Figure 10.6. The torsion Scale Space image of the armchair

To overcome this problem, we propose the *renormalized* TSS image. Let

$$\mathcal{R}(u, \sigma) = (\mathcal{X}(u, \sigma), \mathcal{Y}(u, \sigma), \mathcal{Z}(u, \sigma))$$

and

$$w = \Phi_\sigma(u) = \frac{\int_0^u |\mathcal{R}_v(v, \sigma)| dv}{\int_0^1 |\mathcal{R}_v(v, \sigma)| dv}.$$

Now define

$$\hat{\mathcal{X}}(w, \sigma) = \mathcal{X}(\Phi_\sigma^{-1}(w), \sigma)$$
$$\hat{\mathcal{Y}}(w, \sigma) = \mathcal{Y}(\Phi_\sigma^{-1}(w), \sigma)$$

and

$$\hat{\mathcal{Z}}(w, \sigma) = \mathcal{Z}(\Phi_\sigma^{-1}(w), \sigma).$$

Functions $\hat{\mathcal{X}}(w, \sigma)$, $\hat{\mathcal{Y}}(w, \sigma)$ and $\hat{\mathcal{Z}}(w, \sigma)$ define the renormalized version of each Γ_σ. That is, each evolved curve Γ_σ is reparametrized by its normalized arc length parameter w. Notice that

$$\Phi_\sigma(0) = 0$$

$$\Phi_\sigma(1) = 1$$

and

$$\frac{d\Phi_\sigma(u)}{du} = \frac{|\mathcal{R}_u(u, \sigma)|}{\int_0^1 |\mathcal{R}_v(v, \sigma)| dv} > 0$$

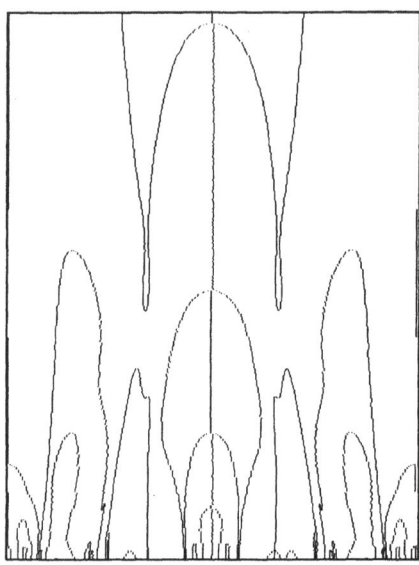

Figure 10.7.   The renormalized torsion Scale Space image of the armchair

at non-singular points. Also,

$$\Phi_0(u) = u.$$

Note that $\Phi_\sigma(u)$ deviates from the identity function

$$\Phi_\sigma(u) = u$$

only to the extent that the arc length parameter on the original curve is distorted by the smoothing process.

The solution of the equation:

$$\tau(w, \sigma) = 0$$

forms the renormalized TSS image of Γ. Figure 10.7 shows the renormalized torsion scale space image of the armchair.

5.   The Resampled Torsion Scale Space Image

It was observed in [188] that as a space curve evolves according to the process defined in section 3, the parametrization of its coordinate functions $x(u)$, $y(u)$ and $z(u)$ does not change. In other words, the function mapping values of the parameter u of the original coordinate functions $x(u)$, $y(u)$ and $z(u)$ to the values of the parameter u of the smoothed coordinate functions $\mathcal{X}(u, \sigma)$, $\mathcal{Y}(u, \sigma)$ and $\mathcal{Z}(u, \sigma)$ is the identity function. It is useful to generalize the definition of evolution so that the mapping function can be different from the identity function. Again let Γ be defined by:

$$\Gamma = \{(x(w), y(w), z(w)) \mid w \in [0, 1]\}.$$

The generalized evolution which maps Γ to Γ_σ is now defined by:

$$\Gamma \longrightarrow \Gamma_\sigma = \{(\mathcal{X}(\mathcal{W}, \sigma), \mathcal{Y}(\mathcal{W}, \sigma), \mathcal{Z}(\mathcal{W}, \sigma)) \mid \mathcal{W} \in [0, 1]\}$$

where

$$\mathcal{X}(\mathcal{W}, \sigma) = x(\mathcal{W}) \otimes g(\mathcal{W}, \sigma)$$

$$\mathcal{Y}(\mathcal{W}, \sigma) = y(\mathcal{W}) \otimes g(\mathcal{W}, \sigma)$$

and

$$\mathcal{Z}(\mathcal{W}, \sigma) = z(\mathcal{W}) \otimes g(\mathcal{W}, \sigma).$$

Note that $\mathcal{W} = \mathcal{W}(w, \sigma)$ and $\mathcal{W}(w, \sigma_0)$ where σ_0 is any value of σ, is a continuous and monotonic function of w. This condition is necessary to ensure physical plausibility since $\mathcal{W}$ is the parameter of the evolved curve Γ_σ. A specially interesting case is when $\mathcal{W}$ always remains the arc length parameter as the curve evolves. When this criterion is satisfied, the evolution of Γ is referred to as *arc length evolution*. An explicit formula for $\mathcal{W}$ can be derived [91]. Let

$$\mathcal{R}(\mathcal{W}, \sigma) = (\mathcal{X}(\mathcal{W}, \sigma), \mathcal{Y}(\mathcal{W}, \sigma), \mathcal{Z}(\mathcal{W}, \sigma)).$$

The Frenet equations for a space curve are given by:

$$\frac{\partial \mathbf{t}}{\partial u} = |\frac{\partial \mathcal{R}}{\partial u}| \kappa \mathbf{n}$$

$$\frac{\partial \mathbf{n}}{\partial u} = -|\frac{\partial \mathcal{R}}{\partial u}| \kappa \mathbf{t} + |\frac{\partial \mathcal{R}}{\partial u}| \tau \mathbf{b}.$$

Let $t = \sigma^2/2$. Observe that

$$\frac{\partial}{\partial t}(|\frac{\partial \mathcal{R}}{\partial u}|^2) = \frac{\partial}{\partial t}(\frac{\partial \mathcal{R}}{\partial u} \cdot \frac{\partial \mathcal{R}}{\partial u}) = 2(\frac{\partial \mathcal{R}}{\partial u} \cdot \frac{\partial^2 \mathcal{R}}{\partial u \partial t}).$$

Note that

$$\frac{\partial \mathcal{R}}{\partial u} = |\frac{\partial \mathcal{R}}{\partial u}| \mathbf{t}$$

$$\frac{\partial \mathcal{R}}{\partial t} = \kappa \mathbf{n}$$

since the Gaussian function satisfies the heat equation. It follows that

$$\frac{\partial}{\partial t}(|\frac{\partial \mathcal{R}}{\partial u}|^2) = 2|\frac{\partial \mathcal{R}}{\partial u}|\mathbf{t}.\frac{\partial}{\partial u}(\kappa \mathbf{n}) = 2|\frac{\partial \mathcal{R}}{\partial u}|\mathbf{t}.(\frac{\partial \kappa}{\partial u}\mathbf{n} - |\frac{\partial \mathcal{R}}{\partial u}|\kappa^2\mathbf{t} + |\frac{\partial \mathcal{R}}{\partial u}|\kappa\tau\mathbf{b})$$

and

$$\frac{\partial}{\partial t}(|\frac{\partial \mathcal{R}}{\partial u}|^2) == -2|\frac{\partial \mathcal{R}}{\partial u}|^2\kappa^2.$$

Therefore

$$2|\frac{\partial \mathcal{R}}{\partial u}|\frac{\partial}{\partial t}|\frac{\partial \mathcal{R}}{\partial u}| = -2|\frac{\partial \mathcal{R}}{\partial u}|^2\kappa^2$$

or

$$\frac{\partial}{\partial t}|\frac{\partial \mathcal{R}}{\partial u}| = -|\frac{\partial \mathcal{R}}{\partial u}|\kappa^2.$$

Let L denote the length of the curve. Now observe that

$$\frac{\partial L}{\partial t} = \int_0^L \frac{\partial}{\partial t}|\frac{\partial \mathcal{R}}{\partial u}|du = -\int_0^L |\frac{\partial \mathcal{R}}{\partial u}|\kappa^2 du = -\int_0^1 \kappa^2 dw.$$

Since the value of the normalized arc length parameter w at any point on the curve, measures the length of the curve from the starting point to that point, it follows that

$$\frac{\partial \mathcal{W}}{\partial t} = -\int_0^W \kappa^2(\mathcal{U}, t)d\mathcal{U}$$

and therefore

$$\mathcal{W}(w, t) = \int_0^t \int_0^W \kappa^2(\mathcal{U}, \mathcal{T})d\mathcal{U}d\mathcal{T} + w. \tag{10.1}$$

Note that,

$$\mathcal{W}(w, 0) = w.$$

Note further that for any given value t_0 of t, $\mathcal{W}(w, t_0)$ is a monotonically increasing function of w. To see this, observe that after changing the order of integration in equation (10.1) and applying the chain rule for derivatives, we obtain

$$\frac{\partial \mathcal{W}(w, t)}{\partial w} = (-\int_0^t \kappa^2(\mathcal{W}, \mathcal{T})d\mathcal{T})\frac{\partial \mathcal{W}(w, t)}{\partial w} + 1$$

or

$$\frac{\partial \mathcal{W}}{\partial w}(1 + \int_0^t \kappa^2(\mathcal{W}, \mathcal{T})d\mathcal{T}) = 1$$

or

$$\frac{\partial \mathcal{W}}{\partial w} = \frac{1}{1 + \int_0^t \kappa^2(\mathcal{W}, \mathcal{T})d\mathcal{T})}$$

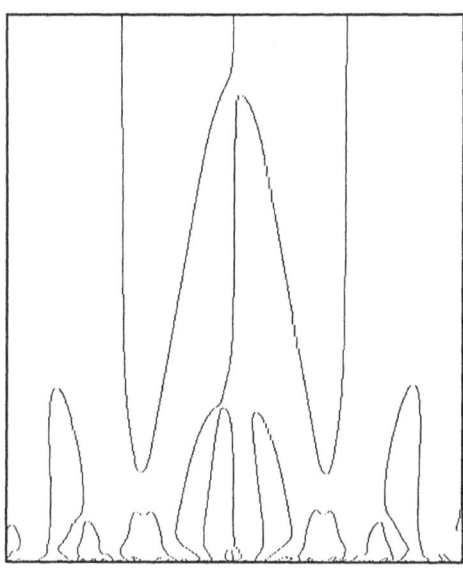

Figure 10.8. The resampled torsion Scale Space image of the armchair

which is always positive for any t. Therefore $\mathcal{W}(w,t)$ is a monotonically increasing function of w for any t.

The solution of the equation:

$$\tau(\mathcal{W},\sigma) = 0$$

forms the resampled TSS image of Γ. Figure 10.8 shows the resampled TSS image of the armchair.

Since the function $\kappa(\mathcal{U},\mathcal{T})$ in equation (10.1) is unknown, $\mathcal{W}(w,t)$ can not be computed directly from equation (10.1). However, the resampled TSS image can be computed as following: A Gaussian filter using a small value of the standard deviation is computed. The curve Γ is parametrized by the normalized arc length parameter and its coordinate functions are convolved with the filter. The resulting curve is reparametrized by its normalized arc length parameter and its coordinate functions are again convolved with the same filter. This process is repeated until the curve has very few torsion zero-crossing points (see theorem 10.12 in section 6). The torsion zero-crossings of each curve are marked in the resampled TSS image.

We will next show that this approximation process converges to solution of equation (10.1). Let ε be the maximum error in the location of any point of Γ when the arc length evolution of Γ is approximated through the process described above using a Gaussian with standard deviation $\Delta\sigma$. Observe that at a point P of Γ

$$\varepsilon \approx |(\mathbf{r} + \kappa\mathbf{n}) - (\mathbf{r} + \Delta\mathbf{r}_g)| = |\kappa\mathbf{n} - \Delta\mathbf{r}_g|$$

where κ and $\mathbf{n}$ are curvature and the principal normal vector at P, $\mathbf{r}$ is the position vector of point P and $\Delta\mathbf{r}_g$ is the amount of change in the position vector of P after Gaussian approximation. Let $\Delta\mathbf{r}_g = \kappa_g\mathbf{n}_g$ where $\mathbf{n}_g$ is a unit vector with the same direction as that of $\mathbf{r}_g$ and κ_g is equal to the length of $\mathbf{r}_g$. $\mathbf{n}_g$ and κ_g can be thought of as the normal vector at P and the curvature of an arc of a circle going through P.

As $\Delta\sigma \to 0$, the curve segment covered by the Gaussian filter can be approximated by a circular segment with constant curvature. It is easily seen that on an arc of a circle, regular evolution causes the same shrinkage rate at every point and is therefore equivalent to arc length evolution. It follows that for a small $\Delta\sigma$, $\kappa_g \to \kappa$ and $\mathbf{n}_g \to \mathbf{n}$. Therefore

$$\varepsilon \approx |\kappa\mathbf{n} - \kappa_g\mathbf{n}_g| \to 0.$$

After j iterations of smoothing, total error is given by $j\varepsilon$ which is small. Hence the approximation process described above converges to solution of equation (10.1).

6.　Evolution Properties of Space Curves

This section contains a number of results on evolution and arc length evolution of space curves as defined in sections 3 and 5. Appendix B contains the proofs of these theorems. The first five theorems express a number of global properties of space curves during evolution and arc length evolution.

THEOREM 10.1 *The order of application of evolution or arc length evolution and a shape preserving transformation (consisting of rotation, uniform scaling and translation) to a space curve does not change the final result.*

This theorem shows that the regular, renormalized and resampled TSS images of a space curve have the *invariance* property. The invariance property is essential since it makes it possible to match a space curve to another of similar shape which has undergone a shape-preserving transformation.

THEOREM 10.2 *A closed space curve remains closed during evolution or arc length evolution.*

THEOREM 10.3 *A connected curve remains connected during evolution or arc length evolution.*

Theorems 10.2 and 10.3 show that connectedness and closedness of a space curve are preserved during evolution or arc length evolution. These theorems show that evolution and arc length evolution of a space curve do not change the physical interpretation of that curve as a contour lying on the surface of a 3-D object. If such a contour is not closed or connected after evolution or arc length evolution, then it can no longer admit a physically plausible interpretation.

THEOREM 10.4 *The center of mass of a space curve is invariant during evolution or arc length evolution.*

THEOREM 10.5 *Let* Γ *be a closed space curve and let* $\mathcal{H}$ *be its convex hull.* Γ *remains inside* $\mathcal{H}$ *during evolution or arc length evolution.*

Theorem 10.4 shows that the center of mass of a space curve does not move as the curve evolves and theorem 10.5 shows that a space curve remains inside its convex hull during evolution or arc length evolution. For studies of convex hull computation, see [96, 97, 127, 161, 258, 257, 259, 319, 338, 14, 32, 31, 73, 51]. Together, theorems 10.4 and 10.5 impose constraints on the physical location of a space curve as it evolves. These constraints become useful whenever the physical location of curves is important. An example is when two or more space curves are used to represent a 3-D object.

The following theorem concerns the uniqueness properties of TSS representations.

THEOREM 10.6 *Let* Γ *be a space curve in* C_3. *Let* $\tau(u)$ *and* $\kappa(u)$ *represent the torsion and curvature functions of* Γ *respectively. The derivatives at a single point on one torsion zero-crossing contour in the TSS image of* Γ *determine function* $\tau^*(u) = \tau(u)\kappa^2(u)$ *uniquely modulo a scale factor (except on a set of measure zero).*

Theorem 10.6 shows that a space curve can be reconstructed modulo the class represented by the function

$$\tau^*(u) = \tau(u)\kappa^2(u)$$

from its TSS image. In a practical shape recognition task, two space curves of different shapes are unlikely to belong to the same class represented by function $\tau^*(u)$ and therefore their TSS images will likely be different. The following theorem makes explicit the conditions under which new torsion zero-crossings will not be observed at the higher scales of TSS images.

THEOREM 10.7 *Let* Γ *be a space curve in* C_3. *If torsion is bounded at every point of* Γ *during evolution and arc length evolution, then all extrema of contours in the regular, renormalized and resampled TSS images of* Γ *are maxima.*

Theorem 10.7 shows that if a space curve remains smooth during evolution and arc length evolution and torsion remains bounded at each of its points, then no new torsion zero-crossings can exist at the higher scales of its TSS images. Theorems 10.8 and 10.9 concern the local behaviour of space curves just before and just after the formation of cusp points during evolution and arc length evolution.

THEOREM 10.8 *Let* $\Gamma = (x(w), y(w), z(w))$ *be a space curve in* C_1 *and let* $x(w)$, $y(w)$ *and* $z(w)$ *be polynomial functions of* w. *Let*

$$\Gamma_\sigma = (\mathcal{X}(\mathcal{W}, \sigma), \mathcal{Y}(\mathcal{W}, \sigma), \mathcal{Z}(\mathcal{W}, \sigma))$$

be an evolved or arc length evolved version of Γ *with a cusp point at* $\mathcal{W}_0$. *There is a* $\delta > 0$ *such that either* $\Gamma_{\sigma-\delta}$ *intersects itself in a neighborhood of point* $\mathcal{W}_0$ *or two projections of* $\Gamma_{\sigma-\delta}$ *intersect themselves in a neighborhood of point* $\mathcal{W}_0$.

THEOREM 10.9 *Let* $\Gamma = (x(w), y(w), z(w))$ *be a space curve in* C_1 *and let* $x(w)$, $y(w)$ *and* $z(w)$ *be polynomial functions of* w. *Let*

$$\Gamma_\sigma = (\mathcal{X}(\mathcal{W}, \sigma), \mathcal{Y}(\mathcal{W}, \sigma), \mathcal{Z}(\mathcal{W}, \sigma))$$

be an evolved or arc length evolved version of Γ *with a cusp point at* W_0, *then either* $\Gamma_{\sigma+\delta}$ *has two new torsion zero-crossings in a neighborhood of* W_0 *or a torsion zero-crossing point exists at* W_0 *on* $\Gamma_{\sigma-\delta}$ *and* $\Gamma_{\sigma+\delta}$.

Theorems 10.8 and 10.9 locally characterize the behaviour of a space curve during evolution and arc length evolution just before and just after the creation of a cusp point. Theorem 10.8 shows that a space curve either intersects itself or two of its projections intersect themselves just before the formation of a cusp point during evolution and arc length evolution, in a neighborhood of the cusp point. These conditions are usually not satisfied during evolution or arc length evolution of a space curve and therefore cusp points are unlikely to occur on space curve during evolution and arc length evolution. Theorem 10.10 defines other conditions under which new torsion zero-crossings can appear.

THEOREM 10.10 *New torsion zero-crossings can appear on a smooth space curve during evolution or arc length evolution in a neighborhood of a point of zero curvature.*

Theorem 10.10 shows that new torsion zero-crossings can indeed occur on a space curve that remains smooth during evolution or arc length evolution at points of zero curvature. Together, theorems 10.9 and 10.10

describe all situations that can lead to creation of new torsion zero-crossings on a space curve during evolution and arc length evolution. This enables one to make inferences about a space curve when new torsion zero-crossings are observed in its TSS image. The last two theorems express important convergence properties of evolution and arc length evolution of space curves.

THEOREM 10.11 *Let* Γ *be a closed space curve and let* Γ_σ *be an evolved or arc length evolved version of* Γ. Γ_σ *tends to a planar curve as* σ *grows large.*

Theorem 10.11 provides information on the limiting shape of a space curve after evolution or arc length evolution.

THEOREM 10.12 *Every closed space curve will reach a state during evolution or arc-length evolution in which new torsion zero-crossings are not created and remains in that state.*

Theorem 10.12 shows that every space curve enters and stays in a stable state after evolution or arc length evolution. Such a state can be used to define the termination point for the computation of the TSS representation of a space curve. That is, when such a state in reached during evolution or arc length evolution, the computation of the TSS representation can terminate.

As argued in the Introduction, representations for 3-D objects based on geometric primitives are not general since they are not suitable for free-form objects with complex surfaces. New techniques for fusion of range images of an object obtained from different viewpoints are available which can construct a complete, detailed model of that object. Such models are stored, in general, as very large polygon meshes which pose problems in terms of storage and processing. Invariant space curves recovered from the surfaces of such objects can serve as compact, efficient representations for them. Such space curves can be defined on manufactured objects as contours formed by surface points where mean curvature has a local maximum, and on complex, free-form objects as contours formed by points with zero Gaussian curvature. This procedure is applicable to a wide variety of 3-D objects and results in rich representations [333]. If necessary, both types of contours can be used together to achieve richer representations. Clearly, several space curves may be used to represent a 3-D object.

7.   Space Curve Matching through the TSS Image

This section explains how an optimal match of two space curves which can differ by an arbitrary amount of rotation (about the x, y and z

axes), uniform scaling and translation and a significant amount of noise can be accomplished through the TSS representations of those curves. The matching algorithm consists of two stages. The first stage makes use of the most significant features in the TSS image (the extrema of the torsion zero-crossing contours) to quickly obtain the best match of the two input TSS images. This match is then used to obtain a *closed-form* least-squares approximation for the parameters of a transformation which would map the first space curve to the second (the second curve is assumed to be a model curve). That approximation is then used as an initial estimate to an optimization technique which iteratively solves for the optimal transformation parameters using the coordinates of the input space curve.

Section 7.1 contains a description of a TSS matching algorithm used to discover the best match of the two input TSS images. Section 7.2 presents a novel technique to obtain a closed-form solution to the transformation parameter estimation problem and discusses the advantages of that technique. Section 7.3 discusses the steps involved in space curve distance computation; a process which is used to verify TSS matches and to optimize the transformation parameters. Section 7.4 describes the details of the optimization technique which finds the optimal transformation parameters.

7.1.   Torsion Scale Space Matching

The fundamental idea behind the TSS matching algorithm is to obtain a coarse-level match using the structural features of the input space curve. Such a match can be found quickly and reliably since at the high levels of TSS images, there are relatively few features to be matched. The actual features used for matching are the extrema of the torsion zero-crossing contours. The reason for using the extrema as features is that they are the most significant points of zero-crossing contours: the TSS coordinates of an extremum convey information on both the location and the scale of the corresponding contour whereas the "body" of the contour is, in general, similar in shape to those of other contours. Furthermore, the extrema are isolated point features and therefore solving the feature correspondence problem is not difficult. This is specially true at the high scales of the TSS image where the extrema are sparse.

The task of the matching algorithm, therefore, is to determine the correspondence between two sets of points: one from each TSS image. The allowed transformation from one set to the other is translation in one direction since this corresponds to uniform scaling, rotation and translation of the input space curve. This matching problem can be solved using a best-first matching strategy [321]. The TSS matching algorithm is therefore as follows:

- For each of the input TSS images, carry out the following: Scan the TSS image starting at the highest scale (where the highest extremum occurs) and ending at the lowest scale chosen for matching and locate the extrema of the torsion zero-crossing contours. Record the coordinates and type (maximum or minimum) of each extremum in a feature list as it is encountered. When the scan ends, this list will be sorted by the scale coordinate of the extrema. Normalize those coordinates so that the horizontal coordinate u varies in the range [0,1].

- Create a number of nodes corresponding to the possible match of the highest-scale extremum of the non-model TSS image and each extremum (of the same type) of the model TSS image. If both input space curves are closed, each extremum matched in a node must have a σ coordinate close to the maximum σ (σ of the highest extremum) of its TSS image. In such a case, discard all nodes which do not satisfy this constraint. Initialize the *cost* of each node to zero.

- For each node created in the previous step, compute a TSS shift parameter α using the following equation:

$$u_m = u_i + \alpha$$

where u_i is the horizontal coordinate of the non-model TSS extremum and u_m is the horizontal coordinate of the model TSS extremum.

- Create two lists for each node created in the second step. The first list will contain the non-model TSS extrema matched within that node and the second list will contain the corresponding model TSS extrema matched within that node. Initialize the first list of each node to contain the highest-scale non-model TSS extremum. Initialize the second list of each node to contain the corresponding model TSS extremum determined in the second step. Expand each node created in the second step using the procedure described in the next step.

- To expand a node, select the highest-scale, non-model TSS extremum unmatched within that node using its first list and apply that node's TSS transformation parameters computed in the third step to map that extremum to the model TSS image (If all non-model extrema have been matched, select a *null* (zero height) extremum at the same location as any unmatched model extremum). Locate the nearest model extremum of the same type which is unmatched within that node using its second list and match the two extrema (If no such model extremum can be found, match the non-model extremum to a null extremum at the same location). The cost of match is defined as the Euclidean distance in the model TSS image between the two extrema. Add the match cost to the node cost. Update the two lists associated with the node.

- Select the lowest-cost node. If there are no more model or non-model extrema that remain unmatched within that node, then return that node as the lowest-cost node. Otherwise, go to the previous step and expand the lowest-cost node.

7.2.   Solving for the Transformation Parameters

Once the best TSS match has been determined, it is possible to obtain a set of points on each of the two curves in order to compute an approximation for the transformation parameters since the correspondence between arc length values on the two space curves is known. A number of iterative as well as closed-form solutions using corresponded points have been developed for this problem. Techniques that lead to closed-form solutions are more efficient and include *singular value decomposition* [22], *polar decomposition* [115], *unit quaternions* [114], and *dual quaternions* [314]. A related problem is fitting of 3-D models to 2-D image data. Iterative techniques such as Newston's method [167] have been utilized to solve this problem. The drawbacks of these techniques are their complexity and computational cost. Furthermore, one also needs to deal with the stability and parameter initialization problems. This section presents a simpler, efficient, robust and general procedure which avoids the complexity of the closed-form solutions and the computational cost of the iterative techniques by converting the problem into a linear one. It is applicable to the *3-D to 3-D* as well as the *3-D to 2-D* case. It is assumed that the transformation to be solved for is a general 3-D linear transformation (and therefore the method is also applicable to objects similar under a 3-D affine transform). It is then shown that a closed-form solution to a least-squares estimation of the parameters can be derived which does not suffer from the complexity and high computational cost of the iterative techniques. Furthermore, the solution does not need to be initialized and does not encounter any stability problems. Any sensitivity to noise (due to the fact that the orthonormality constraints associated with the rotation matrix have not been accounted for) is remedied by applying a transformation parameter optimization technique described in section 7.4 to the least-squares estimate. Experiments showed that a small number of iterations of the optimization technique produced very accurate parameters. Let

$$\mathbf{X} = (x_j, y_j, z_j)$$

be a set of η points on the model curve and let

$$\Psi = (\mu_j, \nu_j, \xi_j)$$

be the set of corresponding points on the other input curve. The parameters of the following transformation:

$$x_j = a\mu_j + b\nu_j + c\xi_j + d$$

$$y_j = e\mu_j + f\nu_j + g\xi_j + h \qquad (10.2)$$

$$z_j = m\mu_j + n\nu_j + p\xi_j + q$$

must be solved for. A *least-squares estimation* procedure is used to estimate values of the transformation parameters. Let the *dissimilarity measure* Ω which measures the difference between the model curve and the transformed curve be defined by:

$$\Omega = \sum_{j=1}^{\eta} \left(x_j^t - x_j^c\right)^2 + \left(y_j^t - y_j^c\right)^2 + \left(z_j^t - z_j^c\right)^2$$

where (x_j^c, y_j^c, z_j^c) is the closest point on the model curve to transformed curve point (x_j^t, y_j^t, z_j^t). Using equations (10.2) to eliminate x_j^t, y_j^t and z_j^t yields:

$$\begin{aligned}
\Omega = \sum_{j=1}^{\eta} & (a\mu_j + b\nu_j + c\xi_j + d - x_j^c)^2 + (e\mu_j + f\nu_j + g\xi_j + h - y_j^c)^2 \\
& + (m\mu_j + n\nu_j + p\xi_j + q - z_j^c)^2.
\end{aligned}$$

Let $\mathcal{P} = (a, b, c, d, e, f, g, h, m, n, p, q)$ be the vector defined by the transformation parameters. The solution of

$$\frac{\partial \Omega}{\partial \mathcal{P}} = 0$$

is the least-squares estimate of those parameters. To compute that estimate, determine the partial derivative of Ω with respect to each of the transformation parameters and set those partial derivatives to zero. The result is a linear system of 12 equations in 12 unknowns which is solved explicitly as following:

$$a = \frac{\mathcal{A}}{\mathcal{B}}$$

where

$$\begin{aligned}
\mathcal{A} &= \mathcal{F}(x_j^c, \mu_j)\mathcal{F}^2(\xi_j, \xi_j)\mathcal{F}(\nu_j, \nu_j) \\
&\quad -\mathcal{F}(x_j^c, \mu_j)\mathcal{F}(\xi_j, \xi_j)\mathcal{F}^2(\nu_j, \xi_j) \\
&\quad -\mathcal{F}(\mu_j, \xi_j)\mathcal{F}(x_j^c, \xi_j)\mathcal{F}(\nu_j, \nu_j)\mathcal{F}(\xi_j, \xi_j) \\
&\quad -\mathcal{F}(\mu_j, \nu_j)\mathcal{F}^2(\xi_j, \xi_j)\mathcal{F}(x_j^c, \nu_j) \\
&\quad +\mathcal{F}(\mu_j, \nu_j)\mathcal{F}(\xi_j, \xi_j)\mathcal{F}(\nu_j, \xi_j)\mathcal{F}(x_j^c, \xi_j) \\
&\quad +\mathcal{F}(\mu_j, \nu_j)\mathcal{F}(\xi_j, \xi_j)\mathcal{F}(\nu_j, \xi_j)\mathcal{F}(\mu_j, \xi_j) \\
&\quad +\mathcal{F}(\mu_j, \xi_j)\mathcal{F}(\nu_j, \xi_j)\mathcal{F}(x_j^c, \nu_j)\mathcal{F}(\xi_j, \xi_j) \\
&\quad -\mathcal{F}^2(\mu_j, \xi_j)\mathcal{F}^2(\nu_j, \xi_j)
\end{aligned}$$

$$\begin{aligned}
\mathcal{B} &= \mathcal{F}(\mu_j, \mu_j)\mathcal{F}^2(\xi_j, \xi_j)\mathcal{F}(\nu_j, \nu_j) - \mathcal{F}(\mu_j, \mu_j)\mathcal{F}(\xi_j, \xi_j)\mathcal{F}^2(\nu_j, \xi_j) \\
&\quad -\mathcal{F}^2(\mu_j, \xi_j)\mathcal{F}(\nu_j, \nu_j)\mathcal{F}(\xi_j, \xi_j) + \mathcal{F}^2(\mu_j, \xi_j)\mathcal{F}^2(\nu_j, \xi_j) \\
&\quad -\mathcal{F}^2(\mu_j, \nu_j)\mathcal{F}^2(\xi_j, \xi_j) + \mathcal{F}(\mu_j, \xi_j)\mathcal{F}(\nu_j, \xi_j)\mathcal{F}(\mu_j, \nu_j)\mathcal{F}(\xi_j, \xi_j)
\end{aligned}$$

and function $\mathcal{F}$ is defined as:

$$\mathcal{F}(\alpha_j, \beta_j) = \sum \alpha_j \beta_j - \frac{1}{\eta} \sum \alpha_j \sum \beta_j$$

$$b = \frac{\mathcal{F}_1}{\mathcal{F}_2}$$

where

$$\begin{aligned}
\mathcal{F}_1 &= \mathcal{F}(x_j^c, \nu_j)\mathcal{F}(\xi_j, \xi_j) - \mathcal{F}(\nu_j, \xi_j)\mathcal{F}(x_j^c, \xi_j) \\
&\quad -a(\mathcal{F}(\mu_j, \nu_j)\mathcal{F}(\xi_j, \xi_j) - \mathcal{F}(\nu_j, \xi_j)\mathcal{F}(\mu_j, \xi_j))
\end{aligned}$$

$$\mathcal{F}_2 = \mathcal{F}(\nu_j, \nu_j)\mathcal{F}(\xi_j, \xi_j) - \mathcal{F}^2(\nu_j, \xi_j)$$

$$c = \frac{\mathcal{F}(x_j^c, \xi_j) - a\mathcal{F}(\mu_j, \xi_j) - b\mathcal{F}(\nu_j, \xi_j)}{\mathcal{F}(\xi_j, \xi_j)}$$

$$d = \frac{1}{\eta}\left(\sum x_j^c - a\sum \mu_j - b\sum \nu_j - c\sum \xi_j\right).$$

Note that expressions for e, f, g and h can be obtained by replacing all x_j^c with y_j^c in the expressions for a, b, c and d respectively. Furthermore, expressions for m, n, p and q can be obtained by replacing all x_j^c with z_j^c in the expressions for a, b, c and d respectively. Note that when solving the 3-D to 2-D case, parameters m, n, p and q are zero but the solution remains the same otherwise.

7.3. Measuring Space Curve Distances

Once an estimate of the transformation parameters is available, it is possible to map the input curve to the space of the model curve. It is then useful to measure the input-model curve distance in order to verify the correctness of the TSS match. Furthermore, as explained in the next section, the computation of that distance is an essential step in the optimization of the transformation parameters. The following procedure is used to compute the input-model curve distance:

- Let $k = 1$. Let $\delta = 0.0$.

- Let η = the number of vertices on the input curve.

- Determine the *closest point* on the model curve to vertex k of the input curve.

- Let $\delta = \delta+$ distance to closest point determined in the previous step.

- Let $k = k + 1$.

- If $k > \eta$, then return δ as the total input-model curve distance and STOP.

- Go to the third step.

To speed up the algorithm, only a small neighborhood on the model curve into which vertex k maps is considered in step 3. Note that the closest point can be a vertex of the model curve or it can be a point on the line segment joining two consecutive vertices. Here, only the vertices of the model curve were considered for the sake of efficiency. No adverse effects were observed in the transformation parameter optimization process explained in the next subsection. This is due to the small distance between consecutive vertices on the space curve.

7.4. Optimizing the Transformation Parameters

The least-squares estimate of the transformation parameters is, in general, not the optimal estimate. This is because the point correspondences computed from the TSS match are not precise due to noise and local shape distortions. Nevertheless, it is possible to optimize those parameters using the following precedure. Note that in the first iteration of this procedure, the point correspondences are obtained from the TSS match.

- Let $\mathcal{D}_p = \infty$.

- Compute the least-squares estimate of the transformation parameters using the current set of point correspondences and map the input curve to the model curve.

- Determine a new set of corresponding points on the model curve as described in the previous subsection and compute the new image-model curve distance $\mathcal{D}_n$.

- If $\mathcal{D}_p - \mathcal{D}_n < \varepsilon$, then STOP.

- Let $\mathcal{D}_p = \mathcal{D}_n$ and go to the second step.

Note that the optimal transformation parameters can be estimated very accurately using a small number of iterations of the procedure above.

8.   Experiments, Discussion and Evaluation

This section presents a number of experiments to demonstrate the stability of TSS representations of space curves under various noise conditions and to illustrate the performance of the matching algorithm described in section 7. It also contains a discussion of the regular, renormalized and resampled TSS representations and an evaluation of those representations according to the criteria proposed in section 1.

Experiments were carried out to examine the behaviour of the proposed TSS representations when input curves are corrupted by noise. Figure 10.9 shows the armchair with moderate and then severe random noise added. The direction as well as the magnitude of the noise is random. Figure 10.10(a) shows the superposition of the TSS image of the armchair with moderate noise and the TSS image of the original armchair. It can be observed that despite the addition of a considerable amount of noise, the TSS image retains its basic structure very well. The TSS image of armchair with severe noise no longer matches well with the TSS image of the original armchair. However, the resampled TSS image of the armchair (shown in figure 10.8) and the resampled TSS image of the armchair with severe noise exhibit a very close match as demonstrated by figure 10.10(b) which shows their superposition.

The last figure of this chapter shows the final matching results obtained using the iterative optimization algorithm described in section 7. Figure 10.11(a) shows the armchair with moderate noise matched to the original armchair. The original curve has been drawn using a thick line. As the figure shows, the optimal transformation parameters have been computed. The initial estimate for the optimization algorithm was provided by the TSS match shown in figure 10.10(a). Figure 10.11(b)

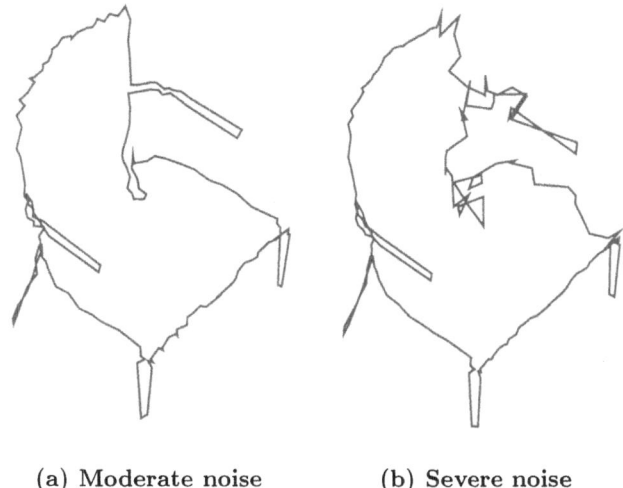

(a) Moderate noise   (b) Severe noise

Figure 10.9. Armchair with added noise

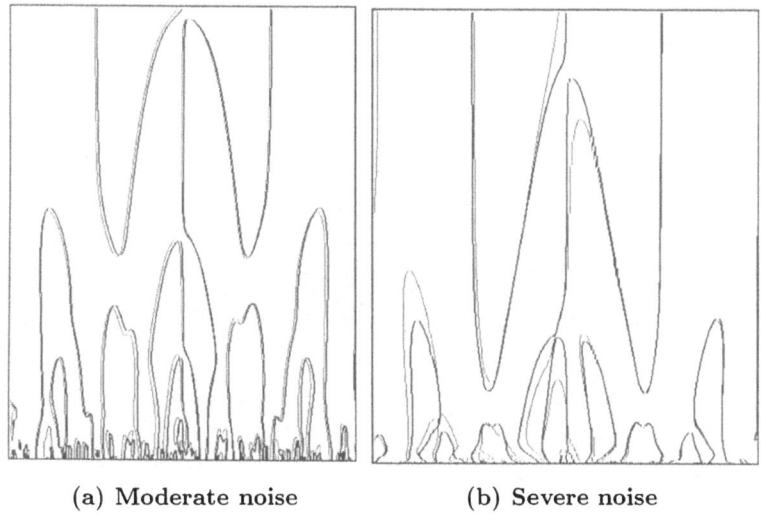

(a) Moderate noise   (b) Severe noise

Figure 10.10. Superposition of the TSS images of the armchair

shows the armchair with severe noise matched to the original armchair (again drawn using a thick line). Despite a severe amount of noise, the optimal parameters have again been computed. The initial estimate for the optimization algorithm was provided by the TSS match shown in figure 10.10(b).

In general, it can be said that the regular TSS image is suitable when the input curve has undergone transformations consisting of uniform

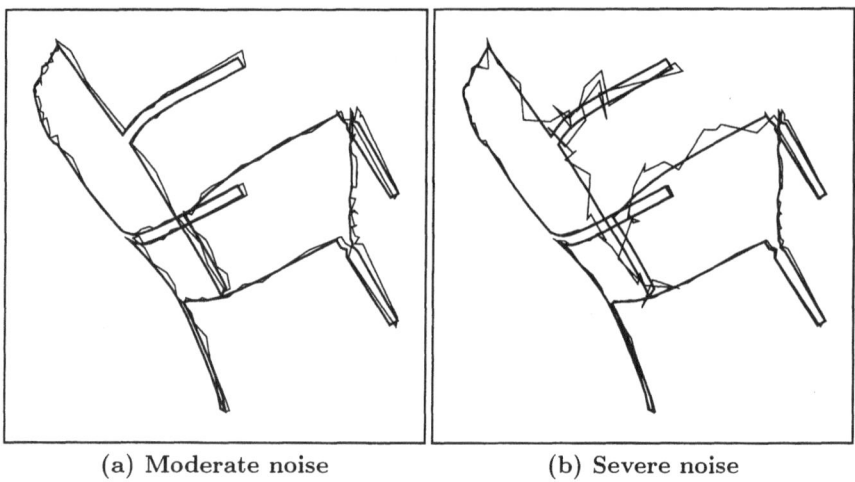

(a) Moderate noise   (b) Severe noise

Figure 10.11. Final match of the armchairs

Representation	Advantages	Disadvantages
The regular TSS Image	Suitable for all shape-preserving transformations as well as uniform noise on the curve.	Non-uniform noise or local difference in shape can cause problems.
The renormalized TSS Image	More suitable when there is non-uniform noise or local shape differences.	Most computationally intensive.
The resampled TSS Image	Most suitable when there is severe non-uniform noise or local shape differences.	De-emphasizes shape differences at fine scales.

Table 10.1. Comparison of regular, renormalized and resampled TSS Images.

scaling, rotation and translation and/or has been corrupted by uniform noise. However, when non-uniform noise and/or local shape changes have altered the shape of the input curve, the renormalized or resampled TSS images are better choices. The resampled TSS image is the most robust one with respect to non-uniform noise and local shape changes and is rather insensitive towards small-scale differences in shape which may be an undesirable feature. The renormalized TSS image is more sensitive towards those small-scale shape differences but requires more computation time. Table 10.1 summarizes the advantages and disadvantages of each representation technique:

The following is an evaluation of the TSS representations according to the criteria proposed in section 1.

Invariance Recall that by invariance, we meant that the representation for the shape of a curve should not change when shape-preserving transformations (rotation, uniform scaling and translation) are applied to that curve. Translation of the curve causes no change in the TSS representations proposed here. rotation causes only a horizontal shift but no structural change in the TSS representations. Uniform scaling causes the TSS representations to undergo uniform scaling as well. If the represented curves are closed, then their TSS representations can be normalized and invariance with respect to uniform scaling will also be achieved. If the represented curves are open, changes due to uniform scaling can be handled during matching.

Uniqueness The uniqueness criterion required that two curves with different shapes be mapped to different representations. As argued earlier, theorem 10.6 shows that a space curve can be reconstructed from any of its TSS representations modulo function $\tau^*(u)$ and therefore the TSS representations partially satisfy the uniqueness criterion. However, TSS images are very rich representations and are therefore believed to be more than sufficient to distinguish space curves with different shapes from each other in practical shape matching tasks. The only arbitrary choice to be made when computing TSS representations is the starting point for parametrization on closed curves. This only causes a horizontal shift in TSS representations but causes no structural change.

Stability The stability criterion requires that a small change in the shape of a curve lead to a small change in its representation and *vice versa*. Theorem 10.3 shows that space curves always remain connected during evolution or arc length evolution and therefore their TSS representations can always be constructed. Furthermore, it can be shown that while a space curve evolves, a small change in the standard deviation of the Gaussian filter always results in a small change in the locations of the torsion zero-crossings on that curve. The experiments of this chapter also show that TSS representations are stable with respect to significant uniform and non-uniform noise on the curves they represent and therefore satisfy the stability criterion.

Local support TSS computations use finite Gaussian filters and finite neighborhoods. Therefore a TSS representation can also be computed for an open curve and, except near the endpoints of the curve, will resemble the corresponding representation for a closed curve of which it is a part. It is therefore believed that the representations also satisfy the local support criterion.

Efficiency All algorithms proposed here are efficient in that their complexities are low order polynomials in the size of the input. The computation of the representations proposed here calls for evaluation of a large number of convolutions. This process can be rendered more efficient using fast Fourier transforms, parallelism, expression of convolutions involving Gaussians of large widths in terms of convolutions involving Gaussians of small widths, and tracking torsion zero-crossing points across multiple scales: when it is known that new torsion zero-crossings will not be created at higher scales, convolutions can be carried out only in a small neighborhood of the existing zero-crossings in order to locate the zero-crossings at the next higher level. Furthermore, TSS representations can be stored very efficiently as encoded binary images. An alternative is to store only a selected subset of points from TSS representations which will be used for matching.

Shape Properties The TSS representations of symmetric space curves are also symmetric since a symmetric space curve also has symmetric torsion zero-crossings across scales. Therefore, the symmetry criterion is satisfied. Furthermore, TSS computations are carried out using finite Gaussian filters and making use of finite-sized neighborhoods. Therefore, a TSS representation can also be computed for an open space curve and, except near the endpoints of the curve, will resemble the corresponding representation for a closed space curve of which it is a part. It is therefore believed that TSS representations also satisfy the part/whole criterion.

Implementation The procedures needed to compute TSS images are not difficult to implement. Convolutions with Gaussian filters are at the heart of the computations. These are standard and well-understood procedures. It follows that this criterion is also satisfied.

9. Conclusions

This chapter introduced a novel shape representation technique for space curves and proposed a number of necessary criteria that any useful shape representation scheme should satisfy. Those criteria are: *invariance, uniqueness, stability, local support, efficiency, shape properties* and *implementation*. Three different ways of computing the representation were described. Each method relies on extracting features of the curve that are invariant under shape preserving transformations at varying scales. These methods result in: the torsion scale space image, the renormalized torsion scale space image and the resampled torsion scale space images. It was shown that each of those representations is suitable under specific circumstances. A two-stage matching algorithm consisting of TSS matching followed by transformation parameter optimization was used to demonstrate the usefulness of the representations in recognition tasks. A number of theoretical properties of those representations were also investigated. These properties together provide a sound foundation for the proposed representations. It was also shown that the proposed representations satisfy the criteria introduced earlier.

Chapter 11

MULTI-SCALE REPRESENTATIONS FOR FREE-FORM 3-D SURFACES

The final chapter is concerned with the generalization of the CSS representation to free-form 3-D surfaces.

A novel technique for multi-scale curvature estimation on a smoothed 3-D surface is presented. This is achieved by convolving local parametrizations of the surface with 2-D Gaussian filters iteratively. In our technique, semigeodesic coordinates are constructed at each vertex of the mesh which becomes the local origin. A geodesic from the origin is first constructed in an arbitrary direction such as the direction of one of the incident edges. The smoothing eliminates surface noise and small surface detail gradually, and results in gradual simplification of the object shape. The surface Gaussian and mean curvature values are estimated accurately at multiple scales together with curvature zero-crossing contours. The curvature values are then mapped to colours/greylevels and displayed directly on the surface. Furthermore, maxima of Gaussian and mean curvatures are also located and displayed on the surface. These features have been utilized by later processes for robust surface matching and object recognition. Our technique is independent of the underlying triangulation and is also more efficient than volumetric diffusion techniques since 2-D rather than 3-D convolutions are employed. Another advantage is that it is applicable to incomplete surfaces which arise during occlusion or surfaces with holes.

1. Introduction

Curvature estimation is an important task in 3-D object description and recognition. Surface curvature provides a unique viewpoint invariant description of local surface shape. Differential geometry [95] provides several measures of curvature, which include Gaussian and mean curvatures. Combination of these curvature values enable the local surface type to be categorized.

This chapter introduces a new technique for multi-scale curvature computation on a smoothed 3-D surface. Complete triangulated models of 3-D objects are constructed and using a local parametrization technique, are then smoothed using a 2-D Gaussian filter. The technique considered here is a generalization of earlier multi-scale representation theories proposed for 2-D contours [222] and space curves [197]. More details of the diffusion technique as well as literature survey appear in [213].

In our approach, diffusion of the surface is achieved through convolutions of local parametrizations of the surface with a 2-D Gaussian filter [213, 214, 331]. *Semigeodesic coordinates* [95] are utilized as a natural and efficient way of locally parametrizing surface shape. The most important advantage of our method is that unlike other diffusion techniques such as volumetric diffusion [146, 144] or level set methods [282], it has *local support* and is therefore applicable to partial data corresponding to surface-segments. This property makes it suitable for object recognition applications in presence of occlusions. It is also more efficient than those techniques since 2-D rather than 3-D convolutions are employed.

This chapter includes examples of simple and complex 3-D objects with their Gaussian and mean curvature values estimated [95]. To visualize these curvature values on the surface, they are then mapped to colours/greylevels. Colour/greylevel mapping is a scalar visualization technique provided in a software package referred to as the Visualization Toolkit (VTK) [278]. Once surface curvatures are estimated, then curvature zero-crossing contours are recovered and displayed on the surface. Finally, maxima of Gaussian and mean curvatures are also located and displayed on the surface.

The organization of this chapter is as follows. Section 2 gives a brief overview of previous work on 3-D object representations. Section 3 describes the relevant theory from differential geometry and explains how a multi-scale shape description can be computed for a free-form 3-D surface. Section 4 is on evolution properties of 3-D surfaces. Section 5 covers the computation of Gaussian and mean curvatures as well as their zero-crossing contours and maxima. Section 6 presents results on 3-D surface smoothing as well as curvature estimation. Section 7 is on the estimation of error in curvature computation. Section 8 describes a robust free-form 3-D object recognition system, and section 9 presents the concluding remarks.

2.   Literature Survey

As the most important application of our work is 3D surface representation and object recognition, this section presents a survey of previous work in those areas. Many object recognition systems rely on restrictions imposed on the geometry of the object. However, complex free-form surfaces may not be modelled easily using volumetric primitives. A free-form surface is a surface such that the surface normal is defined and continuous everywhere, except at sharp corners and edges [33]. Discontinuities in the surface normal or curvature may be present anywhere on a free-form object. The curves that connect these points of discontinuity may meet or diverge smoothly. Recognition of free-form objects is essential in inspection of arbitrary curved surfaces and path planning for robot navigation.

Sinha and Jain [288] provide an overview of geometry based representations derived from range data of objects. Comprehensive surveys of 3-D object recognition systems are presented by Besl and Jain [34], and Chin and Dyer [55]. Some representation schemes for 3-D objects have adopted some form of surface or volumetric parametric models to characterize the shape of the objects. Current volumetric representations rely on representing objects in terms of generalized cylinders, superquadrics, set-theoric combinations of volume primitives as in constructive solid geometry (CSG) or spatial occupancy [251, 291, 53, 271]. However, it may not be possible to express objects with free-form surfaces using for example, superquadric primitives. Surface-based representations describe an object in terms of the surfaces bounding the object and their properties [78, 84], and are employed for recognition. Although there are several methods available to model a surface, triangular meshes are the simplest and most effective form of polygons to cover a free-form surface. The common types of polygonal meshes include the triangular mesh [108, 107] and the four sided spline patches. Triangular meshes have been utilized in our work.

Polyhedral approximations [80] fit a polyhedral object with vertices and relatively large flat faces to a 3-D object. Their disadvantage is that the choice of vertices can be quite arbitrary which renders the representation not robust. Smooth 3-D splines [298] can also be fitted to 3-D objects. Their shortcomings are that the choice of knot points is again arbitrary and that the spline parameters are not invariant. *Generalized cones* or *cylinders* as well as *geons* [255] approximate a 3-D object using globally parametrized mathematical models, but they are not applicable to detailed free-form objects. *multi-view* representations [280] are based on a large number of views of a 3-D object obtained from different viewpoints, but difficulties can arise when a non-standard view is encountered. In *volumetric diffusion* [144] or level set methods [282], an object is treated as a filled area or volume. The object is then blurred by subjecting it to the diffusion equation. The boundary of each blurred object can then be defined by applying the Laplacian operator to the smoothed area or volume. The major shortcoming of these approaches is lack of local support. In other words, the entire object data must be available. This problem makes them unsuitable for object recognition in presence of occlusion. A form of 3-D surface smoothing has been carried out in [302, 304] but this method has drawbacks since it is based on weighted averaging using neighbouring vertices and is therefore dependent on the underlying triangulation. Problems can arise when the same surface is triangulated in different ways (for example as a result of the application of decimation algorithms) resulting in different meshes. As a result, the outcome for each mesh will be different, although they all correspond to the same surface. The smoothing of 3-D surfaces is a result of the diffusion process [306]. For parameterization of a 3-D surface other methods have also been studied, such as the asymptotic coordinates [149], isothermic coordinates [95, 54] and global coordinates [43] used for closed, simply connected objects.

Global representations such as the Extended Gaussian Image [133, 162] describe 3-D objects in terms of their surface normal distributions on the unit sphere with appropriate support functions. However, arbitrary curved objects have to be either approximated by planar patches or divided into regions based on Gaussian curvature. Another approach for specifying a 3-D object is the view-centred representations. The graph approach [145] attempts to group a set of infinite 2-D views of a 3-D object into a set of meaningful cluster of appearances. Murase and Nayar [238] and Swets [300] also exploit photometric information to describe and recognise objects. A major drawback of view-centred representations is lack of complete information. Part based representations capture structure in object descriptions [68], but there is a lack of agreement in deciding the general set of part primitives to be used which would be sufficient and appropriate. Furthermore, computation of parts from a single view of an object is difficult.

Recent approaches using splash and super polygonal segments [296] and algebraic polynomials [256] have addressed the issue of representing complex curved free-form surfaces. However, there are limitations relating to object segmentation issues, restricting objects to be topologically equivalent to a sphere and sensitivity to noise when low-level surface features are used. Note that a substantial amount of work has also been carried out on curvature estimation from range images [42, 299, 163, 312, 83]. However, a range image corresponds to just one view of a 3-D object. It is essentially a 2-D function with known global parameterization, whereas a 3-D mesh is substantially more complex. Finally, mesh simplification techniques [56, 59, 103, 112, 293] are not suitable for object recognition applications since the locations of mesh vertices chosen at higher scales tend to be arbitrary and not robust to noise or local variations of shape.

A number of matching topics have been recognised by researchers as important in 3-D object recognition [310, 85]. These are related to object shape complexity, rigid and flexible objects and occlusion. The success of existing object recognition systems is because of the restrictions they impose on the classes of geometric objects. However, few systems can handle arbitrary surfaces with very few restrictive assumptions about their geometric shapes.

Object recognition is achieved by matching features derived from the scene with stored object model representations. Efficient algorithms were developed for the recognition of flat rigid objects based on the geometric hashing technique in [153, 152]. The technique was also extended to the recognition of arbitrary rigid 3-D objects from single 2-D images [154]. Stein and Medioni [296] and Flynn and Jain [85] have also employed geometric hashing for 3-D object recognition. In a geometric hashing technique the model information is indexed into a hash table using minimal transformation feature points. This technique determines for a given scene's minimal feature set a corresponding feature set on one of the models, by considering only the other scene features which *vote* for the correct interpretation. Other efficient model based object recognition techniques are the Hough (pose) clustering [164, 297], the alignment technique [121] and relational structures [323]. 3-D objects have also been modelled as superquadrics with local and global deformations for recognition purposes [291].

Recent patch-based techniques including point signatures [57] and spin images [130] perform well on scenes containing clutter and occlusion. However, these systems have been designed for single range images, and do not generalize to more general 3-D surfaces which can be obtained by merging two or more range images. In other words, their effectiveness is limited by the use of information in only one range image.

3. Semigeodesic and Geodesic Polar Parametrisation on a 3-D surface

A crucial property of 2-D contours and space curves (or 3-D contours) is that they can be parametrized globally using the arclength parameter. However, free-form 3-D surfaces are more complex. As a result, no global coordinate system exists on a free-form 3-D surface which could yield a natural parametrization of that surface. Indeed, studies of local properties of 3-D surfaces are carried out in differential geometry using local coordinate systems called *curvilinear coordinates* or *Gaussian coordinates* [95]. Each system of curvilinear coordinates is introduced on a patch of a regular surface referred to as a *simple sheet*. A simple sheet of a surface is obtained from a rectangle by stretching, squeezing, and bending but without tearing or gluing together. Given a parametric representation

$$\mathbf{r} = \mathbf{r}(u, v)$$

on a local patch, the values of the parameters u and v determine the position of each point on that patch.

3.1.   Geodesic Lines

Before the semigeodesic and geodesic polar coordinates can be described, it is necessary to define geodesic lines on a regular 3-D surface. The following definitions are useful [95, 106]:

Definition. A geodesic line or a geodesic of a surface is a curve whose *geodesic curvature* is zero at every point. Geodesic curvature is the magnitude of the *vector of geodesic curvature*.

Definition. The vector of geodesic curvature of a curve C lying on a surface S at a point P on C is obtained by projecting the *curvature vector* of C at P on the tangent plane to S at P.

Definition. The curvature vector of a curve C at point P is of the same direction as the principal normal vector at P and of length equal to the curvature of the curve at P.

Definition. The principal normal vector of a curve C at point P is perpendicular to C at P and lies in the *osculating plane* at P. The plane with the highest possible order of contact with the curve C at point P is called the osculating plane at P.

The following crucial property of geodesic lines is actually utilized to construct geodesics on 3-D triangular meshes:

Minimal property of geodesics: An arc of a geodesic line C passing through a point P and lying entirely in a sufficiently small neighborhood of a point P of a surface S of class C_2 is the shortest join of P with any other point of C by a curve lying in the neighborhood.

3.2.   Semigeodesic Coordinates

Semigeodesic coordinates can be constructed in the following way at a point P on a surface S of class C_2:

- Choose a geodesic line C through point P in an arbitrary direction.

- Denote by v the arclength parameter on C, such that P corresponds to the value $v = 0$.

- Take further through every point of C the geodesic line L perpendicular to C at the corresponding point.

- Denote by u the arclength parameter on L.

The two parameters u and v determine the position of each point in the domain swept out by these geodesic lines. It can be shown that in a sufficiently small neighborhood of the point P, semigeodesic coordinates can always serve as curvilinear coordinates in a regular parametric representation of S [95]. The orthogonal cartesian coordinates in the plane are a special case of semigeodesic coordinates on a flat surface.

3.3. Geodesic Polar Coordinates

Geodesic polar coordinates can be constructed at a point P on a surface S of class C_2 in the following way:

- Choose an arbitrary direction **w** on S at point P.

- Take all geodesic lines emanating from point P.

- Denote by v the arclength parameter on each geodesic in the previous step.

- Denote by u the angle between **w** and the tangent vector of each geodesic in step 2 at point P.

Again the two parameters u and v determine the position of each point in the domain swept out by these geodesic lines. It can be shown that in a sufficiently small neighborhood of point P (with P itself deleted), geodesic polar coordinates can always serve as curvilinear coordinates. Point P is a singular point of this parametrization since its coordinates are not uniquely defined. The polar coordinates in the plane are a special case of geodesic polar coordinates on a flat surface.

3.4. Gaussian Smoothing of a 3-D Surface

The procedures outlined above can be followed to construct semigeodesic coordinates or geodesic polar coordinates at every point of a 3-D surface S. In case of semigeodesic coordinates, local parametrization yields at each point P:

$$\mathbf{r}(u, v) = (x(u, v),\ y(u, v),\ z(u, v)).$$

The new location of point P is given by:

$$\mathbf{R}(u, v, \sigma) = (\mathcal{X}(u, v, \sigma),\ \mathcal{Y}(u, v, \sigma),\ \mathcal{Z}(u, v, \sigma))$$

where

$$\mathcal{X}(u, v, \sigma) = x(u, v) \otimes G(u, v, \sigma)$$
$$\mathcal{Y}(u, v, \sigma) = y(u, v) \otimes G(u, v, \sigma)$$
$$\mathcal{Z}(u, v, \sigma) = z(u, v) \otimes G(u, v, \sigma)$$

$$G(u, v, \sigma) = \frac{1}{2\pi\sigma^2}\, e^{-\frac{(u^2+v^2)}{2\sigma^2}}$$

and $\otimes$ denotes convolution. In case of geodesic polar coordinates, the Gaussian function becomes one-dimensional. As a result, each of the 2-D convolutions above can be expressed as a series of 1-D convolutions.

In both cases, this process is repeated at each point of $\mathcal{S}$ and the new point positions after filtering define the smoothed surface. Since the coordinates constructed are valid locally, the Gaussian filters always have $\sigma = 1$, and a filter size of 9.

3.5.   Multi-Scale Description of a 3-D Surface

In order to achieve a multi-scale description of a 3-D surface $\mathcal{S}$, it is smoothed according to the process described earlier. The smoothed surface is then considered as the input to the next stage of smoothing. This procedure is then iterated many times to obtain a multi-scale description of $\mathcal{S}$ [217]. This process is equivalent to diffusion smoothing

$$\frac{\partial \mathcal{S}}{\partial t} = H\,\mathbf{n}$$

since the Gaussian function satisfies the heat equation. In the equation above, t is time, H is mean curvature, and $\mathbf{n}$ is the surface normal vector. t can be regarded as the number of iterations.

3.6.   Implementation on a 3-D Triangular Mesh

The theory explained earlier must be adapted to a 3-D triangular mesh. Both semigeodesic and geodesic polar coordinates involve construction of geodesic lines. Clearly the segment of a geodesic that lies on any given triangle is a straight line. Two situations must be considered:

- Extension of a geodesic when it intersects a triangle edge

- Extension of a geodesic when it intersects a triangle vertex

The following theorem addresses the first situation.

THEOREM 11.1 *Suppose a geodesic intersects an edge e shared by triangles T_1 and T_2. The extension of this geodesic beyond e is obtained by rotating T_2 about e so that it becomes co-planar with T_1, extending the geodesic in a straight line on T_2, and rotating T_2 about e back to its original position (figure C.1).*

Note that the construction above extends to several triangles as long as they remain in a local neighborhood. The next theorem addresses the second case.

THEOREM 11.2 *Suppose a geodesic arrives at a vertex V of the mesh. Define the normal vector **n** at V as the average of the surface normals of all the triangles incident on V weighted by the incident angle. Let Q be the plane formed by the geodesic incident on V and **n**. The extension of this geodesic beyond V is found by intersecting Q with the mesh (figure C.2).*

3.6.1   Implementation of Semigeodesic Coordinates

Semigeodesic coordinates are constructed at each vertex of the mesh which becomes the local origin. The following procedure is employed:

- Construct a geodesic from the origin in an arbitrary direction such as the direction of one of the incident edges.

- Construct the other half of that geodesic by extending it through the origin in the reverse direction using the procedure outlined in theorem 11.2.

- Parametrize that geodesic by the arclength parameter at regular intervals to obtain a sequence of sample points.

- At each sample point on the first geodesic, construct a perpendicular geodesic and extend it in both directions.

- Parametrize each of the geodesics constructed in the previous step by the arclength parameter at regular intervals.

Figure 11.1 shows the complete semigeodesic coordinates on a triangular mesh.

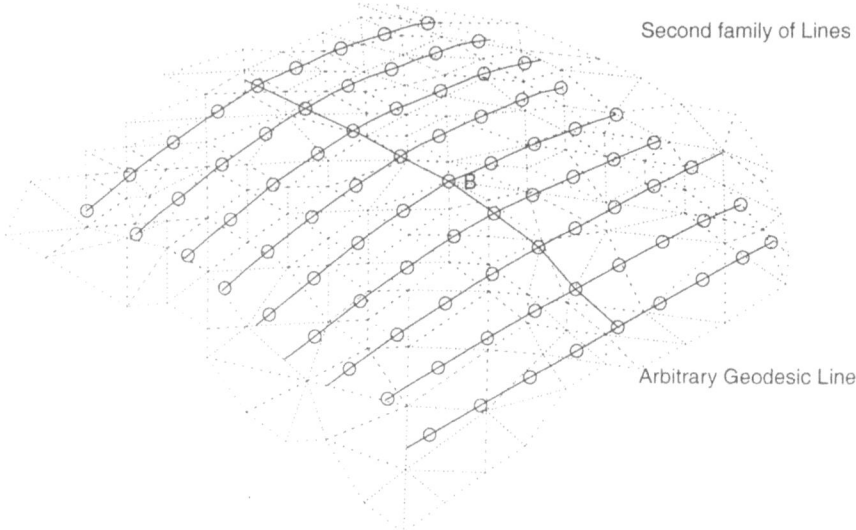

Where "o"s are the semigeodesic coordinates and "B" is the current vertex.

Figure 11.1.   Semigeodesic coordinates on a triangular mesh

3.6.2   Semigeodesic Coordinates on Open Surfaces

Quite often, due to occlusion or complex object shape, it is not possible to construct complete and closed surfaces. As a result, the algorithm described above should be modified to make it also applicable to open surfaces. The algorithm for smoothing an open surface is defined in the following way:

1 Grid construction and smoothing at internal vertices is carried out as on closed surfaces. Any geodesic line that reaches the boundary will stop. The last sample point at or near the boundary will be duplicated until the grid is filled. Likewise, if some geodesic lines can not be constructed, the last geodesic line near the boundary will be duplicated until the grid is filled.

2 If the vertex V of triangle T resides on the boundary, measure the angle α between the two edges of T that are incident on V. Choose the first geodesic line as the bisector of α. Only half of the first geodesic line is constructed because the other half falls outside the surface boundary.

3 At the same vertex, construct another geodesic line perpendicular to the first one.

4 One of those geodesic lines might soon intersect the boundary, so compare the lengths of those lines and choose the longer one. This allows the maximum size grid to be constructed.

5 Construct the second family of geodesic lines as perpendicular to the longer geodesic line determined above.

6 As before, any geodesic line that reaches the boundary will stop, and the last sample point at or near the boundary will be duplicated until the grid is filled.

3.6.3   Implementation of Geodesic Polar Coordinates

Geodesic polar coordinates are also constructed at each vertex of the mesh which again becomes the local origin. The following procedure is used:

- Construct a geodesic from the origin in an arbitrary direction such as the direction of one of the incident edges.

- Let N be the normal plane at the origin defined by the geodesic constructed in the previous step and the normal vector $\mathbf{n}$ (defined in theorem 11.2).

- Rotate N about $\mathbf{n}$ by angle α and intersect it with the mesh to obtain the next geodesic emanating from the origin.

- Repeat the previous step until N is back in its original position.

- Parametrize each of the constructed geodesics by the arclength parameter at regular intervals to obtain a sequence of sample points on each geodesic.

4.   Evolution Properties of 3-D surfaces

This section presents a number of theorems covering fundamental properties of evolution of 3-D surfaces [219]. The proofs of these theorems are given in appendix C.

THEOREM 11.3 *The order of application of evolution and a shape preserving transformation to a surface does not change the final result.*

This result demonstrates that multi-scale description of a free-form 3-D surface is not affected by shape-preserving transformations. This is essential since it makes it possible to match a planar curve to another of similar shape that has undergone a shape-preserving transform.

THEOREM 11.4 *Let S be a closed surface and let $\mathcal{H}$ be its convex hull. S remains inside $\mathcal{H}$ during evolution.*

This result imposes constraints on the physical location of a 3-D surface as it evolves. This constraint becomes useful when the physical location of surfaces in space or their locations with respect to each other is important.

THEOREM 11.5 *Iterative Gaussian filtering of a surface converges to the solution of the heat diffusion equation.*

This theorem presents an efficient, convenient mechanism for the implementation of the heat diffusion process.

THEOREM 11.6 *Let S be a 3-D surface in C_2. Let S_σ be an evolved version of S with a cusp point at P. There is a $\delta > 0$ such that $S_{\sigma-\delta}$ intersects itself in a neighborhood of point P.*

This theorem shows that the formation of a cusp point on a 3-D surface during evolution is preceded by the self-intersection of that surface in a neighborhood of the cusp point.

THEOREM 11.7 *Simple (not self-intersecting) surfaces remain simple during evolution.*

This result is very significant. Simple surfaces usually represent the boundaries of 3-D objects. The evolved versions of those surfaces can only have physical plausibility if they are also simple.

5.   Curvature Estimation

This section presents techniques for accurate estimation of Gaussian and mean curvatures at multiple scales on smoothed free-form 3-D surfaces, and is derived from [330, 332]. Differential geometry provides several measures of curvature, which include Gaussian and mean curvatures [95]. Consider a local parametric representation of a 3-D surface

$$\mathbf{r} = \mathbf{r}(u, v)$$

with coordinates u and v, where

$$\mathbf{r}(u, v) = (x(u, v),\ y(u, v),\ z(u, v))$$

Gaussian curvature K exists at regular points of a surface of class C_2. When $\mathbf{r}(u, v)$ corresponds to semigeodesic coordinates, K is given by:

$$K = \frac{b_{uu}b_{vv} - b_{uv}^2}{x_v^2 + y_v^2 + z_v^2} \tag{11.1}$$

where subscripts denote partial derivatives, and

$$b_{uu} = \frac{Ax_{uu} + By_{uu} + Cz_{uu}}{\sqrt{A^2 + B^2 + C^2}}$$

$$b_{vv} = \frac{Ax_{vv} + By_{vv} + Cz_{vv}}{\sqrt{A^2 + B^2 + C^2}}$$

$$b_{uv} = \frac{Ax_{uv} + By_{uv} + Cz_{uv}}{\sqrt{A^2 + B^2 + C^2}}$$

where

$$A = y_u z_v - z_u y_v$$

$$B = x_v z_u - z_v x_u$$

$$C = x_u y_v - y_u x_v$$

Mean curvature H also exists at regular points of a surface of class C_2. Again, when $\mathbf{r}(u, v)$ corresponds to semigeodesic coordinates, H is given by:

$$H = \frac{b_{vv} + (x_v^2 + y_v^2 + z_v^2)b_{uu}}{2(x_v^2 + y_v^2 + z_v^2)} \tag{11.2}$$

Both Gaussian and mean curvature values are direction-free quantities. Gaussian and mean curvatures are invariant to arbitrary transformation of the (u,v) parameters as well as rotations and translations of a surface. Combination of these curvature measures enable the local surface type to be categorized. On smoothed surfaces of 3-D objects, the procedure for estimating the Gaussian and mean curvatures are as follows. For each point of the surface,

$$p(x(u, v), \ y(u, v), \ z(u, v))$$

the corresponding local neighbourhood data is convolved with the partial derivatives of the Gaussian filter $G(u, v, \sigma)$:

$$x_u = x \otimes \frac{\partial G}{\partial u}, \quad y_u = y \otimes \frac{\partial G}{\partial u}, \quad z_u = z \otimes \frac{\partial G}{\partial u}$$

$$x_v = x \otimes \frac{\partial G}{\partial v}, \quad y_v = y \otimes \frac{\partial G}{\partial v}, \quad z_v = z \otimes \frac{\partial G}{\partial v}$$

$$x_{uu} = x \otimes \frac{\partial^2 G}{\partial u^2}, \quad y_{uu} = y \otimes \frac{\partial^2 G}{\partial u^2}, \quad z_{uu} = z \otimes \frac{\partial^2 G}{\partial u^2}$$

$$x_{vv} = x \otimes \frac{\partial^2 G}{\partial v^2}, \quad y_{vv} = y \otimes \frac{\partial^2 G}{\partial v^2}, \quad z_{vv} = z \otimes \frac{\partial^2 G}{\partial v^2}$$

$$x_{uv} = x \otimes \frac{\partial^2 G}{\partial u \partial v}, \quad y_{uv} = y \otimes \frac{\partial^2 G}{\partial u \partial v}, \quad z_{uv} = z \otimes \frac{\partial^2 G}{\partial u \partial v}$$

where $\otimes$ denotes convolution. Finally, curvature values on a 3-D surface are estimated by substituting these values into equations (11.1) and (11.2), respectively.

5.1.   Curvature Zero-Crossing Contours

Having computed curvature values at each vertex of a smoothed 3-D surface, one can locate curvature zero-crossing contours where curvature functions K or H defined by equations (11.1) and (11.2), are equal to zero. Curvature zero-crossing contours can be useful for segmenting a smoothed 3-D surface into regions. The process of recovery of the curvature zero-crossing contours is identical for Gaussian and mean curvatures. Every edge e of the smoothed surface is examined in turn. If the vertices of e have the same signs of curvature, then there is no curvature zero-crossing point on e. However, if the vertices of e have different signs of curvature, then there exists a point on e at which curvature goes to zero. The zero-crossing point is assumed to be at the midpoint of e. The other two edges of the triangle to which e belongs will then be checked since there will be another zero-crossing point on one of those edges. When that zero-crossing is found, it is connected to the previously found zero-crossing. The curvature zero-crossing contour is tracked in this fashion until one arrives back at the starting point [216].

5.2.   Local Curvature Maxima

Local maxima of Gaussian and mean curvatures are significant and robust feature points on smoothed surfaces since noise has been eliminated from those surfaces. The process of recovery of the local maxima is identical for Gaussian and mean curvatures. Every vertex V of the smoothed surface is examined in turn. The neighbours of V are defined as vertices which are connected to V by an edge. If the curvature value of V is higher than the curvature values of all its neighbours, V is marked as a local maximum of curvature. Curvature maxima can be utilized by later processes for robust surface matching and object recognition with occlusion [219].

5.3.   Torsion Maxima on Zero-Crossing Contours

Once Gaussian and mean curvatures have been determined at each point of a 3-D surface, zero-crossing contours of those curvatures are recovered from that surface. In general, these contours are space curves. Torsion is computed at each point of those contours (see chapter 10), and the maxima of torsion are then recovered. These points are added to the set of feature points recovered from the surface [332].

6.   Results and Discussion

This section presents some results on free-form surface smoothing as well as curvature estimation.

6.1.   Diffusion

The smoothing routines were implemented entirely in C++. Complete triangulated models of some 3-D objects used for our experiments were constructed at CVSSP [108]. In order to experiment with our techniques, both simple and complex 3-D objects with different numbers of triangles were used. Each iteration of smoothing of a surface with 1000 vertices takes about 0.5 second of CPU time on an *UltraSparc 170E*.

The first test object was a cube with 98 vertices and 192 triangles. The smoothing results using semigeodesic coordinates (with filter size equal to 9) are shown in Figure 11.2 (top row). The original cube evolves into a sphere after five iterations. The experiment was repeated using geodesic polar coordinates (with 9 polar lines), and the smoothing results are shown in Figure 11.2 (bottom row). These results indicate that smoothing using semigeodesic coordinates and geodesic polar coordinates produce similar results. The second test object was a foot with 2898 triangles and 1451 vertices. The smoothing results are shown in Figure 11.3. The third test object was a chair with 3788 triangles and 1894 vertices as shown in Figure 11.4. Figure 11.5 shows the fourth test object which was a telephone handset with 11124 triangles and 5564 vertices. These examples show that our technique is effective in eliminating surface noise as well as removing surface detail. The result is gradual simplification of object shape.

The fifth test object was a dinosaur with 2996 triangles and 1500 vertices as shown in Figure 11.6. The object becomes smoother gradually and the legs, tail and ears are removed after 10 iterations. The sixth test object was a cow with 3348 triangles and 1676 vertices as shown in Figure 11.7. The surface noise is eliminated iteratively with the object becoming smoother gradually where after 12 iterations the legs, ears and tail are removed, as was seen for the dinosaur.

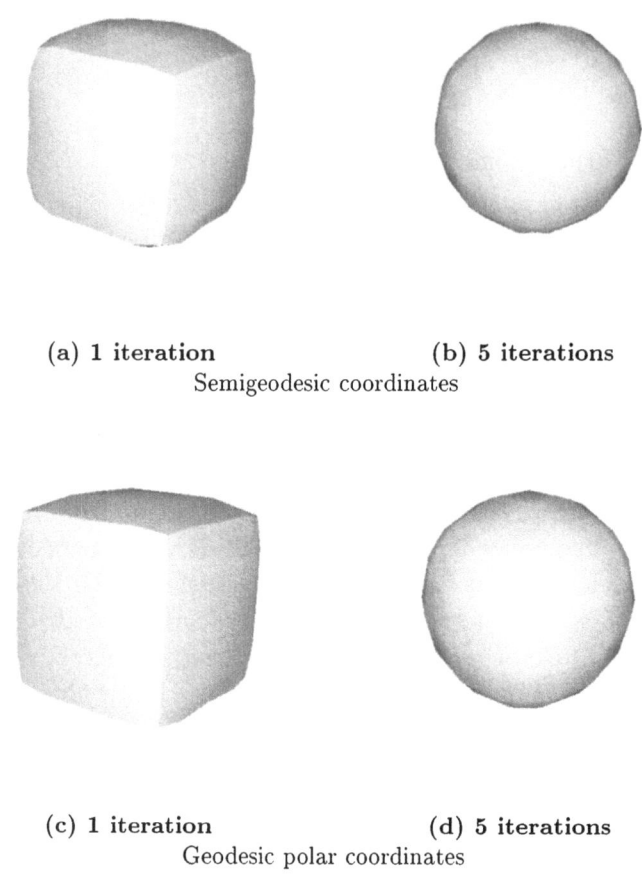

(a) **1 iteration**   (b) **5 iterations**

Semigeodesic coordinates

(c) **1 iteration**   (d) **5 iterations**

Geodesic polar coordinates

Figure 11.2. Smoothing of the Cube

Our smoothing technique was also applied to a number of open or incomplete surfaces. Figure 11.8 shows the results obtained on a part of the telephone handset. This object also has a triangle removed in order to generate an internal hole. Figure 11.9 shows the smoothing results obtained on the lower part of a chair object. Figure 11.10 shows smoothing results obtained on a partial rabbit. The object is smoothed iteratively and the ears disappear eventually.

Note that techniques which apply smoothing in the normal direction need to first estimate curvature in order to displace surface points by a distance proportional to the curvature value. However, curvature estimation itself poses a problem for these methods. Our technique overcomes

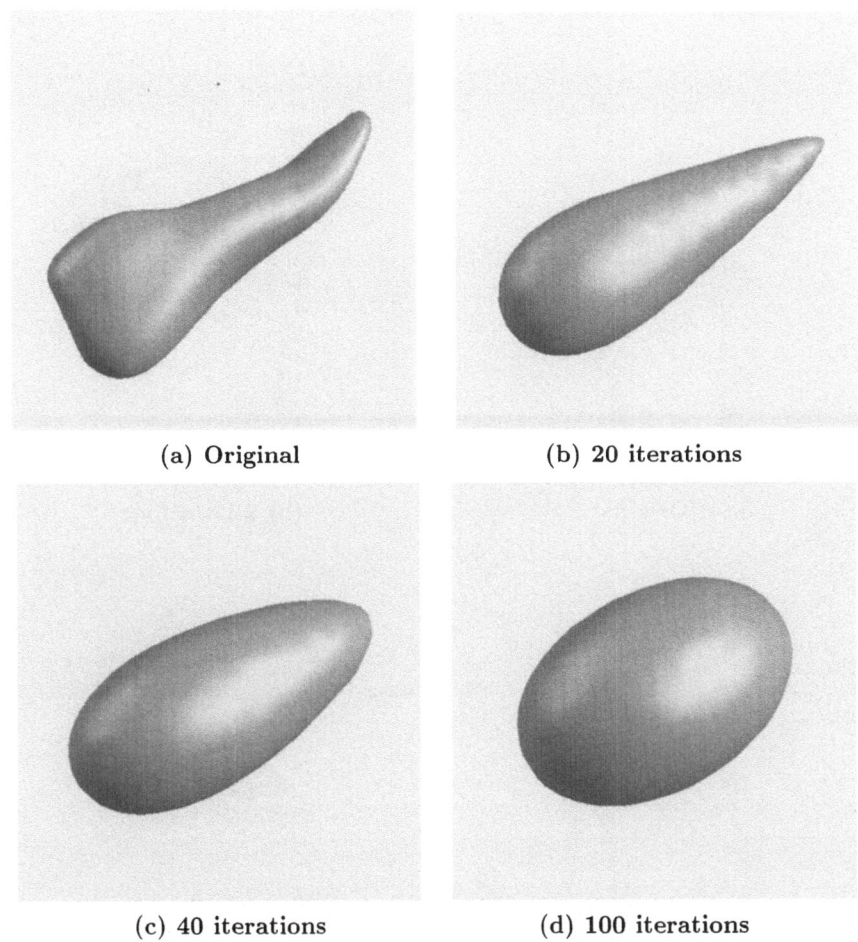

(a) Original (b) 20 iterations

(c) 40 iterations (d) 100 iterations

Figure 11.3. Diffusion of the Foot

this problem by integrating surface smoothing and curvature estimation into one unified formalism. This technique was subjected to further testing using noisy meshes. The noise has been introduced by the range-finder and by the process of range image fusion to build 3D models. The results can be observed in [213] and [331].

This technique is not volume preserving in its current form. We do not consider this to be important. However, if it is important in a given application, the object size can be adjusted after each iteration of smoothing to ensure that its volume is preserved.

It is possible to produce disconnected components of a particular initial object during evolution (figure 11.4(e) and (f)). This is a natural outcome of the smoothing process and occurs at points where the ob-

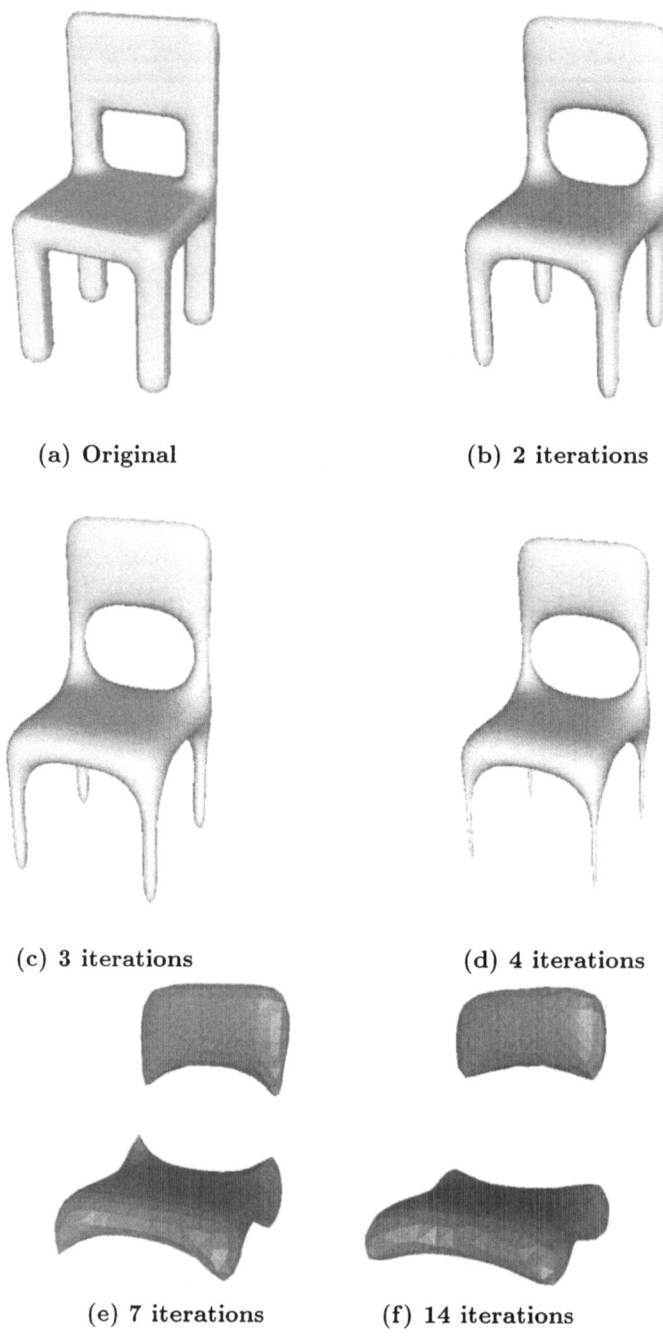

(a) Original   (b) 2 iterations

(c) 3 iterations   (d) 4 iterations

(e) 7 iterations   (f) 14 iterations

Figure 11.4.   Smoothing of the Chair

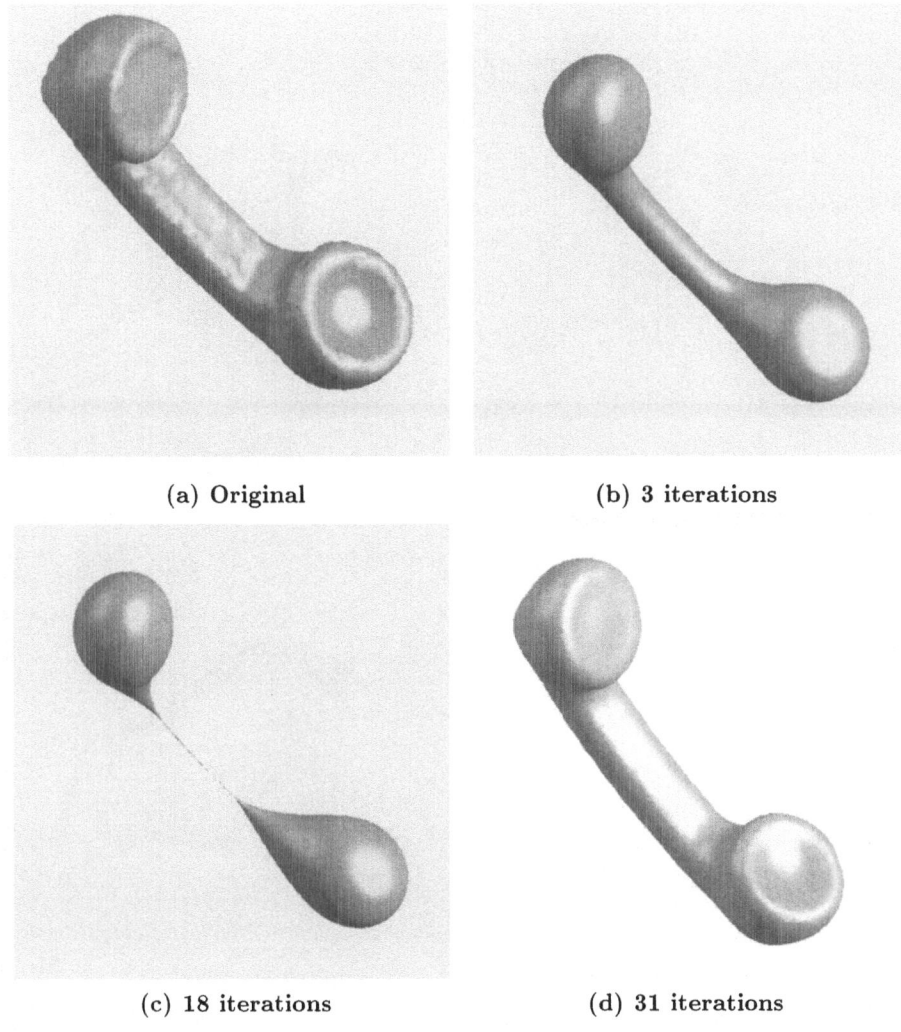

(a) Original   (b) 3 iterations

(c) 18 iterations   (d) 31 iterations

Figure 11.5.   Diffusion of the telephone handset

ject happens to be thin. A decimation algorithm [113] is applied after each iteration of smoothing to remove odd triangles. The same process can segment the object at points where it becomes very thin [136]. We believe that the disconnection of data is desirable when objects become very thin at specific points on their surfaces. This phenomenon can be considered a natural consequence of the multi-scale description of a 3-D object: at larger scales, the smaller/thinner parts of the object tend to disappear, and only the major parts survive. Therefore, this outcome should not be avoided.

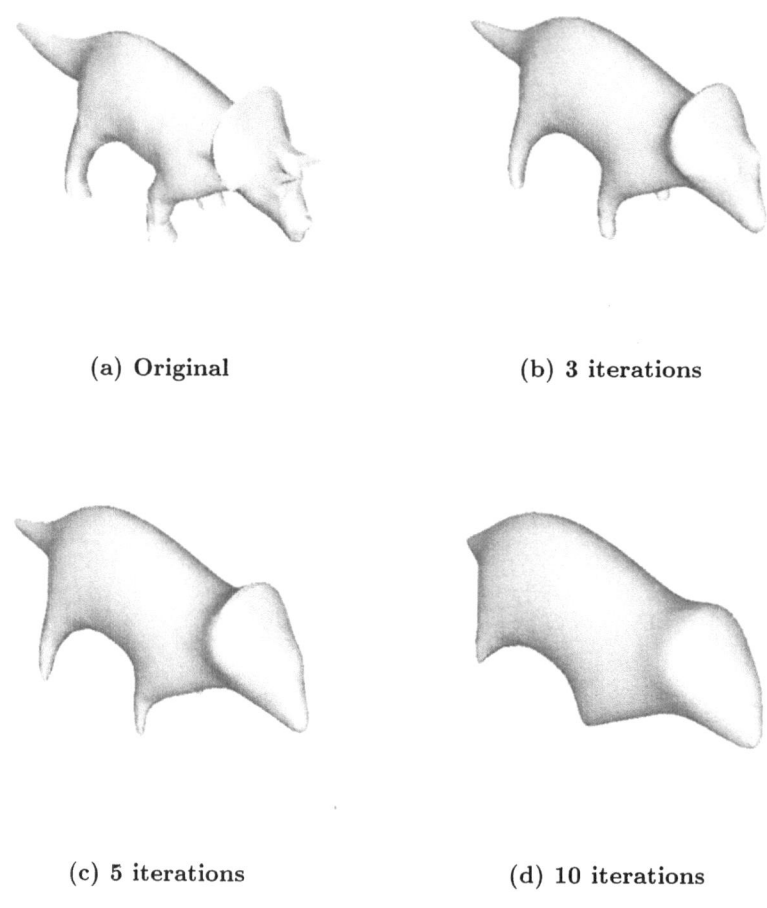

(a) Original   (b) 3 iterations

(c) 5 iterations   (d) 10 iterations

Figure 11.6. Smoothing of the dinosaur

Animation of surface diffusion can be observed at the web site:
http://www.ee.surrey.ac.uk/Research/VSSP/demos/CSS3d/.

6.2.   Curvature Estimation

This section presents the results of application of our curvature estimation techniques to 3-D objects using methods described in section 5.
The diffusion results for other 3-D surfaces were given in [213]. The first
example is a phone handset. After smoothing the object, the Gaussian
curvatures of all vertices were estimated. To visualize these curvature

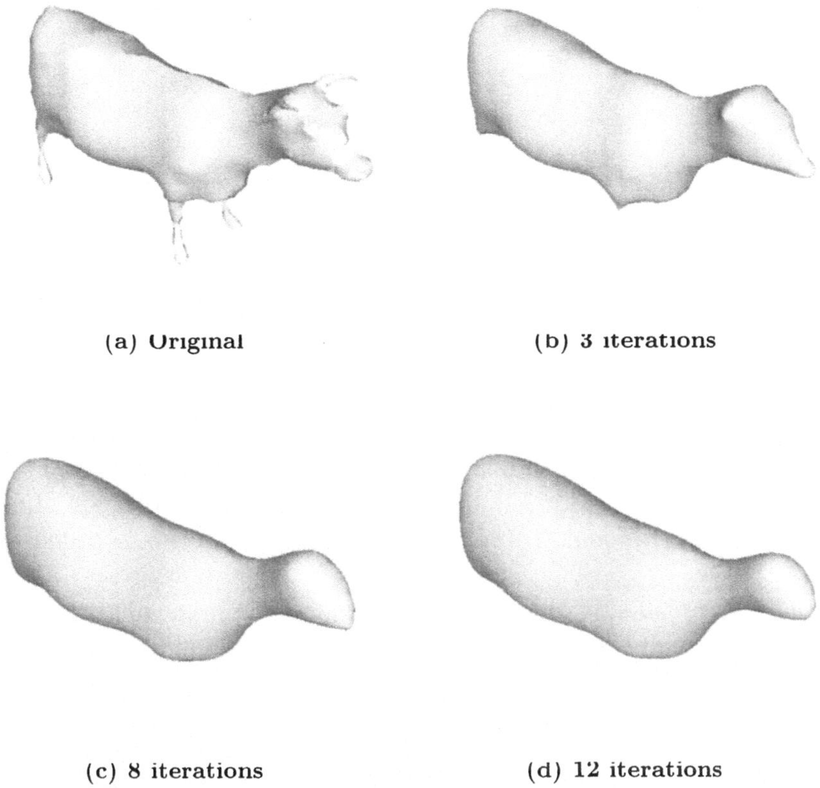

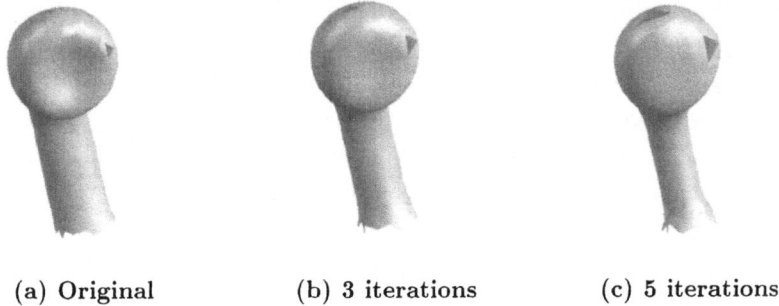

(a) Original   (b) 3 iterations

(c) 8 iterations   (d) 12 iterations

Figure 11.7. Smoothing of the cow

(a) Original   (b) 3 iterations   (c) 5 iterations

Figure 11.8. Diffusion of the partial telephone handset

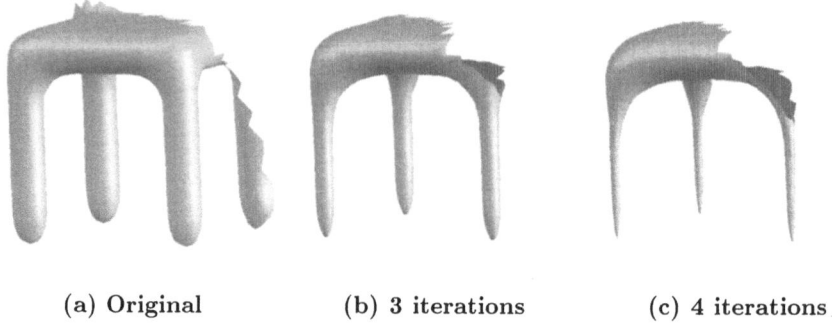

(a) Original	(b) 3 iterations	(c) 4 iterations

Figure 11.9.   Smoothing of the partial chair

values on the surface, they are then mapped to colours/greylevels using the Visualization Tool Kit (VTK) [278], and the results are shown in Figures 11.11(a) and (b). Surface curvature colours are coded as follows:
red = high, blue = low and other colours designate non-extreme values. All convex corners of the phone handset are red, indicating high curvature values, whereas the concave corners are blue indicating low curvature values and flat areas are green since their curvature values are close to zero. The same experiment was repeated to estimate the mean curvatures of the phone handset and the results are shown in Figures 11.11(c) and (d). This indicates that mean curvature values for the edges are different than those for flat areas, as expected.

The next object was a chair. Its Gaussian curvature values were estimated and results are shown in Figures 11.12(a) and (b). These results again confirm that the curvature values are high and low at convex and concave corners, respectively, and close to zero on flat regions such as the seat and the back. The mean curvatures of the chair were also estimated and the results are shown in Figures 11.12(c) and (d). Gaussian and mean curvatures were also estimated for more complex objects, and Figures 11.13, 11.14 and 11.15 show the results for a rabbit, a dinosaur and a cow, respectively,

Next, the curvature zero-crossing contours of these surfaces were found and displayed on the surface using VTK. Curvature zero-crossing contours can be used for segmenting surfaces into regions. Figures 11.16(a) and (b) show Gaussian curvature zero-crossing contours for the smoothed phone handset. Figures 11.16(c) and (d) show mean curvature zero-crossing contours for the same object.

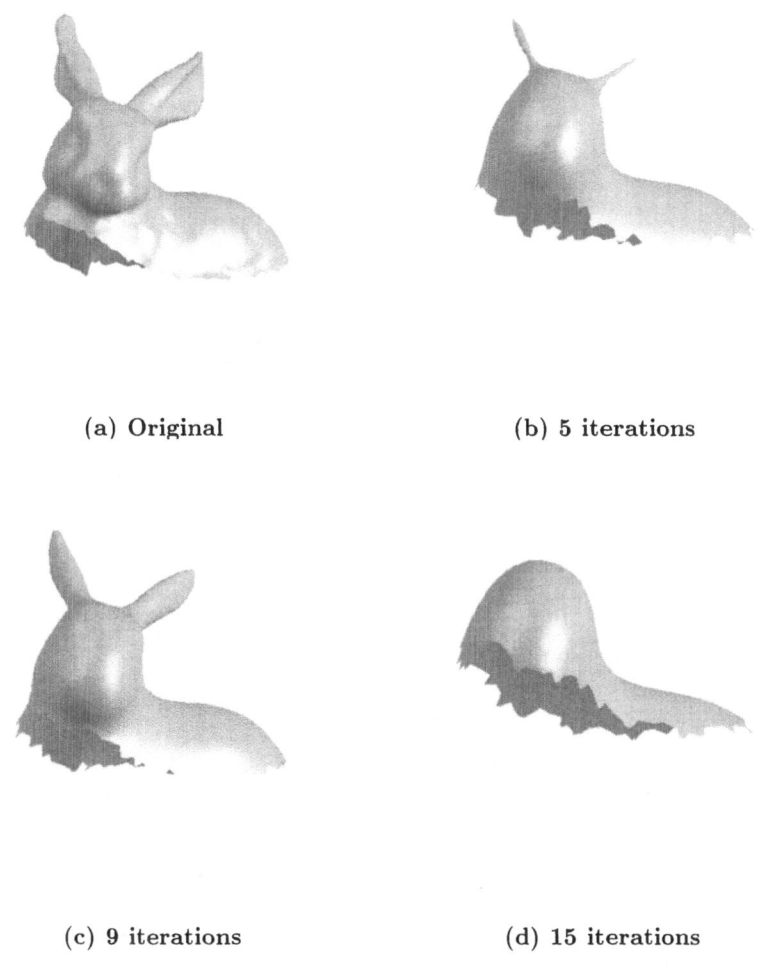

(a) Original   (b) 5 iterations

(c) 9 iterations   (d) 15 iterations

Figure 11.10. Smoothing of the partial rabbit

Figures 11.17(a) and (b) show the Gaussian curvature zero-crossing contours for the smoothed chair, and figures 11.17(c) and (d) show the mean curvature zero-crossing contours for the same object. Figure 11.18 shows Gaussian and mean curvature zero-crossing contours for the smoothed rabbit. The same experiments were also repeated for the dinosaur and cow, and these results are shown in Figure 11.19 and Figure 11.20, respectively. Notice that the number of curvature zero-crossing contours are reduced, as the object is smoothed iteratively.

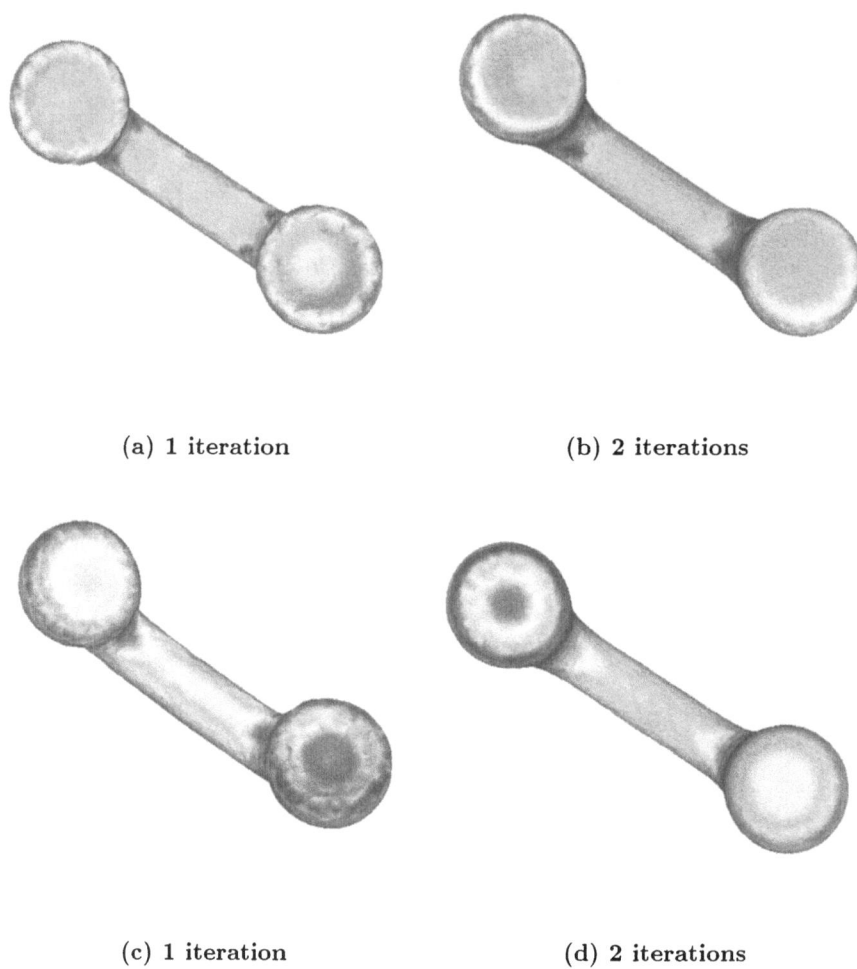

(a) 1 iteration   (b) 2 iterations

(c) 1 iteration   (d) 2 iterations

Figure 11.11. Gaussian (top row) and mean (bottom row) curvatures on the phone handset

Finally, the local curvature maxima for the smoothed telephone hand-set were identified. The local maxima of Gaussian curvature are dis-played on the surface as shown in Figure 11.21(a). Figure 11.21(b) shows the local maxima of mean curvature for the same object. Figure 11.22(a) shows the local maxima of Gaussian curvature for the smoothed chair, and Figure 11.22(b) shows the local maxima of mean curvature for the same object. Note that some of the curvature maxima are not visible due to self-occlusion.

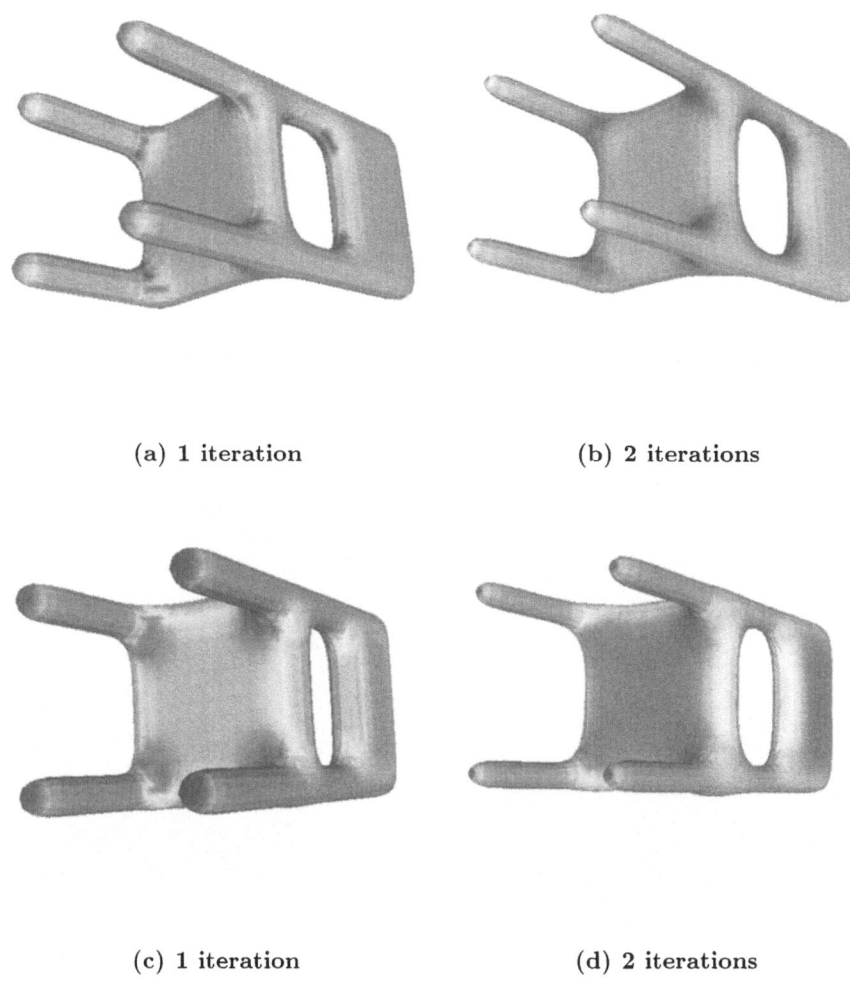

(a) 1 iteration · · · · · · · · · · · · · · · · · · (b) 2 iterations

(c) 1 iteration · · · · · · · · · · · · · · · · · · (d) 2 iterations

Figure 11.12. Gaussian (top row) and mean (bottom row) curvatures on the chair

Figures 11.23 (a) and (b) show the local maxima of Gaussian and mean curvatures for the rabbit. The local maxima of Gaussian and mean curvatures for the dinosaur and cow are also shown in Figure 11.24 and Figure 11.25, respectively. All curvature maxima are shown after one iteration. These features can be utilized by later processes for robust surface matching and object recognition with occlusion. As these examples show, we believe that multi-scale feature recovery is more robust than feature extraction at a single scale [192].

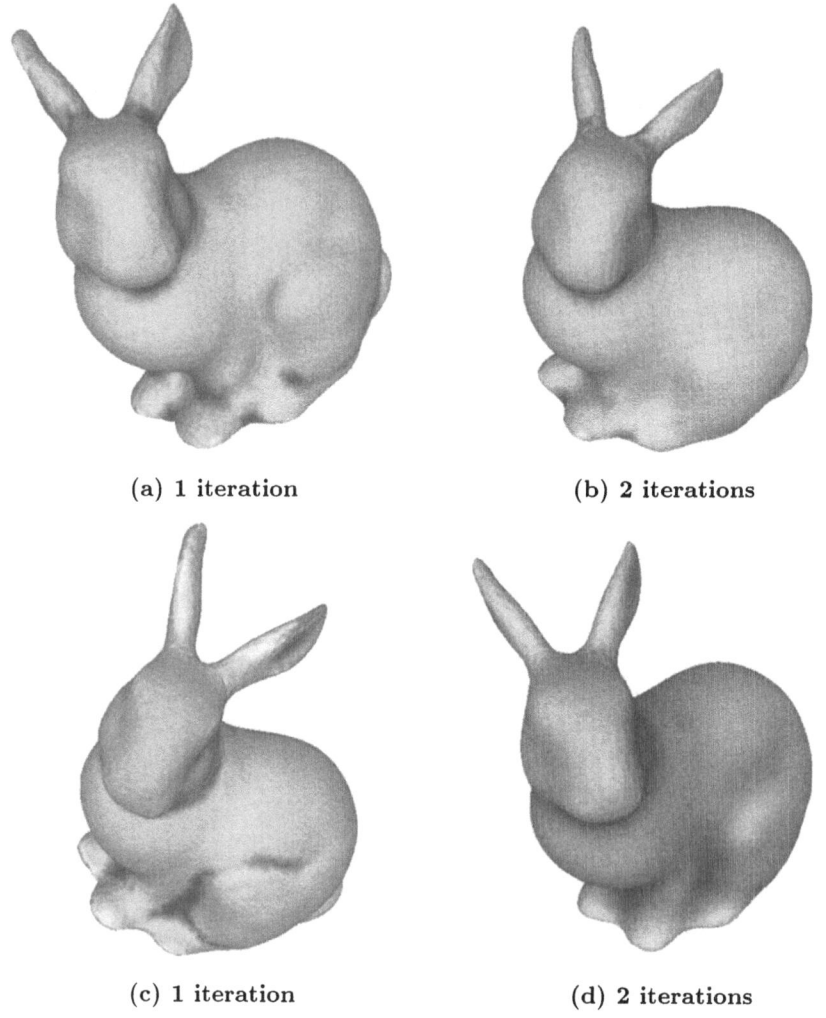

(a) 1 iteration   (b) 2 iterations

(c) 1 iteration   (d) 2 iterations

Figure 11.13.   Gaussian (top row) and mean (bottom row) curvatures on the rabbit

Next, the torsion maxima of curvature zero-crossing contours which are alternative features that can also be used for matching are determined and displayed on the object. A robust method for torsion estimation on free-form space curves was presented in chapter 10. Figures 11.26(a) and (b) show the torsion maxima of curvature zero-crossing contours of the phone handset for Gaussian and mean curvatures, respectively. Figures 11.27 and 11.28 show the results for the dinosaur and cow, respectively.

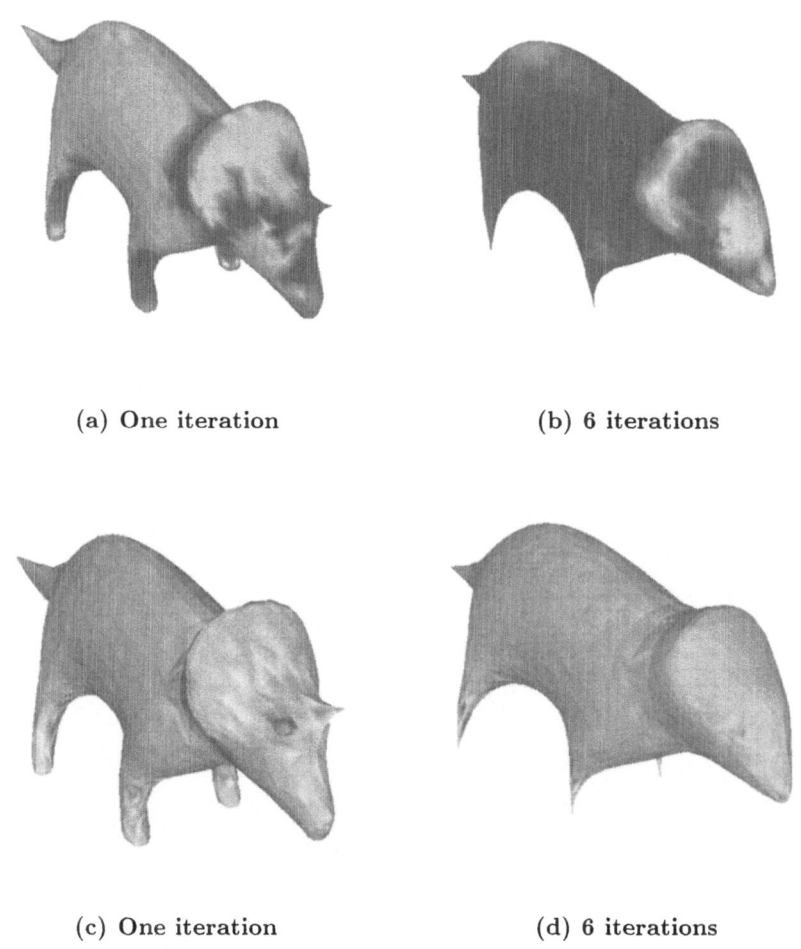

(a) One iteration   (b) 6 iterations

(c) One iteration   (d) 6 iterations

Figure 11.14. Gaussian (top row) and mean (bottom row) curvatures on the dinosaur

Table 11.1 contains a comparison of our multi-scale technique for free-form 3-D surface description to other relevant methods.

7.   Estimation of Error in Curvature Computation

Gaussian and mean curvatures can be defined in terms of principal curvatures k_1 and k_2 as following:

$$K = k_1 k_2$$

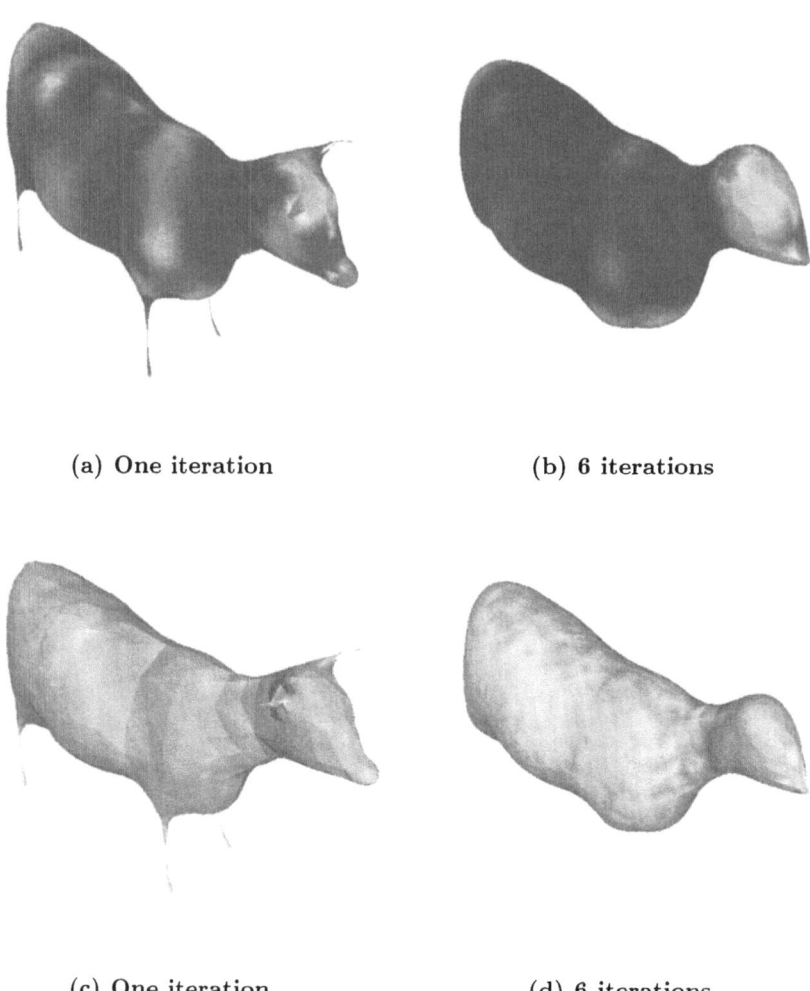

(a) One iteration   (b) 6 iterations

(c) One iteration   (d) 6 iterations

Figure 11.15.   Gaussian (top row) and mean (bottom row) curvatures on the cow

and

$$H = \frac{k_1 + k_2}{2}$$

Now if $k_1 = k_1 + \varepsilon$ and $k_2 = k_2 + \varepsilon$, where ε represents error ($\varepsilon \ll 1$), then Gaussian curvature is given by:

$$K = k_1 k_2 + \varepsilon(k_1 + k_2) + \varepsilon^2 \tag{11.3}$$

and mean curvature H is given by:

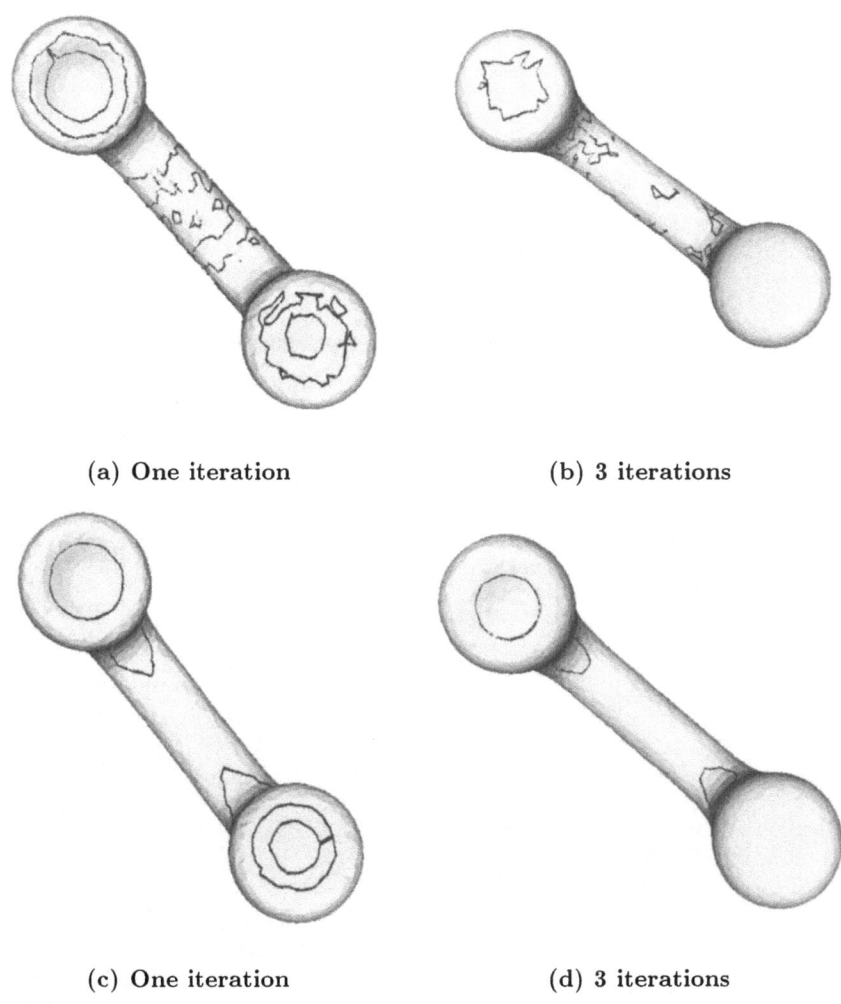

(a) One iteration   **(b) 3 iterations**

(c) One iteration   **(d) 3 iterations**

Figure 11.16. Gaussian (top row) and mean (bottom row) curvature zero-crossing contours on the phone handset

$$H = \frac{k_1 + k_2}{2} + \varepsilon \qquad (11.4)$$

Since k_1 and k_2 are very small values (object sizes are quite large to avoid numerical problems) and

$$\varepsilon(k_1 + k_2) + \varepsilon^2 \ll \varepsilon$$

it follows that for an error in the values of principal curvatures k_1 and k_2, the error introduced in Gaussian curvature is expected to be smaller than

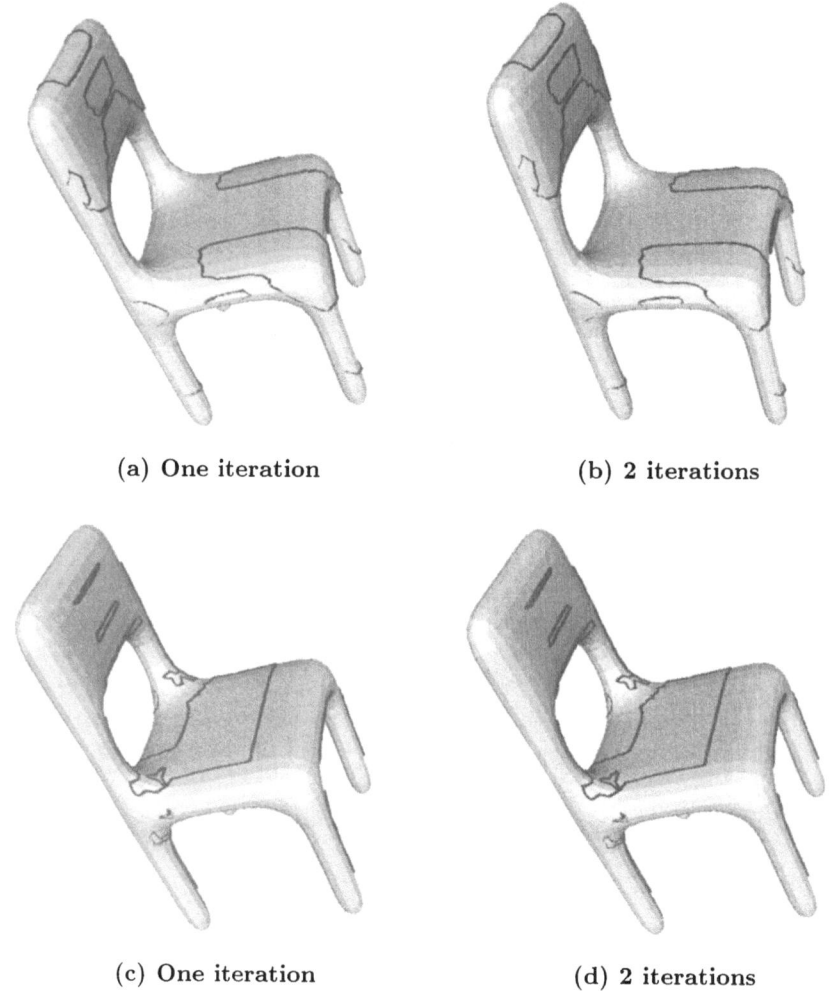

(a) One iteration   (b) 2 iterations

(c) One iteration   (d) 2 iterations

Figure 11.17. Gaussian (top row) and mean (bottom row) curvature zero-crossing contours on the chair

that of mean curvature. Note that Gaussian and mean curvatures have different units. The point of this analysis was to demonstrate that the numerical value of the estimation error in Gaussian curvature is smaller than that of mean curvature. On smoothed surfaces of 3-D objects, the procedure for curvature estimation error is as follows:

For each vertex the curvature values are computed for all possible directions of the first geodesic line, and the average curvature value is then used as the correct value of curvature for that vertex. Then, the error in direction i is given by:

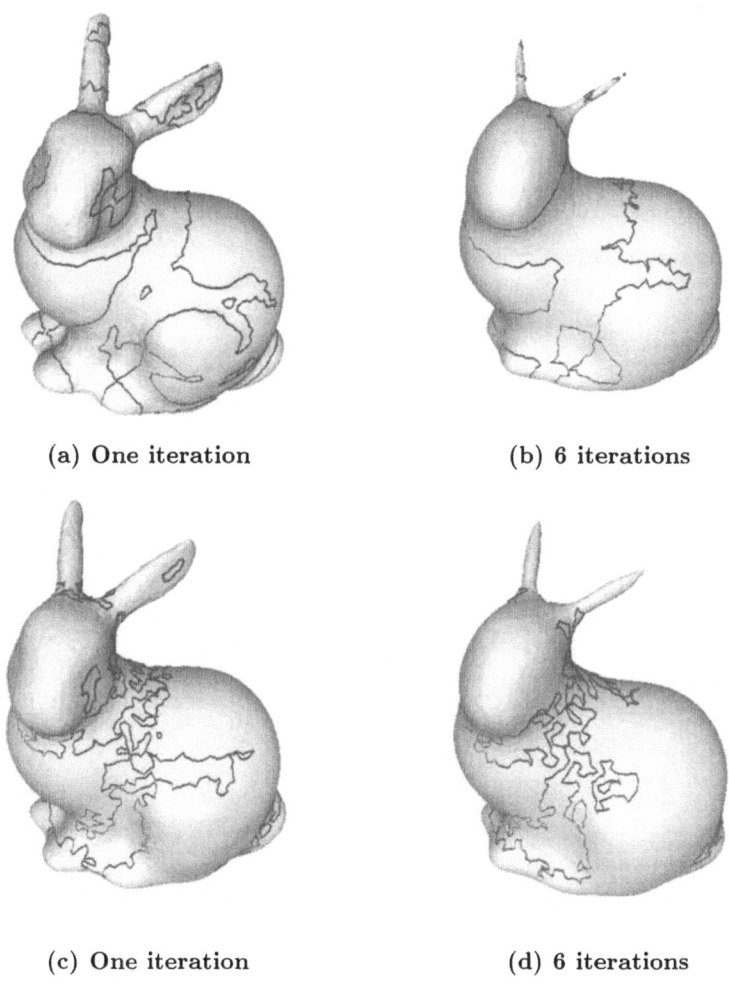

(a) One iteration   (b) 6 iterations

(c) One iteration   (d) 6 iterations

Figure 11.18. Gaussian (top row) and mean (bottom row) curvature zero-crossing contours on the rabbit

$$error_i = \frac{|\bar{k} - k_i|}{|\bar{k}|} \tag{11.5}$$

where k_i is the curvature values for direction i, and $\bar{k}$ is the average curvature value.

We first applied our curvature estimation technique to a surface with known curvature values. Our method was tested on a 3-D mesh representing a sphere. It was confirmed that Gaussian and mean curvature values estimated at the vertices were approximately equal. The errors

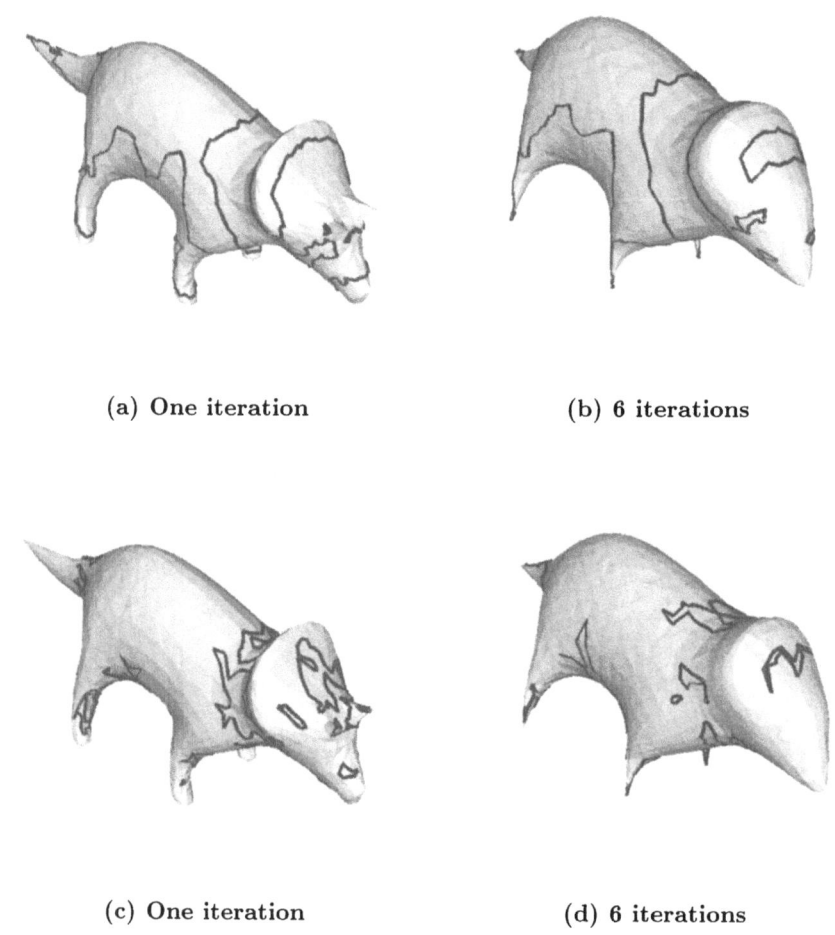

(a) One iteration   (b) 6 iterations

(c) One iteration   (d) 6 iterations

Figure 11.19.   Gaussian (top row) and mean (bottom row) curvature zero-crossing contours on the dinosaur

observed were in agreement with expectations (see below). However, we believe that the use of objects with known curvatures is not a satisfactory method of testing our method. Our technique was designed for curvature estimation on noisy, free-form meshes with unknown curvature values. Objects with known curvatures usually have simple shapes. As a result, it is not difficult to devise techniques for curvature estimation on those objects which will perform well by exploiting constraints about object shape. It should be noted that the best evidence in favor

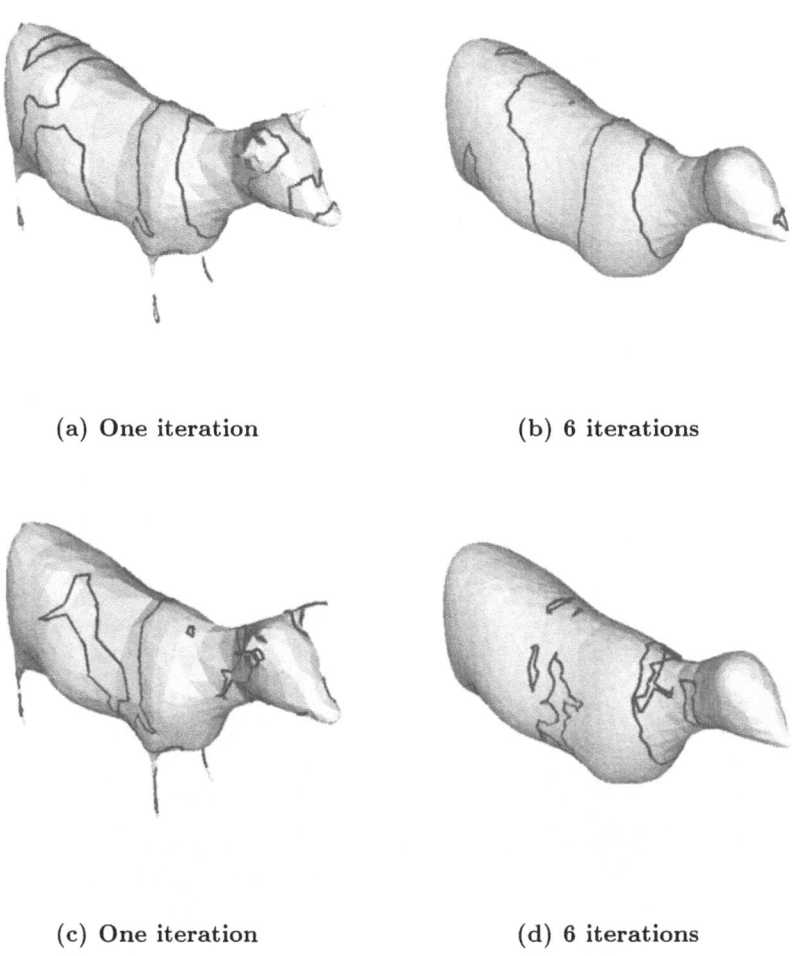

(a) One iteration   (b) 6 iterations

(c) One iteration   (d) 6 iterations

Figure 11.20. Gaussian (top row) and mean (bottom row) curvature zero-crossing contours on the cow

of the accuracy of our method is the comprehensive work on 1-D and 2-D approaches to numeric curvature estimation and their comparative performance assessment [75, 76, 63].

As mentioned earlier, in our method the direction of the first geodesic line is randomly selected. We will therefore examine curvature estimation error for *all* possible directions. The smoothing procedure was repeated where for each vertex all possible directions for the first geodesic line were constructed and different step sizes were considered. Specifically, the direction of each incident edge was defined as a possible di-

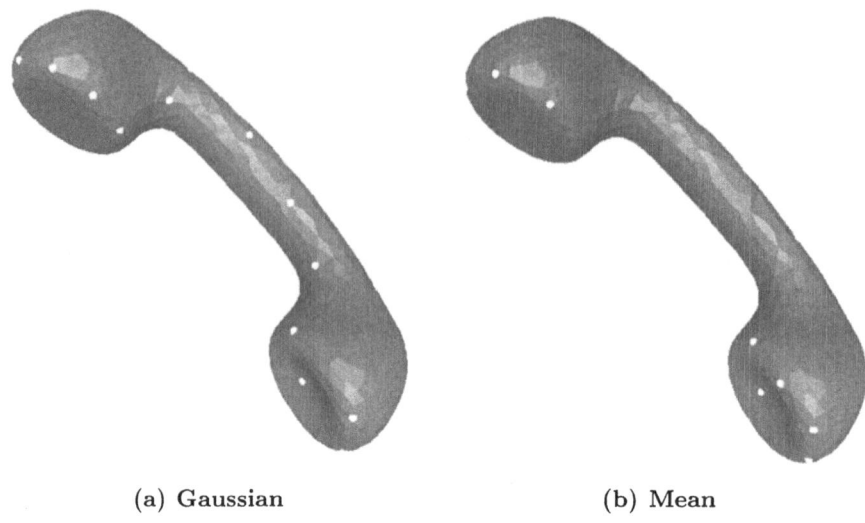

(a) Gaussian (b) Mean

Figure 11.21. Curvature maxima of the phone handset

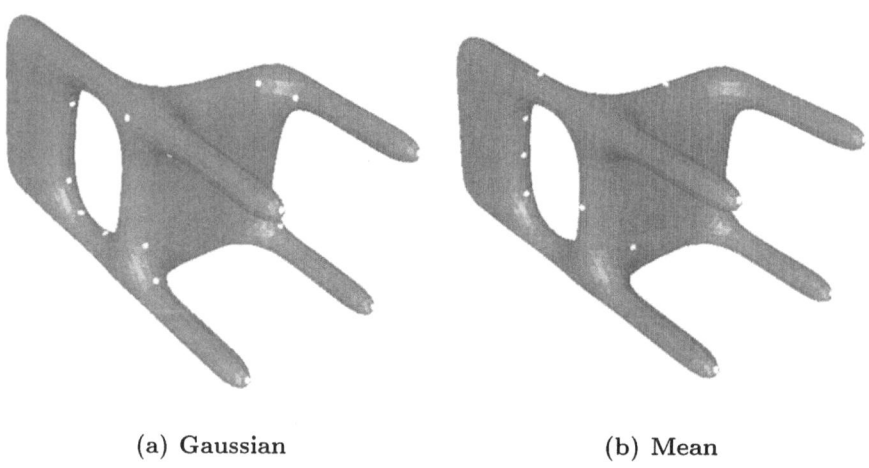

(a) Gaussian (b) Mean

Figure 11.22. Curvature maxima of the chair

rection. After the object was smoothed, for each vertex the curvature values were computed for all directions and the curvature errors were estimated using equation (11.5). Then, the maximum, minimum and average values of the curvature errors were computed. The first test object was the foot. Figure 11.29(a) shows the maximum error distribution for estimating Gaussian curvature with the step size varying from 500 to

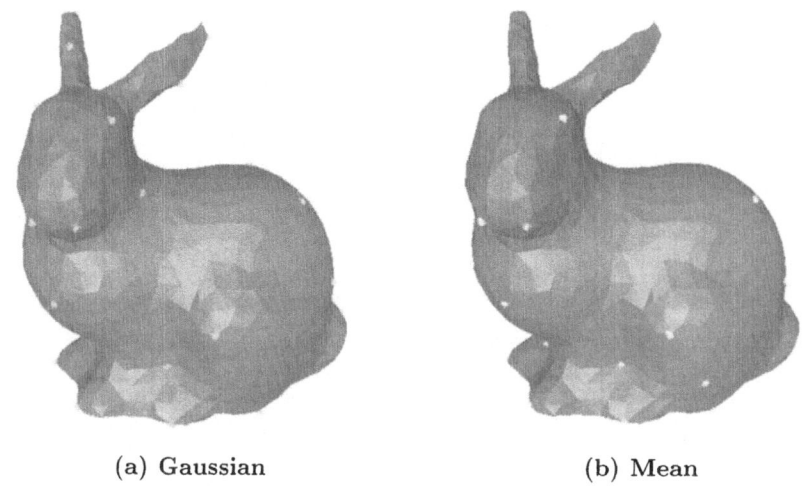

(a) Gaussian   (b) Mean

Figure 11.23. Curvature maxima of the rabbit

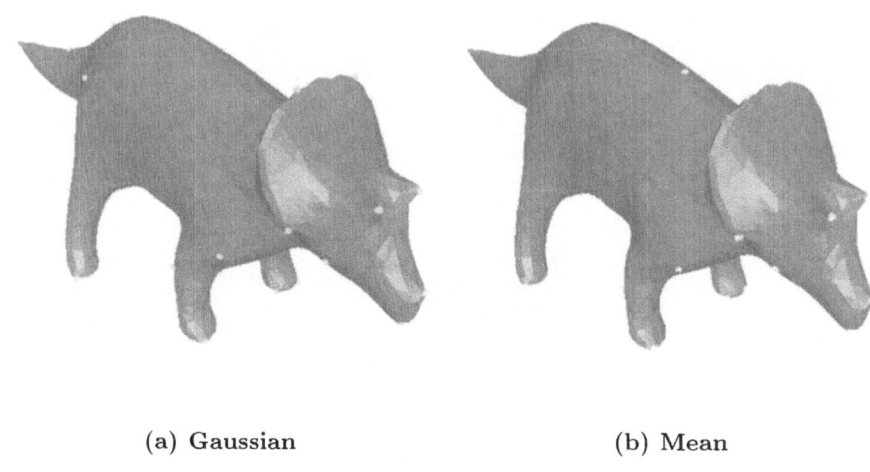

(a) Gaussian   (b) Mean

Figure 11.24. Curvature maxima of the dinosaur

3000. These results indicate that for the step size between 1000 to 2000 the error is reduced to about 1.0% after one iteration. We then repeated the experiment such that for each vertex the minimum error in curvature values of all possible directions were computed and results are shown in Figure 11.29(b). These indicate that for the step size between 1000 to 2000 the error is reduced to about 0.5% after one iteration. When aver-

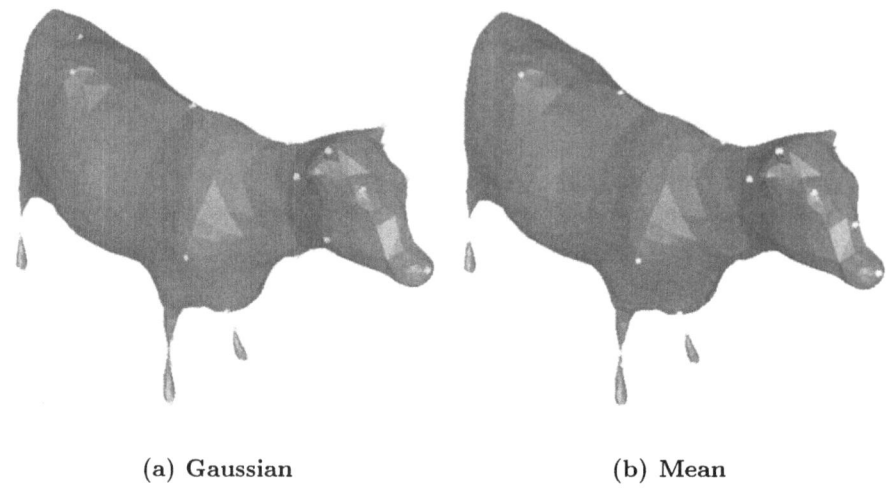

(a) Gaussian   (b) Mean

Figure 11.25.   Curvature maxima of the cow

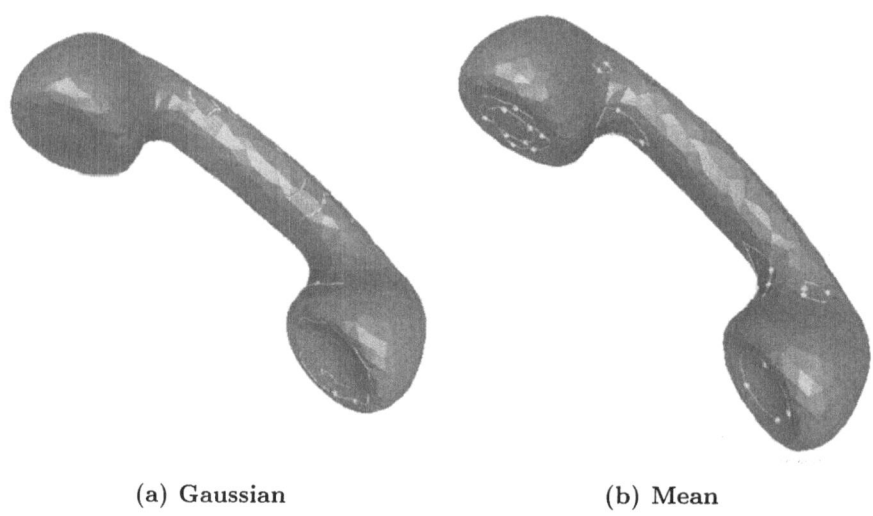

(a) Gaussian   (b) Mean

Figure 11.26.   Torsion maxima of curvature zero-crossing contours of the phone

age curvature value of all possible directions was calculated the average curvature error was about 0.75% after one iteration, as shown in Figure 11.29(c). However, as the surface becomes smooth iteratively, the errors reduce as shown in Figure 11.29. After 100 iterations the errors in maximum, minimum and average curvature values are reduced to 0.55%, 0.35% and 0.45%, respectively.

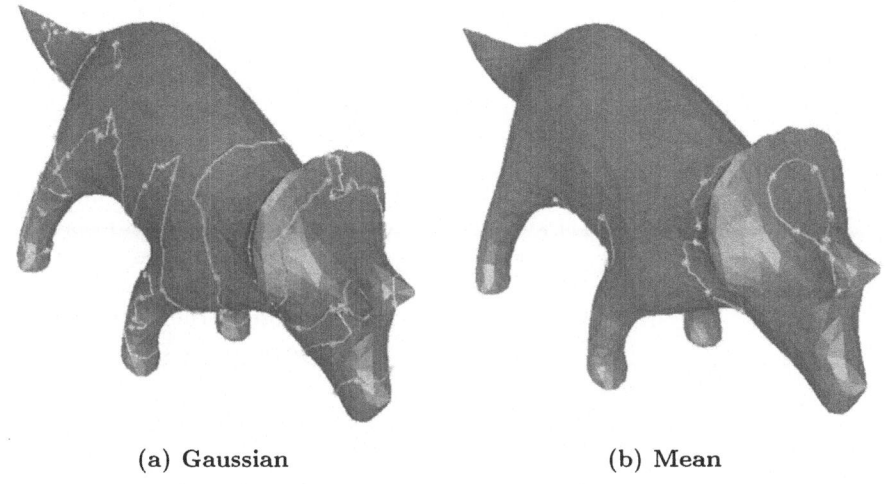

(a) Gaussian   (b) Mean

Figure 11.27. Torsion maxima of curvature zero-crossing contours of the dinosaur

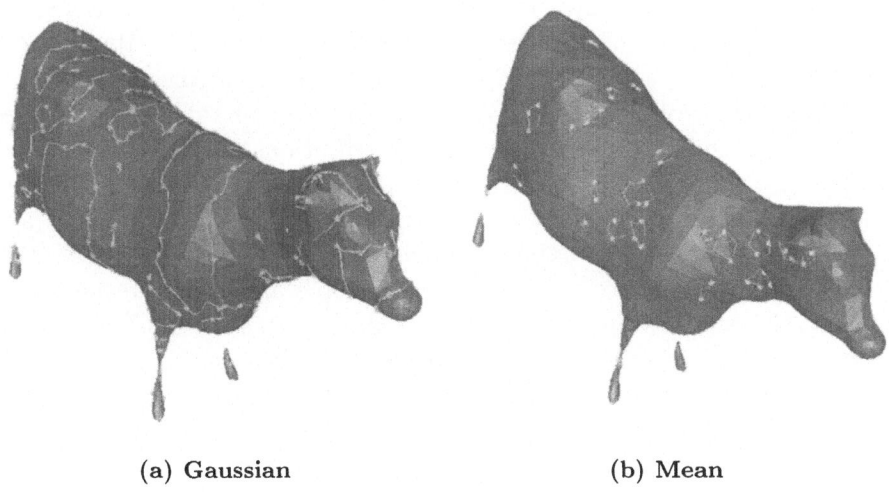

(a) Gaussian   (b) Mean

Figure 11.28. Torsion maxima of curvature zero-crossing contours of the cow

The above procedures were also repeated for estimation of mean curvatures. Figures 11.30(a) and (b) show the error distributions for the estimation of mean curvatures. Again, the errors are reduced for the step sizes between 1000 to 2000, which are about 3.0% and 1.0% for the maximum and minimum errors in mean curvatures, respectively. The average value of error in mean curvature is about 2.0% as shown

Technique	Short Analysis
Range image analysis	Applies only to a single-view range image but not a free-form 3-D mesh structure.
Mesh neighborhood weighted averaging	(1) Outcome depends on how the surface is triangulated. (2) Does not provide a way for curvature estimation.
Level set methods and volumetric diffusion	(1) They lack local support so not suitable for incomplete surfaces / object recognition with occlusion. (2) Inefficient since problem dimensionality is increased.
Mesh simplification	(1) Mesh vertices not robust to noise/shape distortions. (2) Does not provide a way for curvature estimation.
Proposed method	(1) Combines multi-scale description and curvature estimation. (2) Applicable to noisy, free-form, incomplete meshes. (3) Efficient and robust to mesh geometry.

Table 11.1. Comparison to other techniques

in Figure 11.30(c). As the surface becomes smooth iteratively, the errors are reduced as shown in Figure 11.30 and after 100 iterations, the maximum, minimum and average values of error drop to about 2.2%, 0.5% and 1.4%, respectively. Notice that the error in the estimation of Gaussian curvatures are lower than that of mean curvature as discussed earlier. Figure 11.31 shows the error distributions for estimating Gaussian curvatures of the rabbit when all possible directions were selected. Again the errors were reduced for step sizes between 1000 to 2000 and for one iteration, the maximum, minimum and average values of error in Gaussian curvatures are about 1.23%, 0.96% and 1.15%, respectively. After 24 iterations these errors reduce to about 0.92%, 0.5% and 0.7%, respectively. Figure 11.32 also shows the error distributions for estimating mean curvature of the rabbit and these results also indicate that the maximum, minimum and average values of error are about 2.85%, 2.7% and 2.8%, respectively. After 24 iterations these errors reduce to about 2.1%, 1.6% and 1.9%, respectively.

Figure 11.33 shows the error distribution for estimating Gaussian curvatures of the dinosaur when all possible directions were selected. The errors were reduced for step sizes between 1000 to 2500 and for one iteration, the maximum, minimum and average values of error in Gaussian curvatures are about 0.94%, 1.23% and 1.10%, respectively. After 17 iterations these errors reduce to about 0.52%, 0.92% and 0.7%, respectively. Figure 11.34 also shows the error distributions for estimating mean curvature of the dinosaur and these results also indicate that the maximum, minimum and average values of error are about 2.60%, 2.72%

and 2.65%, respectively. After 17 iterations these errors reduce to about 2.0%, 1.9% and 1.92%, respectively. Figure 11.35 shows the error distributions for estimating Gaussian curvatures of the cow when all possible directions were selected. Again the errors were reduced for step sizes between 1000 to 2000 and for one iteration, the maximum, minimum and average values of error in Gaussian curvatures are about 1.29%, 1.0% and 1.1%, respectively. After 15 iterations these errors reduce to about 1.0%, 0.54% and 0.76%, respectively. Figure 11.36 further shows the error distribution for estimating mean curvature of the cow and these results also indicate that the maximum, minimum and average values of error are about 2.70%, 2.57% and 2.62%, respectively. After 15 iterations these errors reduce to about 2.07%, 1.96% and 2.0%, respectively.

Our experiments indicate that estimation of Gaussian and mean curvatures on smoothed surfaces is very accurate and not affected by the arbitrary direction of the first geodesic line when constructing semi-geodesic coordinates. smoothing is necessary to remove noise from the surface before curvature can be estimated reliably. While smoothing might cause a displacement of features, since scale changes slowly, the resulting displacement is small, and therefore, it is not difficult to determine the correspondence between the features across scales.

8. Robust Free-Form 3-D Object Recognition

This section presents a system for the recognition of free-form 3-D objects using 3-D models under different viewing conditions based on geometric hashing and global verification [218].

The matching stage of the algorithm uses the hash table prepared in the off-line stage. Given a scene of feature points, one tries to match the measurements taken at scene points to those memorized in the hash table. The techniques described earlier in this chapter were used for feature recovery from 3-D surfaces. Smoothing is used to remove noise and reduce the number of feature points to add to the efficiency and robustness of the system. The local maxima of Gaussian and mean curvatures are selected as feature points. Furthermore the torsion maxima of the zero-crossing contours of Gaussian and mean curvatures are also selected as feature points. Recognition results are demonstrated for rotated and scaled as well as partially occluded objects. In order to verify match, 3D translation, rotation and scaling parameters are used for verification and results indicate that our technique is invariant to those transformations. Our technique for smoothing and feature extraction is more suitable than level set methods or volumetric diffusion for object recognition applications since it is applicable to incomplete surface data that

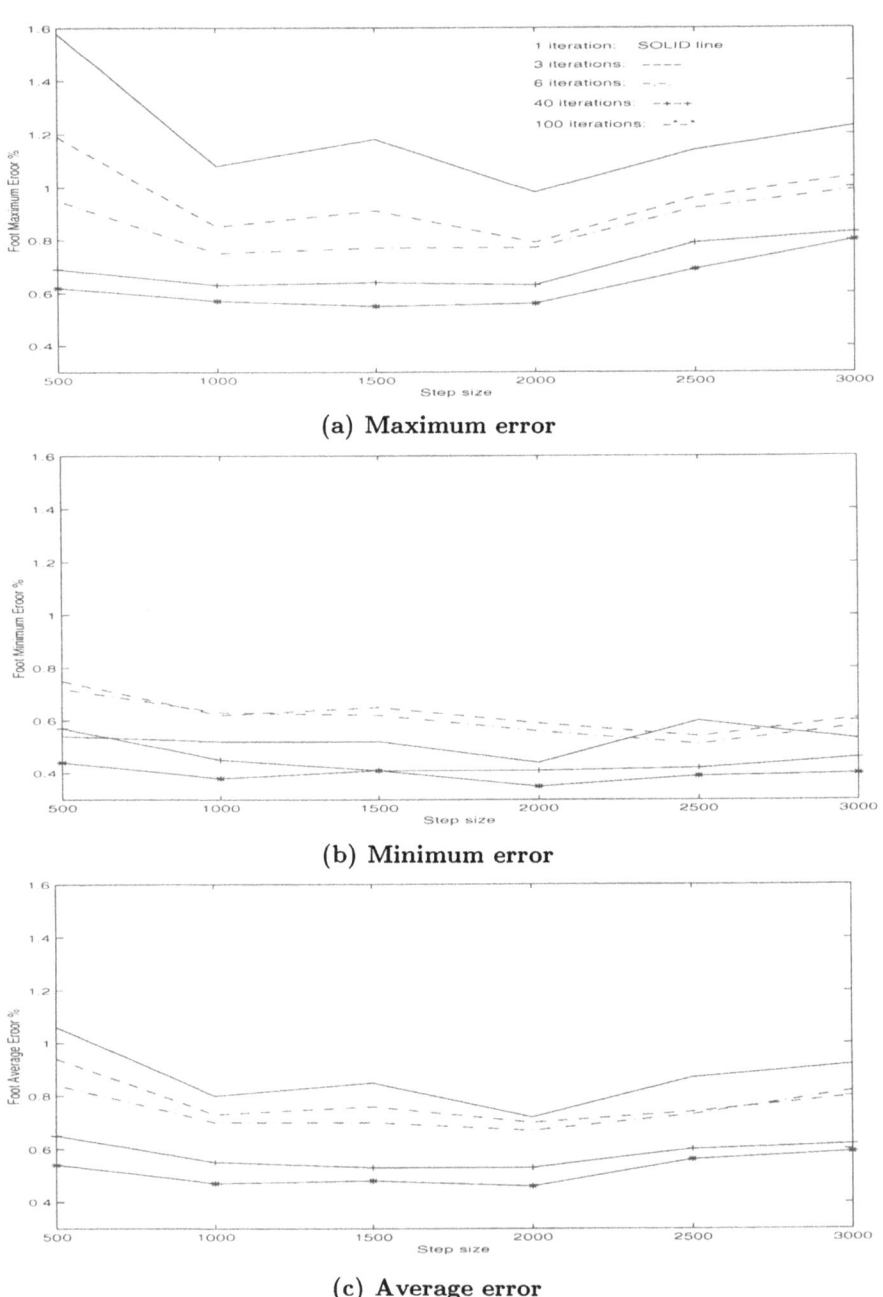

(a) Maximum error

(b) Minimum error

(c) Average error

Figure 11.29.   Gaussian curvature error distribution of the foot

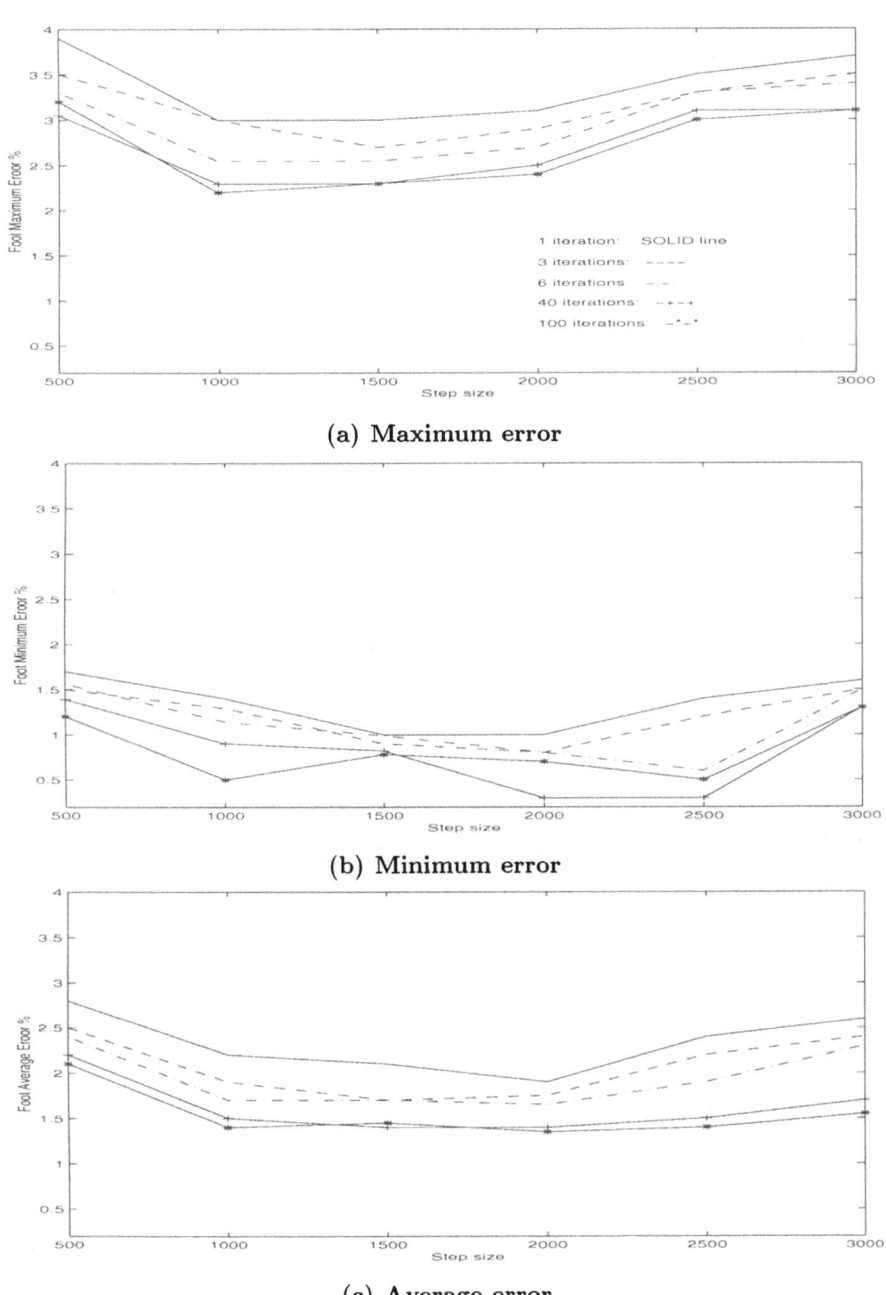

(a) **Maximum error**

(b) **Minimum error**

(c) **Average error**

Figure 11.30. Mean curvature error distribution of the foot

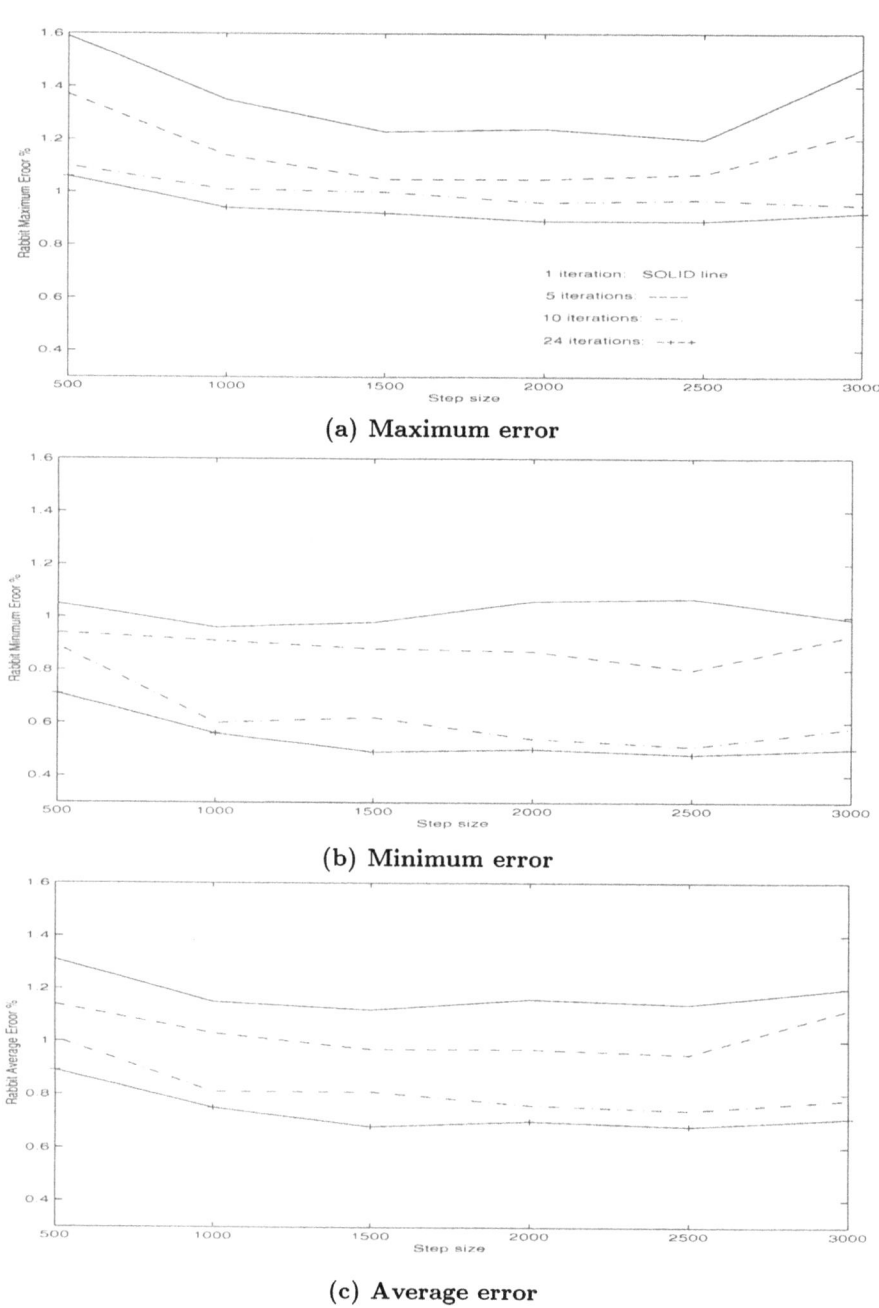

(a) Maximum error

(b) Minimum error

(c) Average error

Figure 11.31. Gaussian curvature error distribution of the rabbit

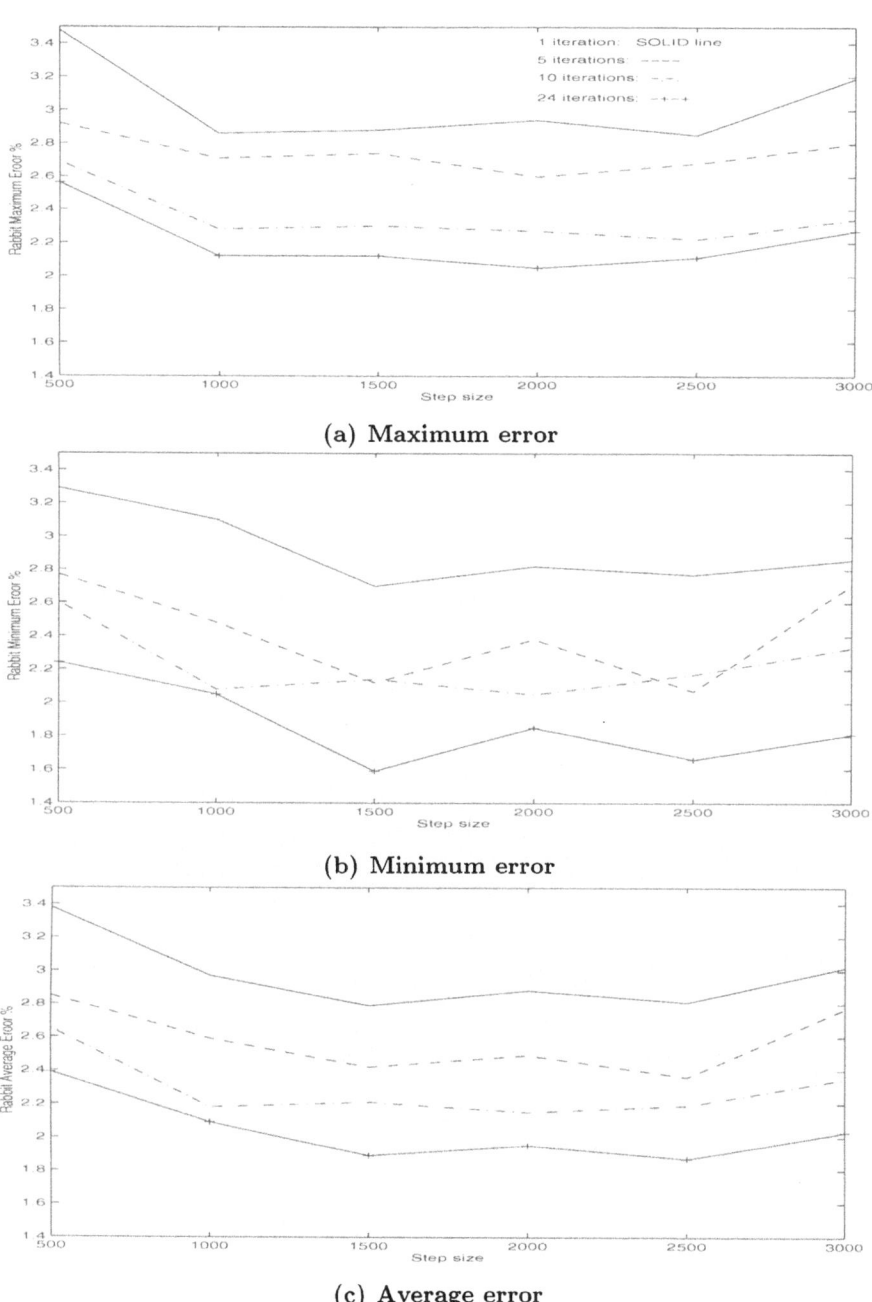

(a) **Maximum error**

(b) **Minimum error**

(c) **Average error**

Figure 11.32.   Mean curvature error distribution of the rabbit

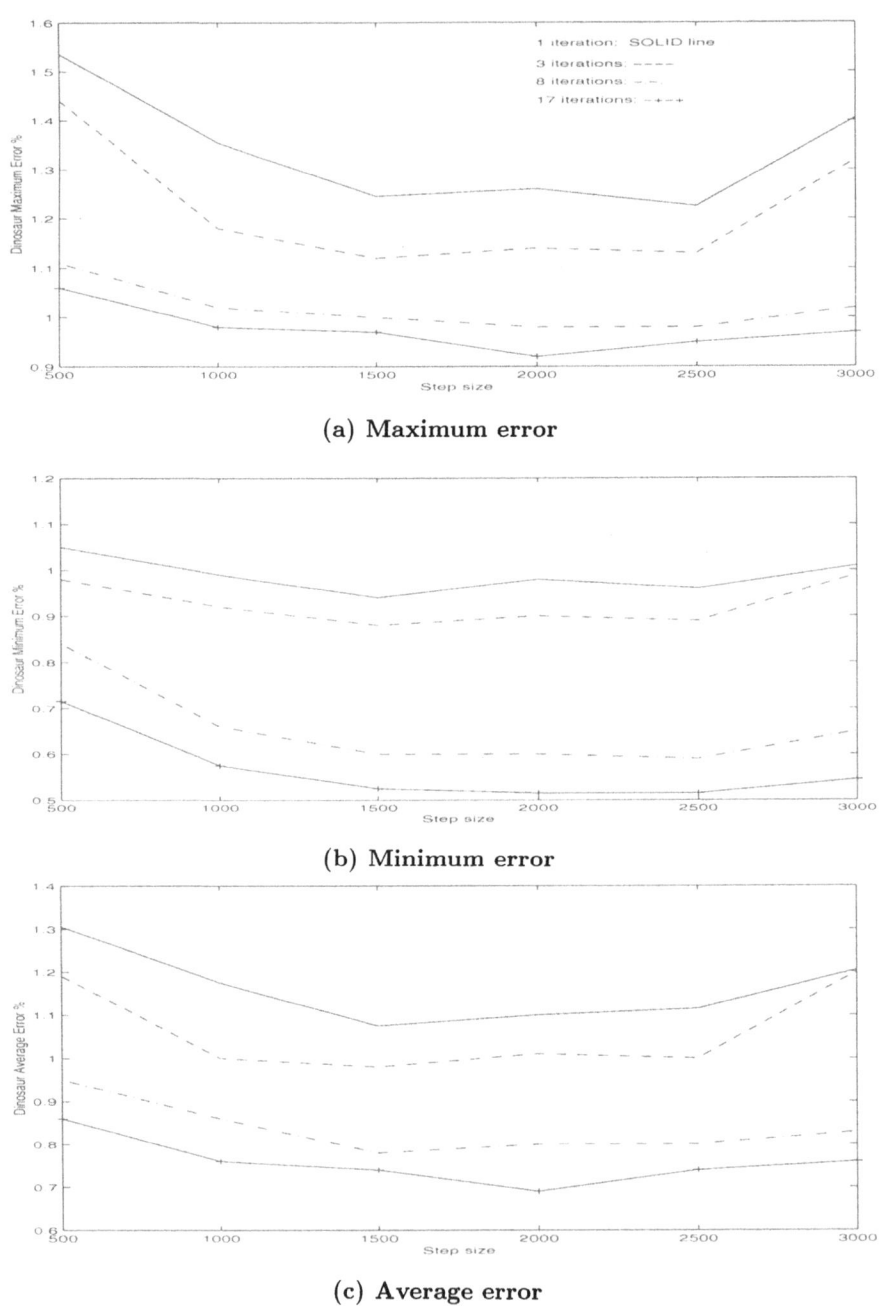

(a) Maximum error

(b) Minimum error

(c) Average error

Figure 11.33.   Gaussian curvature error distribution of the dinosaur

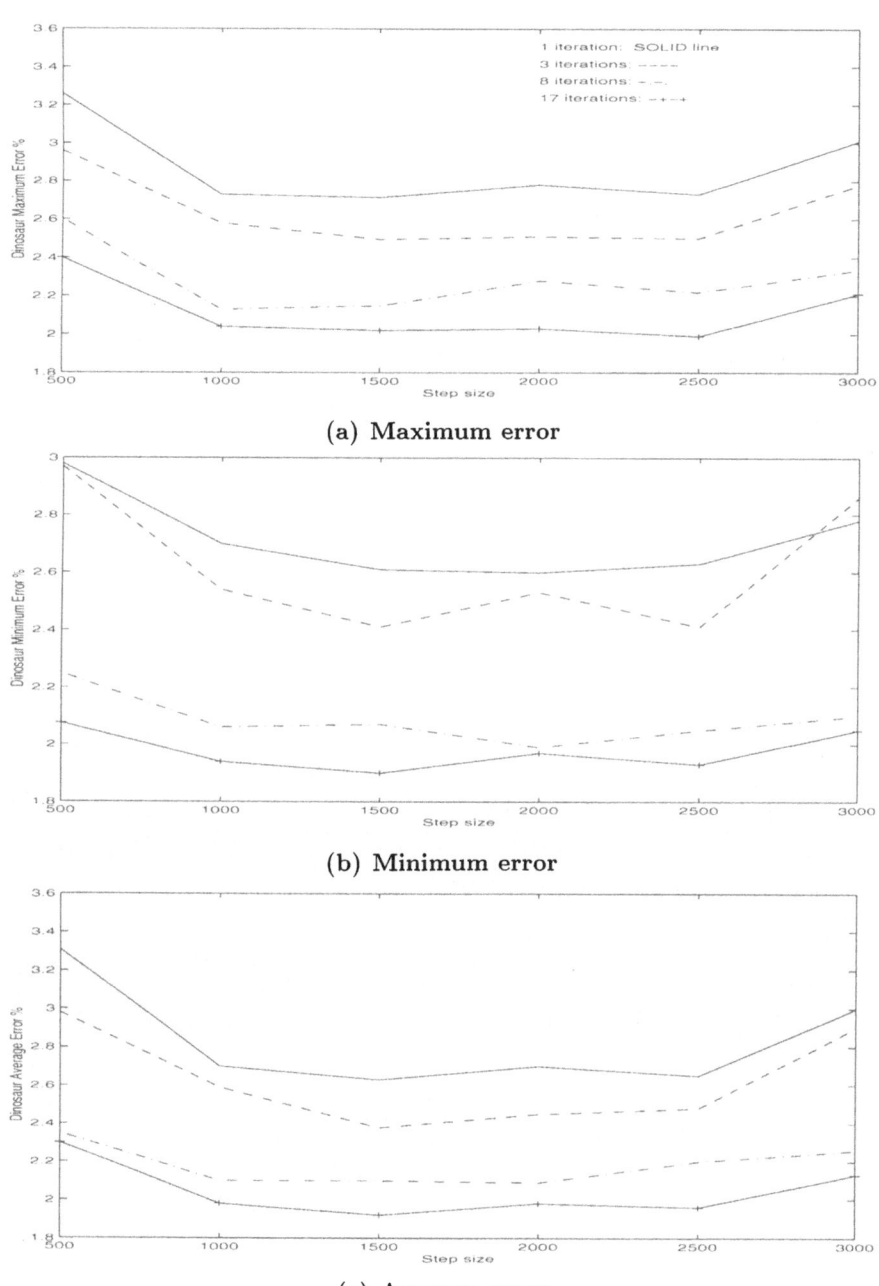

(a) Maximum error

(b) Minimum error

(c) Average error

Figure 11.34.   Mean curvature error distribution of the dinosaur

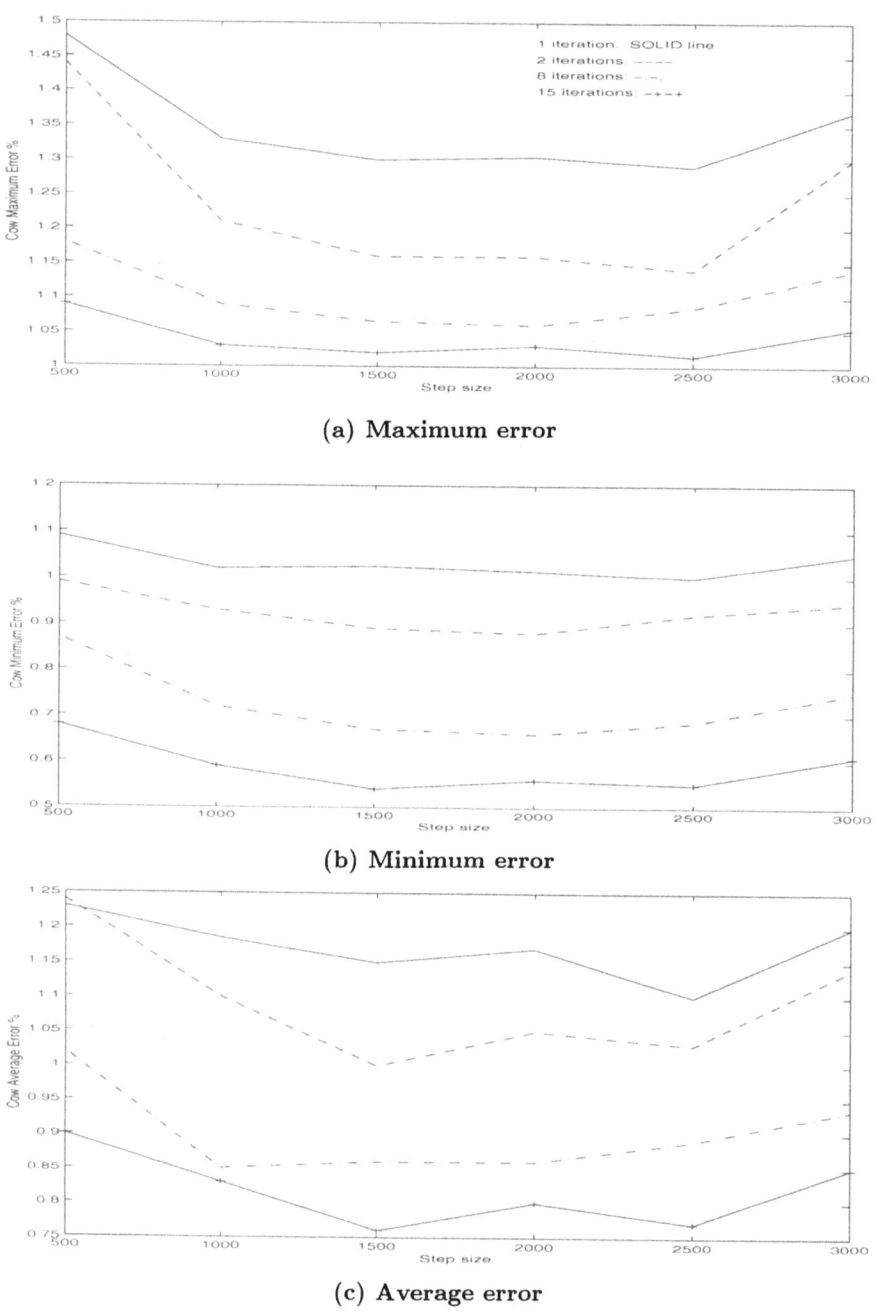

(a) Maximum error

(b) Minimum error

(c) Average error

Figure 11.35.   Gaussian curvature error distribution of the cow

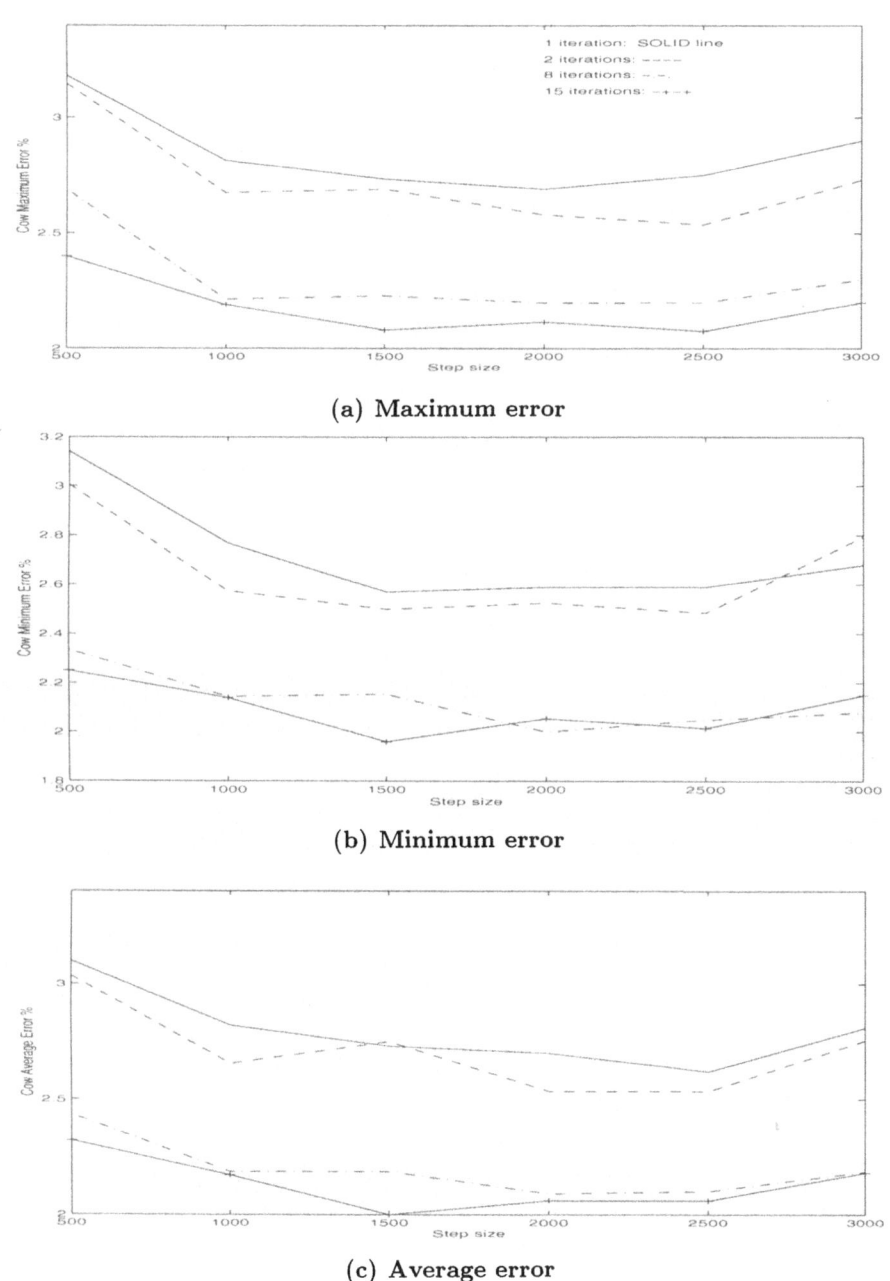

(a) **Maximum error**

(b) **Minimum error**

(c) **Average error**

Figure 11.36. Mean curvature error distribution of the cow

arises during occlusion. It is also more efficient and allows for accurate estimation of curvature values.

Object recognition is a major task in computer vision. Surface curvature provides a unique viewpoint invariant description of local surface shape. Differential geometry [95] provides several measures of curvature, which include Gaussian and mean curvatures. Combination of these curvature values enable the local surface type to be categorized.

In this section the recognition of free-form 3-D objects using 3-D models based on the geometric hashing technique and global verification is addressed. This technique is useful for partially occluded objects. The model information is indexed into a hash-table using minimal transformation invariant features. Some interest features are extracted from the objects, so that both the model object and the observed scene can be represented by sets of these interest features. The recognition of a partially occluded object in a scene amounts to the discovery of a match between a subset of the scene interest features and a subset of the interest features of some model object. The recognition time depends directly on the complexity of the scene to be recognized.

The section is concluded with examples showing 3-D objects with their local curvature maxima as well as the maxima of torsion of zero-crossing contours of Gaussian and mean curvatures, located and displayed on the surface. Furthermore, matching results for 3-D surfaces with arbitrary shapes in a database will be presented.

8.1.   The Geometric Hashing Algorithm

Geometric hashing technique for model based object recognition was introduced in [152, 154]. Geometric hashing has also been employed for 3-D object recognition in [296, 85]. In a model based object recognition system, one has to address representation and matching problems. The representation should be rich enough to allow reliable distinction between the different objects in the database as well as efficient matching. A major factor in a reliable representation scheme is its ability to deal with partial occlusion. The objects are represented as sets of geometric features such as points, and their geometric relations are encoded using minimal sets of such features under the allowed object transformations.

The matching stage of the algorithm uses the hash table prepared in the off-line stage. Given a scene of feature points, one tries to match the measurements taken at scene points to those memorized in the hash table. On smoothed surfaces of 3-D objects, the procedure for indexing data into the hash table is as follows. For each 3-D object in the database:

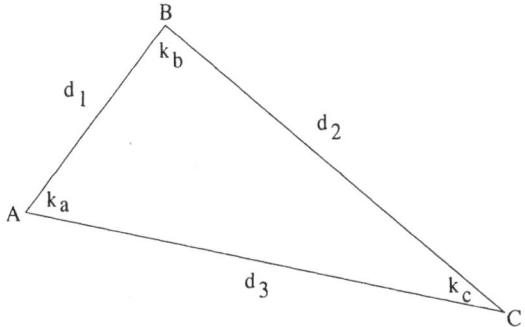

Figure 11.37. Triplet of non-collinear points A, B and C

1 The local maxima of Gaussian curvature are selected as feature points. Furthermore, local maxima of mean curvature and the torsion maxima of the zero-crossing contours of Gaussian and mean curvatures are also selected as feature points.

2 Choose an arbitrary ordered triplet of non-collinear points A, B and C to form a triangle ABC. Denote the curvature values of points A, B and C by k_a, k_b and k_c, and the edge lengths AB, BC and AC as d_1, d_2 and d_3, respectively, see Figure 11.37.

Select the maximum curvature value and edge length. Let k_a and d_1 be maximum curvature value and edge length. Calculate the indexed value IV for the hash table as:

$$IV = \frac{k_a}{k_b}\frac{k_a}{k_c}\frac{d_2 + d_3}{d_1} \tag{11.6}$$

3 Go back to step (2) and repeat the procedure for different triplets of feature points. All possible triplets of feature points which are not co-linear are considered.

We now have produced a hash table with all the data indexed into its memory from a given database. Given a scene of feature points from a 3-D object, we try to match the index value IV as well as the individual ratios to those memorized in the hash table. Notice that the input 3-D object can either be complete or incomplete. Thus, given a 3-D object in a scene, the matching procedure is as follows:

4 Repeat steps (1) to (3) above, and then for each indexed value IV check the appropriate entry in the hash-table. Tally a vote for each model which appears at that location.

5 If several objects score large numbers of votes close to each other, then the most likely candidate will be chosen using global verification applied at next stage.

The voting is carried out simultaneously for all models in the hash table. The overall recognition time is dependent on the number of feature points in the scene. Our aim is to significantly reduce the number of feature points that are due to noise. This is achieved through the smoothing process which also removes noise. In order to find the optimal smoothing scale for each object, we employed the following multi-scale procedure:

- Apply one iteration of smoothing and count the number of feature points recovered from the object.

- Repeat the first step until several iterations of smoothing have been carried out. Construct a graph of the number of feature points vs. the number of iterations.

- Smooth this graph, and find the point where the slope is minimized. This indicates that the features have become stable at the corresponding scale which is used as the optimal scale for smoothing.

8.2.   Global Verification

In general the voting scheme may yield more than one candidate solution with very close scores from the geometric hashing stage. In this case we use a threshold to select the most likely models.

Global verification requires the estimation of the 3-D transformation parameters for the surviving models. It is possible to make use of closed form solution techniques [115, 114, 307, 80] to obtain these parameters. However, these techniques are quite complex to implement and relatively inefficient. Considering that the estimation procedure must be repeated many times, it is advantageous to use a method which is as efficient as possible. We have developed a relatively simple and efficient technique which generates approximate solutions. We found that this was quite satisfactory for our application [215].

From the 3 points of the scene model selected for matching, as discussed in the previous section, another point is also determined which is the centre of gravity of all 3 points. Let $P_1(x_{1p}, y_{1p}, z_{1p})$, $P_2(x_{2p}, y_{2p}, z_{2p})$ and $P_3(x_{3p}, y_{3p}, z_{3p})$ be the 3 non-collinear points selected from the model object and $P_4(x_{4p}, y_{4p}, z_{4p})$ be the centre of gravity of P_1, P_2 and P_3. We then form a plane P from these points. The same procedure is also applied to the object in the scene. Let $Q_1(x_{1q}, y_{1q}, z_{1q})$, $Q_2(x_{2q}, y_{2q}, z_{2q})$, $Q_3(x_{3q}, y_{3q}, z_{3q})$ and $Q_4(x_{4q}, y_{4q}, z_{4q})$ be the scene points. Point Q_4 is the centre of gravity of points Q_1, Q_2 and Q_3, thus a plane Q is also formed. We linearize the problem to simplify the solution, hence, the linear equations for the transformation, mapping model points to scene points are given by [197]:

$$
\begin{bmatrix}
x_{1p} & y_{1p} & z_{1p} & 1 \\
x_{2p} & y_{2p} & z_{2p} & 1 \\
x_{3p} & y_{3p} & z_{3p} & 1 \\
x_{4p} & y_{4p} & z_{4p} & 1
\end{bmatrix}
\cdot
\begin{bmatrix}
a & e & m \\
b & f & n \\
c & g & p \\
d & h & q
\end{bmatrix}
=
\begin{bmatrix}
x_{1q} & y_{1q} & z_{1q} \\
x_{2q} & y_{2q} & z_{2q} \\
x_{3q} & y_{3q} & z_{3q} \\
x_{4q} & y_{4q} & z_{4q}
\end{bmatrix}
\tag{11.7}
$$

Note that this approach is employed in order to obtain a quick and approximate solution which is sufficient for verification. From the set of linear equations (11.7), one can solve for the twelve parameters a, b, c, d, e, f, g, h, m, n, p and q. In order to verify match, 3-D translation, rotation and scaling will be used to determine global consistency. The translation parameters are (d, h, q). Let γ, β and α be the angles in the x, y and z directions for the rotation of the plane P in 3-D space. The 3-D rotation matrices about x-axis, y-axis and z-axis denoted $R_x(\gamma)$, $R_y(\beta)$ and $R_z(\alpha)$, respectively, are given by:

$$
R_x(\gamma) =
\begin{bmatrix}
1 & 0 & 0 \\
0 & \cos\gamma & -\sin\gamma \\
0 & \sin\gamma & \cos\gamma
\end{bmatrix}
\tag{11.8}
$$

$$
R_y(\beta) =
\begin{bmatrix}
\cos\beta & 0 & \sin\beta \\
0 & 1 & 0 \\
-\sin\beta & 0 & \cos\beta
\end{bmatrix}
\tag{11.9}
$$

$$
R_z(\alpha) =
\begin{bmatrix}
\cos\alpha & -\sin\alpha & 0 \\
\sin\alpha & \cos\alpha & 0 \\
0 & 0 & 1
\end{bmatrix}
\tag{11.10}
$$

The columns (and the rows) of matrices $R_x(\gamma)$, $R_y(\beta)$ and $R_z(\alpha)$ are mutually perpendicular unit vectors and they have determinants of 1, so they are orthogonal. Therefore the rotation parameters (γ, β, α) can be obtained from the product $R_z(\alpha).R_y(\beta).R_x(\gamma)$ and the solution of equation (11.7):

$$
\begin{bmatrix} a \\ e \\ m \\ n \\ p \end{bmatrix} = \begin{bmatrix} cos\alpha \; cos\beta \\ sin\alpha \; cos\beta \\ -sin\beta \\ cos\beta \; sin\gamma \\ cos\beta \; cos\gamma \end{bmatrix} \tag{11.11}
$$

To compute the scaling factor $\Re$, the distances from the centre of gravity points P_4 and Q_4 to their corresponding 3 points are measured and the shortest distances for each object are selected. Let r_1 and r_2 be the shortest distances selected from the model object and the scene object, then the scaling factor $\Re$ is equal to their ratio:

$$
\Re = \frac{r_1}{r_2}
$$

A number of model objects with close high scores are selected for the global verification stage. The hash table yields many candidate matches for each selected model. For each of these candidates, seven global transform parameters are estimated, using the procedure described earlier. The candidates are compared and if their corresponding parameters are compatible, they are clustered together. The largest cluster then indicates the largest group of globally consistent matches for each model. The model objects with the largest clusters are then chosen as the most likely objects present in the scene.

Our clustering algorithm is quite efficient since it avoids the creation of an explicit high-dimensional parameter space. The following is a step-by-step description of our clustering algorithm:

- Create a cluster for one of the points in the multi-dimensional parameter space. Consider that point as the centre of the cluster.

- Find another point which is closer than a threshold to the centre of the cluster, and add that point to the cluster. If no points are added, go to step 4.

- Compute the new centre of the cluster as the centre of mass of the points already in the cluster. Go to step 2.

- Repeat this procedure for all points which are not already in a cluster.

8.3.   Recognition System Results and Discussion

This section presents the matching results of the system applied to free-form 3-D surfaces in an object database. Given a 3-D surface in a scene, the aim is to match the measurements taken at scene points to those memorized in the hash table. There are 20 different objects in our database. All of those are shown in Figure 11.38. It should be pointed out that most of the objects in the database correspond to real range data. They were created by merging range images of real objects obtained from different viewpoints. The matching system was implemented entirely in C++ and ran on an UltraSparc 170E. The system was quite fast with matching times not exceeding 2-3 CPU seconds in each case.

Note that while some object recognition systems have employed grey-level images as their input [280, 121], others have made use of range images [57, 130] to achieve recognition. These systems have been designed for feature extraction and matching based on range images, and can not cope with more general 3-D surfaces. Our system goes further by accepting 3-D surfaces that are more general than range images. They can be formed by merging two or more range images obtained from different viewpoints. This makes it possible to obtain more information about the objects to be recognized, and achieve more reliable recognition.

Once the local maxima of Gaussian and mean curvatures of each object are obtained, they are then indexed into the hashing table as explained earlier.

Note that there is no exact correspondence between model vertices and input object vertices. This is because the model and input objects are subjected to different levels of smoothing followed by decimation which modifies the vertex structure on those objects.

The first experiment consisted of applying arbitrary amounts of scaling and 3-D rotation to the database objects, and to determine whether they can be recognised correctly by the system. All objects were in fact recognized correctly by the system. In the first stage, geometric hashing was applied to the input object. If one of the models $\mathcal{M}$ received a vote count that was substantially higher than the vote counts for all other objects, then $\mathcal{M}$ was selected as the correct object and the system terminated. Otherwise, two or more models received high vote counts that were relatively similar. In this case, the system applied global verification only to the surviving models in order to select one of them. Table 11.2 shows the results of geometric hashing for some of the rotated/scaled objects. The numbers shown are the percentages of triangles on each input object which received votes. When the number shown is greater than 100, some triangles received more than one vote. Table 11.3 shows the results of global verification. The numbers shown are the number of points in the largest clusters for each input object. Blanks indicate that no verification was considered necessary for the corresponding object.

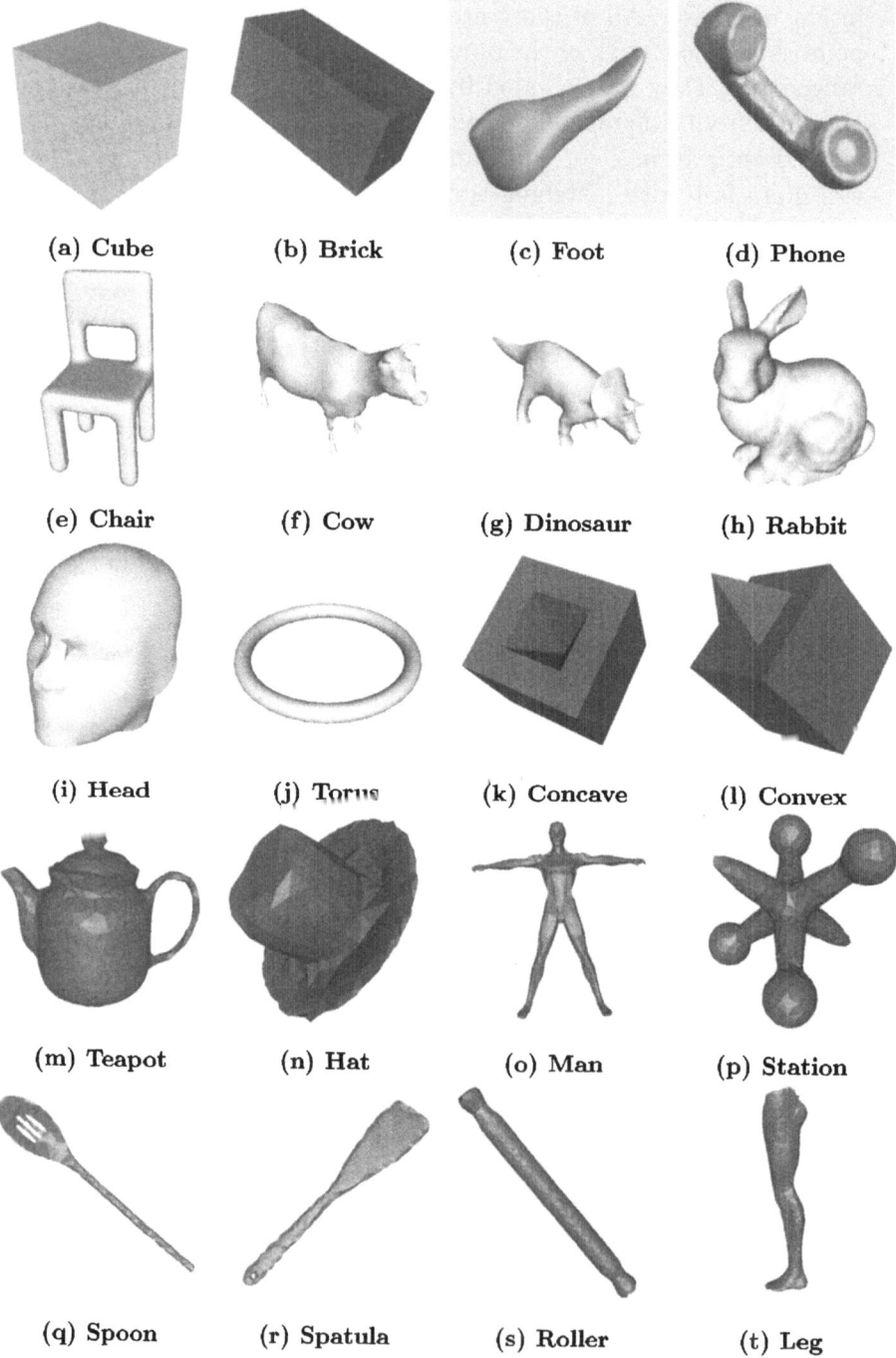

(a) Cube   (b) Brick   (c) Foot   (d) Phone

(e) Chair   (f) Cow   (g) Dinosaur   (h) Rabbit

(i) Head   (j) Torus   (k) Concave   (l) Convex

(m) Teapot   (n) Hat   (o) Man   (p) Station

(q) Spoon   (r) Spatula   (s) Roller   (t) Leg

Figure 11.38. Free-form 3-D objects used for matching experiments

input/model	chair	cow	dino	foot	head	leg	man	phone	rabbit
chair	68	76	66	1	33	27	52	50	47
cow	21	85	49	1	19	24	41	29	33
dino	17	54	83	1	21	21	35	29	32
foot	20	63	57	91	31	37	45	46	37
head	21	56	59	2	87	21	39	39	44
leg	21	72	58	2	20	87	55	50	34
man	16	48	43	1	17	26	59	30	27
phone	37	96	90	5	50	65	78	116	65
rabbit	19	55	53	1	28	21	37	0	87

Table 11.2. Results of geometric hashing for some rotated/scaled objects

input/model	chair	cow	dino	foot	head	leg	man	phone	rabbit
chair	357	43	22				19	33	13
cow									
dino		277	3665						
foot		6		35					
head			24		1309				
leg		123	28			1029			
man		224	76				2081	130	
phone		327	180					9395	
rabbit									

Table 11.3. Results of global verification for the rotated/scaled objects

The second experiment made use of incomplete surfaces which were again subjected to arbitrary amounts of scaling and 3-D rotation. In order to obtain incomplete surfaces, up to 60% of connected vertices were removed from database objects. The procedure utilized for vertex removal was as following: a vertex chosen randomly was removed from a given surface. Following that, all of its neighboring vertices were removed from that surface, and all the neighbors of those vertices, etc. This process continued until the desired number of vertices had been removed from the surface.

As in the previous experiment, for each input object, the system applied geometric hashing to all database models followed by global verification to the surviving models. Again, all input objects were correctly recognized by the system. This experiment shows that incomplete surfaces can also be matched successfully to the database models by the

input/model	chair	cow	dino	foot	head	leg	man	phone	rabbit
chair	70	0	59	1	34	23	41	51	46
cow	15	46	40	1	18	20	29	32	25
dino	13	40	68	0	22	14	28	24	30
foot	25	66	60	38	28	26	49	35	40
head	19	56	62	0	65	21	54	38	42
leg	0	0	50	0	0	50	0	0	0
man	19	0	47	1	24	23	48	34	33
phone	38	89	87	5	58	56	67	118	68
rabbit	20	43	46	1	25	21	30	41	52

Table 11.4. Results of geometric hashing for rotated/scaled incomplete objects

input/model	chair	cow	dino	foot	head	leg	man	phone	rabbit
chair	196	11	9				6	19	6
cow		28	6		5	6	8	17	5
dino		41	1296						
foot	7	4	8				4		2
head	4	4			24		4	6	2
leg			1			1			
man	35	17			6	15	370	61	11
phone	12							781	
rabbit	8	5			5		13	12	312

Table 11.5. Results of global verification for rotated/scaled incomplete objects

system. Table 11.4 shows the results of geometric hashing for a number of rotated/scaled incomplete objects, and table 11.5 shows the results of global verification.

In the third experiment, three complex scenes were created each consisting of two or more objects. Figure 11.39 shows the *kitchen-ware* scene and some of the features recovered from that scene. This scene contains a dish, a kettle, a spatula and a roller. Figure 11.40 shows the *bull-rider* scene and the corresponding recovered features. This scene contains a cow and a rider. Figure 11.41 shows the *space-station* scene and its recovered features. This scene contains a space-station and a space-ship attached to it.

As in the earlier experiments, the system applied geometric hashing to all database models followed by global verification to the surviving models. In the *kitchen-ware* scene, *dish* scored highest with other

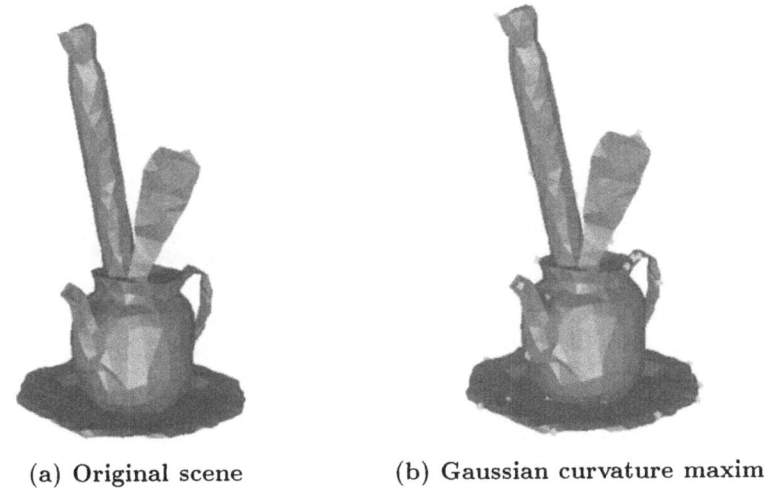

(a) **Original scene**   (b) **Gaussian curvature maxima**

Figure 11.39. Kitchen-ware scene and extracted features

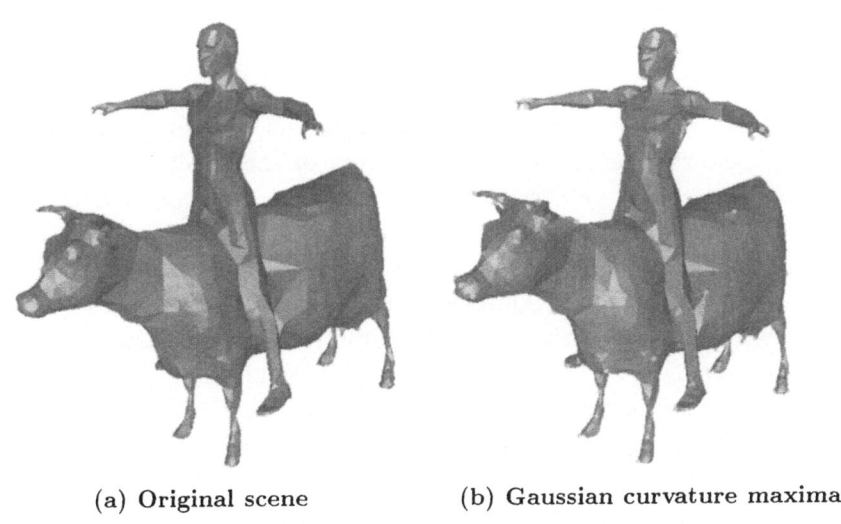

(a) **Original scene**   (b) **Gaussian curvature maxima**

Figure 11.40. Bull-rider scene and extracted features

scene objects also receiving high scores. In the *bull-rider* scene, *cow*, *dinosaur* and *rider* received the three highest scores respectively. In the *space-station* scene, the station itself received the highest score. This experiment shows that scenes depicting occlusion can also be recognized satisfactorily by the system.

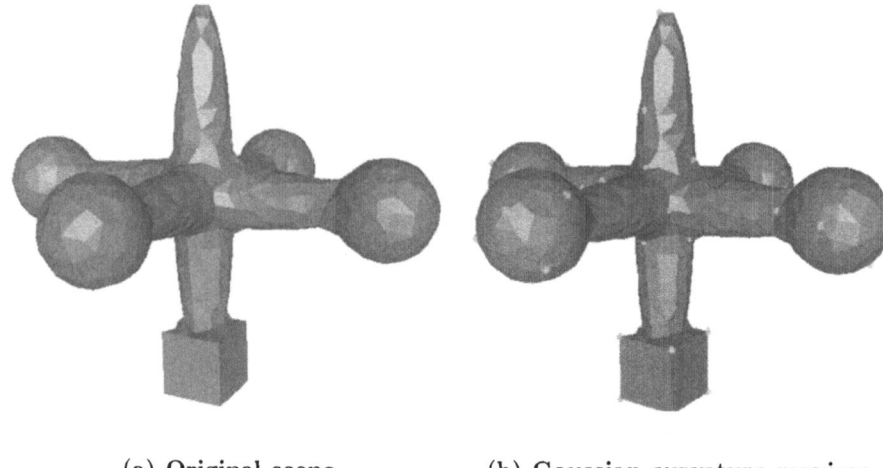

(a) Original scene   (b) Gaussian curvature maxima

Figure 11.41.   Space-station scene and extracted features

9.   Conclusions

A novel technique for multi-scale free-form 3-D surface description and curvature estimation was presented. In our technique semigeodesic coordinates are constructed at each vertex of the mesh which becomes the local origin. A geodesic from the origin is first constructed in an arbitrary direction such as the direction of one of the incident edges. During the diffusion process, 3-D surfaces were also sampled locally using different step sizes. Complete triangulated models of 3-D objects are constructed and using a local parametrization technique, are then smoothed using a 2-D Gaussian filter. The smoothing eliminated the surface noise and small surface detail gradually, and resulted in gradual simplification of object shape. The surface Gaussian and mean curvatures were also estimated. To visualize these curvature values on the surface, they are then mapped to colours/greylevels, and shown directly on the surface using the Visualization Toolkit. All convex corners of the surface indicated high Gaussian curvature values, whereas the concave corners indicated low Gaussian curvature values and the curvature values of flat areas were close to zero.

Next, the Gaussian and mean curvature zero-crossing contours were also recovered and displayed on the surface. Results indicated that as the surface is smoothed iteratively, the number of curvature zero-crossing contours were reduced. Curvature zero-crossing contours can be used for segmenting surfaces into regions. Furthermore, the local maxima of Gaussian and mean curvatures were also located and displayed on the surface. Next, the torsion maxima of the zero-crossing contours of Gaussian and mean curvatures were also selected as feature points.

A robust system for the recognition of free-form 3-D objects using 3-D models under different viewing conditions based on the geometric hashing algorithm and global verification was also presented. This technique was also shown to be useful for partially occluded objects. In order to verify match, 3-D translation, rotation and scaling parameters were used for global verification and results indicated that our technique is invariant to those transformations.

Appendix A
Proofs of Theorems of Chapter 1

This appendix contains the proofs of the 11 theorems of chapter 1 regarding planar curves.

Proof of Theorem 1.1. We will show that evolution is invariant under a general affine transform which includes transformations consisting of rotation, uniform scaling and translation.

Let $\Gamma = (x(u), y(u))$ be a planar curve and let $\Gamma_\sigma = (X(u, \sigma), Y(u, \sigma))$ be its evolved version. If Γ_σ is transformed according to an affine transform, then the following relationships hold between its old coordinates, $X(u, \sigma)$ and $Y(u, \sigma)$, and its new coordinates, $x_1(u, \sigma)$ and $y_1(u, \sigma)$:

$$x_1(u, \sigma) = aX(u, \sigma) + bY(u, \sigma) + c,$$

$$y_1(u, \sigma) = dX(u, \sigma) + eY(u, \sigma) + f.$$

Now suppose Γ is transformed according to an affine transform and then evolved. The coordinates, $x_2(u)$, $y_2(u)$ of the new curve are:

$$x_2(u, \sigma) = (ax(u) + by(u) + c) \otimes g(u, \sigma),$$

$$y_2(u, \sigma) = (dx(u) + cy(u) + f) \otimes g(u, \sigma).$$

Because the convolution operator is distributive [135], it follows that

$$x_2(u, \sigma) = a(x(u) \otimes g(u, \sigma)) + b(y(u) \otimes g(u, \sigma)) + c = x_1(u, \sigma),$$

$$y_2(u, \sigma) = d(x(u) \otimes g(u, \sigma)) + e(y(u) \otimes g(u, \sigma)) + f = x_1(u, \sigma).$$

Note that this result holds for any convolution operator not just the Gaussian. The proof for arc length evolution is very similar. ◻

Proof of Theorem 1.2. A closed curve has $(x(0), y(0)) = (x(1), y(1))$. It follows that $(X(0, \sigma), Y(0, \sigma)) = (X(1, \sigma), Y(1, \sigma))$. ◻

Proof of Theorem 1.3. Let $\Gamma = (x(u), y(u))$ be a connected planar curve and let $\Gamma_\sigma = (X(u, \sigma), Y(u, \sigma))$ be its evolved version. We will show that Γ_σ is also a connected curve.

Since Γ is connected, $x(u)$, $y(u)$ and therefore $X(u, \sigma)$ and $Y(u, \sigma)$ are continuous functions. Let u_0 be any value of parameter u and let x_0 and y_0 be the values of $X(u, \sigma)$ and $Y(u, \sigma)$ at u_0 respectively. It follows that if u goes through an infinitesimal change, then $X(u, \sigma)$ and $Y(u, \sigma)$ will also go through infinitesimal changes

$$X(u_0, \sigma) \to x_0 + \delta, \quad Y(u_0, \sigma) \to y_0 + \xi.$$

As a result, point $P(x_0, y_0)$ on Γ_σ goes to point $P'(x_0 + \delta, y_0 + \xi)$. Let the distance between P and P' be D. Then

$$D = \sqrt{(x_0 + \delta - x_0)^2 + (y_0 + \xi - y_0)^2} = \sqrt{\delta^2 + \xi^2} \leq \delta\sqrt{2},$$

assuming $|\delta|$ is the larger of $|\delta|$ and $|\xi|$. It follows that an infinitesimal change in parameter u also results in an infinitesimal change in the value of the vector-valued function Γ_σ. Therefore Γ_σ is a connected curve. ⊠

Proof of Theorem 1.4. The proof of this theorem will be given for arc length evolution only. The proof for regular evolution is very similar.

Let M be the center of mass of $\Gamma = (x(w), y(w))$ with coordinates (x_M, y_M). Then

$$x_M = \int_0^1 x(w)\, dw \bigg/ \int_0^1 dw = \int_0^1 x(w)\, dw,$$

$$y_M = \int_0^1 y(w)\, dw \bigg/ \int_0^1 dw = \int_0^1 y(w)\, dw.$$

Let $\Gamma_\sigma = (X(W, \sigma), Y(W, \sigma))$ be an arc length evolved version of Γ with $N = (X_N, Y_N)$ as its center of mass. Observe that

$$X_N = \int_0^1 X(W, \sigma)\, dW$$

$$= \int_0^1 \int_{-\infty}^{\infty} g(v, \sigma)x(W - v)\, dv\, dW$$

$$= \int_{-\infty}^{\infty} g(v, \sigma) \left(\int_0^1 x(W - v)\, dW \right) dv.$$

W covers Γ_σ exactly once. Therefore

$$\int_0^1 x(W - v)\, dW = x_M.$$

Therefore, $X_N = x_M$. Similarly, $Y_N = y_M$. Hence, M and N are the same point. ⊠

Proof of Theorem 1.5. The proof of this theorem will be given for arc length evolution only. The proof for regular evolution is very similar.

Since G is simple and convex, every line L tangent to G contains that curve in the left (or right) half plane it creates. Since Γ is inside G, Γ is also contained in the same half plane. Now, rotate L and Γ so that L becomes parallel to the y axis. L is now described by the equation $x = c$. Since L does not intersect Γ, it follows that $x(w_0) \geq c$ for every point w_0 on Γ. Let Γ_σ be an arc length evolved version of Γ. Every point of Γ_σ is a weighted average of all the points of Γ. Therefore, $X(W_0, \sigma) \geq c$ for every point W_0 on Γ_σ and Γ_σ is also contained in the same half plane. This result holds for *every* line tangent to G; therefore, Γ_σ is contained inside the intersection of all the left (or right) half planes created by the tangent lines of G. It follows that Γ_σ is also inside G. ⟌

Proof of Theorem 1.6. The proof of this theorem for the regular CSS image of a planar curve Γ will be given first. This will be followed by another proof for the resampled CSS image of Γ.

The following is a proof of theorem 1.6 for the regular CSS image. Section A shows that the derivatives at a point on a curvature zero-crossing contour provide homogeneous equations in the moments of the Fourier transforms of the coordinate functions of the curve. Section B shows that the moments are related to the coefficients of expansion of the coordinate functions of the curve in functions related to the Hermite polynomials. Section C shows that the quadratic equations obtained in section A can be converted into a homogeneous linear system of equations which can be solved uniquely for the curvature function of the curve.

A. Constraints from the Curvature zero-crossing Contours

Let $\Gamma = (x(u), y(u))$ be the arc-length parametrization of the curve with Fourier transform $\tilde{\Gamma} = (\tilde{x}(\omega), \tilde{y}(\omega))$. The Fourier transform of the Gaussian filter

$$G(u, t) = \frac{1}{\sqrt{2t}} \, e^{-u^2/4t}$$

is $\tilde{G}(\omega) = e^{-\omega^2 t}$ where $t = \sigma^2/2$.

Let $\Gamma_{t_0} = (x(u, t_0), y(u, t_0))$ be a curve obtained by convolving $x(u)$ and $y(u)$ with $G(u, t_0)$. Since Γ is in C_1, there exists a t_0 such that Γ_{t_0} is in C_∞. The curvature zero-crossings in a neighborhood of t_0 are given by solutions of $\alpha(u, t) = 0$ where

$$\alpha(u, t) = \dot{x}(u, t)\ddot{y}(u, t) - \dot{y}(u, t)\ddot{x}(u, t),$$

where ˙ represents derivative with respect to u. Using the convolution theorem, the term in $\alpha(u, t)$ can be expressed as follows:

$$\dot{x} = \int e^{-\omega^2 t} \, e^{i\omega u} (i\omega)\tilde{x}(\omega) \, d\omega,$$

$$\dot{y} = \int e^{-\omega^2 t} \, e^{i\omega u} (i\omega)\tilde{y}(\omega) \, d\omega,$$

$$\ddot{x} = \int e^{-\omega^2 t} \, e^{i\omega u} (-\omega^2)\tilde{x}(\omega) \, d\omega,$$

$$\ddot{y} = \int e^{-\omega^2 t} e^{i\omega u} (-\omega^2) \tilde{y}(\omega) \, d\omega.$$

The Implicit Function Theorem guarantees that the contours $u(t)$ are C_∞ in a neighborhood of t_0. Let ξ be a parameter of the curvature zero-crossing contour. Then

$$\frac{d}{d\xi} = \frac{du}{d\xi} \frac{\partial}{\partial u} + \frac{dt}{d\xi} \frac{\partial}{\partial t}.$$

On the curvature zero-crossing contour, $\alpha = 0$ and $(dk/d\xi^k)\alpha = 0$ for all integers k. Furthermore, since the curvature zero-crossing contour is known, all the derivatives of u and t with respect to ξ are known as well.

We can now compute the derivatives of α with respect to ξ at (u_0, t_0). The first derivative is given by:

$$
\begin{aligned}
\frac{d\alpha}{d\xi} = {} & \frac{du}{d\xi} \left(\int e^{-\omega^2 t} e^{i\omega u} (i\omega)^3 \tilde{y}(\omega) \, d\omega \int e^{-\omega^2 t} e^{i\omega u} i\omega \tilde{x}(\omega) \, d\omega \right. \\
& \left. - \int e^{-\omega^2 t} e^{i\omega u} (i\omega)^3 \tilde{x}(\omega) \, d\omega \int e^{-\omega^2 t} e^{i\omega u} (i\omega) \tilde{y}(\omega) \, d\omega \right) \\
& + \frac{dt}{d\xi} \left(\int e^{-\omega^2 t} e^{i\omega u} (i\omega)^3 \tilde{x}(\omega) \, d\omega \int e^{-\omega^2 t} e^{i\omega u} (-\omega^2) \tilde{y}(\omega) \, d\omega \right. \\
& + \int e^{-\omega^2 t} e^{i\omega u} (\omega^4) \tilde{y}(\omega) \, d\omega \int e^{-\omega^2 t} e^{i\omega u} (i\omega) \tilde{x}(\omega) \, d\omega \\
& - \int e^{-\omega^2 t} e^{i\omega u} (i\omega) \tilde{y}(\omega) \, d\omega \int e^{-\omega^2 t} e^{-i\omega u} (-\omega^2) \tilde{x}(\omega) \, d\omega \\
& \left. - \int e^{-\omega^2 t} e^{i\omega u} \omega^4 \tilde{x}(\omega) \, d\omega \int e^{-\omega^2 t} e^{i\omega u} (i\omega) \tilde{y}(\omega) \, d\omega \right).
\end{aligned}
\tag{A.1}
$$

Note that the moment of order k of function $f(\omega) = e^{-\omega^2 t} e^{i\omega u} (i\omega) \tilde{x}(\omega)$ is defined by:

$$M_k = \int\limits_{-\infty}^{\infty} (i\omega)^k e^{-\omega^2} e^{i\omega u} (i\omega) \tilde{x}(\omega) \, d\omega$$

and the moment of order k of function $f'(\omega) = e^{-\omega^2 t} e^{i\omega u} (i\omega) \tilde{y}(\omega)$ is defined by:

$$M'_k = \int\limits_{-\infty}^{\infty} (i\omega)^k e^{-\omega^2 t} e^{i\omega u} (i\omega) \tilde{y}(\omega) \, d\omega.$$

As a result, Equation (A.1) can be re-written as:

$$
\begin{aligned}
\frac{d}{d\xi} \alpha(u_0, t_0) = {} & \frac{du}{d\xi} (M'_2 M_0 - M_2 M'_0) \\
& + \frac{dt}{c\xi} (M_2 M'_1 + M'_3 M_0 - M'_2 M_1 - M_3 M'_0).
\end{aligned}
\tag{A.2}
$$

Furthermore, the second derivative is given by:

$$\frac{d^2}{d\xi^2}\alpha(u_0, t_0) = \frac{d^2 u}{d\xi^2}(M_2' M_0 - M_2 M_0')$$

$$+ \frac{d^2 t}{d\xi^2}(M_2 M_1' + M_3' M_0 - M_2' M_1 - M_3 M_0')$$

$$+ \left(\frac{du}{d\xi}\right)^2 (M_3' M_0 + M_1 M_2' - M_3 M_0' - M_1' M_2)$$

$$+ 2\frac{du}{d\xi}\frac{dt}{d\xi}(M_4' M_0 - M_4 M_0')$$

$$+ \left(\frac{dt}{d\xi}\right)^2 (M_4 M_1' + 2M_3' M_2 + M_5' M_0$$

$$- M_4' M_1 - 2M_3 M_2' - M_5 M_0'). \tag{A.3}$$

Since the parametric derivatives along the curvature zero-crossing contours are zero, equations (A.2) and (A.3) are equal to zero. Note that equation (A.2) is a quadratic equation in the first four moments of functions $f(\omega)$ and $f'(\omega)$ and equation (A.3) is a quadratic equation in the first six moments of those functions. In general, the $k+1$st equation, $(d^k/d\xi^k)\alpha(u, t) = 0$, is a quadratic equation in the first $2k + 2$ moments of each of the functions $f(\omega)$ and $f'(\omega)$ or a total of $4k + 4$ moments.

B. The Moments and the Coefficients of Expansion of $\dot{x}(u)$ and $\dot{y}(u)$

This section shows that the moments and the moment-pairs in equations:

$$(d^k/d\xi^k)\alpha(u, t)$$

are related respectively to the coefficients of the expression of the functions $\dot{x}(u)$, $\dot{y}(u)$ and the curvature function of Γ, $\kappa(u)$, in functions related to the Hermite polynomials. Expand function

$$\dot{x}(u) = \frac{d}{du}x(u)$$

as

$$\dot{x}(u) = \sum a_k(\sigma)\phi_k(u, \sigma)$$

in terms of functions $\phi_k(u, \sigma)$ related to the Hermite polynomials

$$H_k(u) = (-1)^k e^{u^2}\frac{d^k}{du^k}e^{-u^2}$$

by

$$\phi_k(u, \sigma) = (-1)^k \frac{\sigma^{k-1}}{(\sqrt{2})^{k+1}\sqrt{\pi}}H_k\left(\frac{u}{\sigma\sqrt{2}}\right).$$

The coefficients $a_k(\sigma)$ of the expansion are given by:

$$a_k(\sigma) = \langle w_k(u, \sigma), \dot{x}(u)\rangle,$$

where $\langle , \rangle$ denotes inner product in L^2 and $\{w_k(u,\sigma)\}$ is the set of functions biorthogonal to $\{\phi_k(u,\sigma)\}$. The $\{\phi_k(u,\sigma)\}$ are given explicitly by

$$\phi_k(u,\sigma) = \frac{\sigma^{2k-1}}{k!\sqrt{2\pi}} e^{u^2/2\sigma^2} \frac{d^k}{du^k} e^{-u^2/2\sigma^2}$$

and the $w_k(u,\sigma)$ by

$$w_k(u,\sigma) = (-1)^k \frac{d^k}{du^k} e^{-u^2/2\sigma^2}.$$

Since

$$\dot{x}(u) = \frac{1}{\sqrt{2\pi}} \int e^{i\omega u}(i\omega)\tilde{x}(\omega)\,d\omega$$

the a_k are given by:

$$a_k(\sigma) = \frac{1}{\sqrt{2\pi}}(-1)^k \int \left\langle \frac{d^k}{du^k} e^{-u^2/2\sigma^2}, e^{i\omega u} \right\rangle (i\omega)\tilde{x}(\omega)\,d\omega.$$

The inner product is just the inverse Fourier transform of $w_k(u,\sigma)$. Therefore

$$a_k(\sigma) = \int (i\omega)^k e^{(-\omega^2\sigma^2)/2}(i\omega)\tilde{x}(\omega)\,d\omega$$

which is equal to M_k modulo a factor $e^{i\omega u}$, since $t = \sigma^2/2$.
Similarly, the function

$$\dot{y}(u) = \frac{d}{du}y(u)$$

can be expanded in terms of the functions $\phi_k(u,\sigma)$ by

$$\dot{y}(u) = \sum b_k(\sigma)\phi_k(u,\sigma)$$

and it again follows that

$$b_k(\sigma) = \int (i\omega)^k e^{(-\omega^2\sigma^2)/2}(i\omega)\tilde{y}(\omega)\,d\omega$$

which is equal to M'_k modulo a factor $e^{i\omega u}$.
   Furthermore, $a'_k(\sigma)$ and $b'_k(\sigma)$, the coefficients of expansion of functions $\ddot{x}(u)$ and $\ddot{y}(u)$ in terms of the functions $\phi_k(u,\sigma)$, can be seen to be related to $a_k(\sigma)$ and $b_k(\sigma)$ according to the following relationships:

$$a'_{k-1}(\sigma) = a_k(\sigma),$$

$$b'_{k-1}(\sigma) = b_k(\sigma).$$

Therefore $\kappa(u)$, the curvature function of Γ can be expressed as:

$$\kappa(u) = \dot{x}(u)\ddot{y}(u) - \dot{y}(u)\ddot{x}(u) = \sum\sum (a_j b_{k+1} - b_j a_{k+1})\phi_j\phi_k.$$

Therefore if the pairs $a_j(\sigma)b_k(\sigma)$, $j,k = 0,\ldots,2n+1$, are all known, the curvature function of Γ can be reconstructed.

C. Reconstructing the Curvature Function

	M_0'	M_1'	M_2'	M_3'	M_4'	M_5'
M_0			+	+	×	×
M_1			+		×	
M_2	+	+		×		
M_3	+		×			
M_4	×	×				
M_5	×					

Table A.1.   The moment-pairs in equations (A.2) and (A.3)

Equations (A.2) and (A.3) can be converted into homogeneous linear equations by assuming that each moment-pair appearing in those equations is a new variable. Table A.1 shows the moment-pairs in equations (A.2) and (A.3). The + signs designate the moment-pairs in equation (A.2) and the + and × signs together designate the moment-pairs in equation (A.3).

Note that all other moment-pairs in table A.1 can be computed from the existing ones using the following relationships:

$$M_i M_j' = \frac{M_{i=1} M_j'. M_i M_{j+1}'}{M_{i-1} M_{j+1}'} = \frac{M_{i+1} M_j'. M_i M_{j+1}'}{M_{i+1} M_{j+1}'}$$

$$= \frac{M_i M_{j-1}'. M_{i-1} M_j'}{M_{i-1} M_{j-1}'} = \frac{M_i M_{j-1}'. M_{i+1} M_j'}{M_{i+1} M_{j-1}'}. \tag{A.4}$$

We proceed to compute the first n derivatives at point (u_0, t_0) on one of the curvature zero-crossing contours. We now obtain $n+1$ homogeneous linear equations in some of the moment-pairs $M_i M_j'$. Since this system is in terms of the first $2n + 2$ moments of functions $f(\omega)$ and $f'(\omega)$, it will contain $O(n^2)$ moment-pairs. Therefore additional equations are required to constrain the system. To obtain those equations, we proceed as follows:

Assume that moments of order higher than $2n + 1$ are zero. Compute derivatives of order higher than n at (u_0, t_0) but set moments of order higher than $2n + 1$ to zero in the resulting equations. If a sufficient number of derivatives are computed at (u_0, t_0), the number of equations obtained will be equal to the number of moment-pairs and our linear system will be constrained. Note that this procedure allows us to use only one point from the curvature scale apace image.

It follows from an assumption of generality that the system will have a unique zero eigenvector and therefore a unique solution modulo scaling. Once the moment-pairs in the system are known, all other moment-pairs can be computed from the known ones using Equation (A.4). Since all the moment-pairs $M_i M_j'$ together determine the curvature function of the curve, it follows that the curve can be determined modulo a rigid motion and constant scaling.

It has been shown in [333] that a 1-D signal can be reconstructed using two points from its scale space image. Note that our result implies that only one point is sufficient for the reconstruction of that signal.

To prove the same result about the resampled curvature scale space of Γ, recall that derivatives at one point (at any scale) on any curvature zero-crossing contour in the curvature scale space of Γ were computed, and it was shown that the resulting equations can be solved for the coefficients of expansion of the curvature function of Γ in functions related to the Hermite polynomials.

As before, we choose a point on a zero-crossing contour at any scale of the resampled curvature scale space image of Γ and compute the necessary derivatives. The value of σ in the resulting equations is then set to zero. Consequently, the arc length evolved curve Γ_σ, where σ corresponds to the scale at which the derivatives were computed, is reconstructed modulus uniform scaling, rotation, and translation.

The next step is to recover the original curve Γ. This is done by applying reverse *arc length evolution* to Γ_σ. Let the arc length evolved curve Γ_σ be defined by

$$\Gamma_\sigma = \{(X(W, \sigma), Y(W, \sigma)) \mid W \in [0, 1]\}.$$

A reverse arc length evolved curve Γ is defined by

$$\Gamma = \{(x(w), y(w)) \mid w \in [0, 1]\},$$

where

$$x(w) = X(w, \sigma) \otimes D_N(w, \sigma),$$
$$y(w) = Y(w, \sigma) \otimes D_N(w, \sigma),$$

where D_N is a deblurring operator defined in [120] and

$$w(W, \sigma) = \int_0^\sigma \int_0^w \kappa^2(w, \sigma) \, dw \, d\sigma.$$

As a result, Γ is recovered modulo uniform scaling, rotation, and translation.

To prove the same result about the renormalized curvature scale space image, evolved curve Γ_σ is again reconstructed. Then, each of its coordinate functions is deblurred by convolving it with the deblurring operator D_N. Once again, Γ is recovered modulo uniform scaling, rotation, and translation.   ⊠

Proof of Theorem 1.7. Since by assumption all evolved and arc length evolved curves Γ_σ are in C_2, the conditions of the implicit function theorem are satisfied on contours $\kappa(u, \sigma)$, $\kappa(w, \sigma)$, and $\kappa(w, \sigma) = 0$ in the regular, renormalized, and resampled curvature scale space images of Γ. Therefore, the proofs are very similar. The following is a proof for the regular CSS image:

On any contour in curvature scale space image,

$$k(u, \sigma) = 0.$$

Since all Γ_σ are in C_2, this is equivalent to:

$$\dot{X}(u, \sigma)\ddot{Y}(u, \sigma) - \ddot{X}(u, \sigma)\dot{Y}(u, \sigma) = 0.$$

To exploit the properties of the heat equation [120], it is convenient to change variables and let

$$t = \frac{1}{2}\sigma^2$$

so we define

$$x(u,t) = X(u,\sigma)$$
$$y(u,t) = Y(u,\sigma)$$

$$\alpha(u,t) = x_u(u,t)y_{uu}(u,t) - x_{uu}(u,t)y_u(u,t). \tag{A.5}$$

The functions $x(u,t)$ and $y(u,t)$ are obtained by convolving

$$(1/\sqrt{4\pi t})e^{-(1/4t)u^2}$$

with the original curve coordinates $x(u)$ and $y(u)$ respectively, and so they satisfy the heat equation:

$$x_{uu}(u,t) = x_t(u,t), \tag{A.6}$$

$$y_{uu}(u,t) = y_t(u,t). \tag{A.7}$$

The implicit function theorem is satisfied on contours

$$\alpha(u,t) = 0$$

because of the assumption that all Γ_σ are smooth. Therefore we can write:

$$t = t(u)$$

and

$$\dot{t}(u) = \frac{dt}{du} = \frac{-\alpha_u}{\alpha_t}.$$

The theorem will be proven if we can show that for all points such that

$$\dot{t}(u) = 0$$

we have

$$\ddot{t}(u) < 0.$$

Observe that

$$\dot{t}(u) = 0 \quad \text{iff} \quad \alpha_u(u,t) = 0. \tag{A.8}$$

At an extremism where (A.8) holds, we have

$$\ddot{t}(u) = \frac{d}{du}\left(\frac{-\alpha_u}{\alpha_t}\right) = \frac{\partial}{\partial u}\left(\frac{-\alpha_u}{\alpha_t}\right) + \frac{\partial}{\partial t}\left(\frac{-\alpha_u}{\alpha_t}\right)\frac{dt}{du} = \frac{-\alpha_{uu}}{\alpha_t}.$$

So we must show that if

$$\alpha(u,t) = \alpha_u(u,t) = 0$$

then

$$\alpha_{uu}/\alpha_t > 0.$$

We shall show that these conditions require

$$\alpha_{uu}/\alpha_t = 1$$

which proves the theorem. From (A.5), (A.6) and (A.7) we have

$$\alpha = x_u y_t - x_t y_u,$$

$$\alpha_u = x_{uu}y_t + x_u y_{ut} - x_{ut}y_u - x_t y_{uu}.$$

But using (A.6)

$$\alpha_u = x_u y_{ut} - x_{ut} y_u. \tag{A.9}$$

Similarly

$$\alpha_{uu} = (x_u y_{tt} - x_{tt} y_u) + (x_t y_{ut} - x_{ut} y_t).$$

If $\alpha = \alpha_u = 0$ then using (A.5) and (A.9)

$$x_t y_{ut} - x_{ut} y_t \;=\; x_t \left(y_{ut} - x_{ut} \frac{y_t}{x_t} \right) = x_t \left(y_{ut} - x_{ut} \frac{y_u}{x_u} \right)$$

$$=\; \frac{x_t}{x_u} (x_u y_{ut} - x_{ut} y_u) = 0$$

so

$$\alpha_{uu} = x_u y_{tt} - x_{tt} y_u.$$

We also have

$$\alpha_t = (x_u y_{tt} - x_{tt} y_u) - (x_t y_{ut} - x_{ut} y_t)$$

so

$$\alpha_t = x_u y_{tt} - x_{tt} y_t$$

and hence

$$\alpha_{uu} = \alpha_t$$

as claimed.   ⋈

Proof of Theorem 1.8. Since the class of polynomial functions is closed under convolution with a Gaussian [120], it follows that $X(u, \sigma)$ and $Y(u, \sigma)$ are also polynomial functions:

$$X(u, \sigma) = a_0 + a_1 u + a_2 u^2 + a_3 u^3 + \cdots,$$

$$Y(u, \sigma) = b_0 + b_1 u + b_2 u^2 + b_3 u^3 + \cdots.$$

Suppose that Γ_σ goes through the origin of the coordinate system at $u_0 = 0$. It follows that $a_0 = b_0 = 0$. Assume further that there is a singularity on Γ_σ at u_0. We have:

$$X_u(u, \sigma) = a_1 + 2a_2 u + 3a_3 u^2 + 4a_4 u^3 + \cdots,$$

$$Y_u(u, \sigma) = b_1 + 2b_2 u + 3b_3 u^2 + 4b_4 u^3 + \cdots.$$

Since $X_u(u, \sigma)$ and $Y_u(u, \sigma)$ are zero at a singular point, it also follows that $a_1 = b_1 = 0$. We will now perform a case analysis of the singular point at u_0 to determine when it corresponds to a cusp point. Since we will examine a small neighborhood of point u_0, we will approximate the curve using the lowest degree terms in $X(u, \sigma)$ and $Y(u, \sigma)$:

$$\Gamma_\sigma = (u^m, u^n).$$

Assume w.l.o.g. that $n > m$. From above we know that $m > 1$. Using

$$\Gamma_u(u, \sigma) = (X_u(u, \sigma), Y_u(u, \sigma)) = (mu^{m-1}, nu^{n-1})$$

it follows that

$$r_u(\epsilon, \sigma) = (m\epsilon^{m-1}, n\epsilon^{n-1}) = \epsilon^{m-1}(m, n\epsilon^{n-m})$$

and

$$r_u(-\epsilon, \sigma) = (m(-\epsilon)^{m-1}, n(-\epsilon)^{n-1}).$$

We can now analyze the singular point in each of the four possible cases:

1. m and n are both even: $m - 1$ and $n - 1$ are both odd. So

$$r_u(-\epsilon, \sigma) = (-m\epsilon^{m-1}, -n\epsilon^{n-1}) = -\epsilon^{m-1}(m, n\epsilon^{n-m}).$$

A comparison of $r_u(\epsilon, \sigma)$ and $r_u(-\epsilon, \sigma)$ shows that an infinitesimal change is the parameter u results in a large change in the direction of the tangent vector. Therefore the singular point *is* also a cusp point in this case.

2. m and n are both odd: $m - 1$ and $n - 1$ are both even. So

$$r_u(-\epsilon, \sigma) = (m\epsilon^{m-1}, n\epsilon^{n-1}) = \epsilon^{m-1}(m, n\epsilon^{n-m}).$$

Comparing $r_u(\epsilon, \sigma)$ to $r_u(-\epsilon, \sigma)$ now shows that the tangent direction does not change with u in a small neighborhood of the singular point. Therefore this singular point is not a cusp point.

3. m is odd and n is even: $m - 1$ is even and $n - 1$ is odd. So

$$r_u(-\epsilon, \sigma) = (m\epsilon^{m-1}, -n\epsilon^{n-1}) = \epsilon^{m-1}(m, n\epsilon^{n-m}).$$

An infinitesimal change in u also results in an infinitesimal change in the tangent direction. Again, this singular point is *not* a cusp point.

4. m is even and n is odd: $m - 1$ is odd and $n - 1$ is even. So

$$r_u(-\epsilon, \sigma) = (-m\epsilon^{m-1}, n\epsilon^{n-1}) = \epsilon^{m-1}(-m, n\epsilon^{n-m}).$$

An infinitesimal change in u now results in a large change in the tangent direction. Therefore this singular point *is* a cusp point.

It follows from the case analysis above that only the singular points in cases 1 and 4 are cusp points. We will now derive analytical expressions for the curve $\Gamma_{\sigma-\delta}$ so that it can be analyzed in a small neighborhood of the cusp point.

To deblur function $f(u) = u^k$, we convolve a rescaled version of that function with the function

$$(2/\sqrt{\pi}) e^{-v^2}(1 - v^2)$$

(an approximation to the deblurring operator derived in [120]) as follows:

$$(D_t f)(u) = \int_{-\infty}^{\infty} \frac{2}{\sqrt{\pi}} e^{-v^2}(1 - v^2) f(u + 2v\sqrt{t}) \, dv,$$

or

$$(D_t f)(u) = \frac{2}{\sqrt{\pi}} \int_{-\infty}^{\infty} e^{-v^2}(1 - v^2) f(u + 2v\sqrt{t})^k \, dv,$$

where t is the scale factor and controls the amount of deblurring. Note that our approximation is good for small t. Solving the integral above yields

$$(D_t f)(u) = \sum_{\substack{p=0 \\ (p \text{ even})}} 1.3 \ldots (p-1)$$

$$\times \frac{(2t)^{p/2} k(k-1) \ldots (k-p+1)}{p!} (1-p) u^{k-p}. \qquad (A.10)$$

The following are four functions of the form $f(u) = u^k$ and their deblurred versions:

    a. $f(u) = u^2$ $(D)tf(u) = u^2 - 2t$
    b. $f(u) = u^3$ $(D_t f)(u) = u^3 - 6tu$
    c. $f(u) = u^4$ $(D_t f)(u) = u^4 - 12tu^2 - 36t^2$
    d. $f(u) = u^5$ $(D_t f)(u) = u^5 - 20tu^3 - 180t^2 u$

We can now analyze the cusp points identified in cases 1 and 4 above. In case 1, the curve Γ_σ is approximated by (u^m, u^n) where m and n are both even numbers. We now deblur the curve to obtain:

$$(D_t x)(u) = u^m - c_1 t u^{m-2} - \cdots - c_{(m-2)/2} t^{(m-2)/2} u^2 - c_{m/2} t^{m/2},$$

$$(D_t y)(u) = u^n - c_1' t u^{n-2} - \cdots - c_{(n-2)/2}' t^{(n-2)/2} u^2 - c_{n/2}' t^{n/2}.$$

Note that all powers of u are even and the constants c_j and c_j' are all positive as follows from an examination of (A.10). It follows that

$$(D_t \dot{x})(u) = mu^{m-1} - (m-2)c_1 t u^{m-3} - \cdots 2c_{(m-2)/2} t^{(m-2)/2} u,$$

$$(D_t \dot{y})(u) = nu^{n-1} - (n-2)c_1' t u^{n-3} - \cdots 2c_{(n-2)/2}' t^{(n-2)/2} u$$

contain only odd powers of u and $(D_t \dot{r})(\epsilon) = -(D_t \dot{r})(-\epsilon)$. Hence there is also a cusp point on the curve $\Gamma_{\sigma-\delta}$ at u_0. This is a contradiction of the assumption that Γ_σ is the fast evolved version of Γ with a cusp at u_0. It follows that Γ_σ can not have a cusp point of this kind at u_0.

    We shall now turn to the cusp points encountered in case 4. Recall that, in that case, the curve Γ_σ is approximated, in a small neighborhood of the cusp point, by (u^m, u^n) where m is even and n is odd. Again we deblur the curve to obtain:

$$(D_t x)(u) = u^m - c_1 t u^{m-2} - \cdots - c_{(m-2)/2} t^{(m-2)/2} u^2 - c_{m/2} t^{m/2},$$

$$(D_t y)(u) = u^n - c_1' t u^{n-2} - \cdots - c_{(n-2)/2}' t^{(n-2)/2} u^2 - c_{n/2}' t^{n/2}.$$

Again note that constants c_j and c_j' are all positive. The deblurred curve intersects itself if there are two values of u, u_1 and u_2, such that

$$x(u_1) = x(u_2), \qquad (A.11)$$

$$y(u_1) = y(u_2). \qquad (A.12)$$

Since $(D_t x)(u)$ contains even powers of u only, it follows from (A.11) that $u_1 = -u_2$. Since $(D_t y)(u)$ contains odd powers of u only, substituting $u_1 = -u_2$ in (A.12), letting $t = \delta$, and simplifying yields:

$$u_1^n - c_1' \delta u_1^{n-2} - c_2' \delta^2 u_1^{n-4} - \cdots - c_{(n-1)/2}' \delta^{(n-1)/2} u_1 = 0. \qquad (A.13)$$

$u_1 = 0$ is one of the roots of this equation. For very small values of u_1, the LHS of (A.13) is negative since the first term will be smaller than each of the other temps (which are negative). As u_1 grows larger, the first term becomes larger than the sum of all other terms and therefore the LHS of (A.13) becomes positive. It follows that there exists a positive value of u_1 at which (A.13) is satisfied. Therefore $\Gamma_{\sigma-\delta}$ is self-intersecting in a small neighborhood of the cusp point of Γ_σ.

It follows that the theorem must also be true of arc length parametrization or close approximations. ⋈

Proof of Theorem 1.9. Assume by contradiction that Γ is a simple curve that intersects itself during arc length evolution. The location vector of each point of Γ is a continuous function of t during arc length evolution; therefore Γ must touch itself at point P before self-intersection. Let Γ_{t_0} be such a curve. Consider two neighborhoods S_1 and S_2 of Γ_{t_0} that have point P in common. S_1 and S_2 correspond to nonoverlapping ranges of the arc length parameter W. Note that S_1 and S_2 have co-linear tangents at P. Let L be the line of those tangents. The tangents exist since it follows from Theorem 8 that P can not be a cusp point on either S_1 or S_2 because Γ_t does not self-intersect for $t \leq t_0$.

Recall that the infinitesimal movement during arc length evolution of each point of S_1 and S_2 is determined by the equation

$$\frac{\partial \mathbf{R}}{\partial t} = \kappa \mathbf{n}.$$

Therefore, during arc length evolution, every point will move in the direction of the normal vector by an amount equal to the curvature at that point. Similarly, during reverse arc length evolution, every point will move in the opposite direction of the normal vector by an amount equal to the curvature at that point.

It follows that if S_1 and S_2 are on opposite sides of L, after an infinitesimal amount of reverse arc length evolution, they will intersect. This is a contradiction of the assumption that Γ was simple before touching itself. Assume then that S_1 and S_2 are on the same side of L. Note that S_1 and S_2 can not be overlapping since they would still be overlapping after an infinitesimal amount of reverse arc length evolution, which is also a contradiction of the assumption that Γ was simple before touching itself. Let S_1 be the segment *inside* S_2, i.e., the tangent to S_2 always has S_1 to the same side. It can be seen that S_1 has a larger curvature at P than S_2. Therefore, after an infinitesimal amount of reverse arc length evolution, point P on S_1 and point P on S_2 will move in the same direction, but point P on S_1 will move by a larger amount. It follows that after an infinitesimal amount of reverse arc length evolution, S_1 and S_2 will intersect, which is, again, a contradiction. It follows that Γ remains simple during arc length evolution. ⋈

Proof of Theorem 1.10. It will be shown that this theorem holds for an arbitrary parametrization of Γ_σ. Therefore, it must also be true of arc length parametrization or close approximations.

Let $(x(u), y(u))$ be an arbitrary parametrization of Γ_σ with a cusp point at u_0. Using a case analysis similar to the one in the proof of Theorem 1.8 to characterize all possible kinds of singularities of Γ_σ at u_0, we can again conclude that only the singular points in cases 1 and 4 are cusp points. In case 1, the curve is approximated by (u^m, u^n) in a neighborhood of u_0, where m and n are both even. This type of cusp point cannot arise on Γ_σ if Γ is in C_1. We now turn to the cusp points of case 4. Recall that in case 4, the curve Γ_σ is approximated, in a neighborhood of u_0, by (u^m, u^n), where m is even, and n is odd. Observe that

$$\dot{x}(u) = mu^{m-1}, \quad \ddot{x}(u) = m(m-1)u^{m-2},$$

$$\dot{y}(u) = nu^{n-1}, \quad \ddot{y}(u) = n(n-1)u^{n-2}$$

and

$$\kappa(u) = \frac{\dot{x}(u)\ddot{y}(u) - \dot{y}(u)\ddot{x}(u)}{(\dot{x}(u)^2 + \dot{y}(u)^2)^{3/2}}$$

$$= \frac{mn(n-1)u^{m+n-3} - m(m-1)nu^{m+n-3}}{(m^2u^{2m-2} + n^2u^{2n-2})^{3/2}}$$

since nm, $\kappa(u)$ is always positive on either side of the cusp point in a neighborhood of u_0. Therefore, no curvature zero crossings exist in that neighborhood on Γ_σ.

We now derive analytical expressions for $\Gamma_{\sigma+\delta}$ so that it can be analyzed in a neighborhood of u_0. To blur function $f(u) = u^k$, we convolve a resealed version of that function with the function $(1/\sqrt{\pi})\,e^{-x^2}$, which is the blurring operator, as follows:

$$F(u) = \int_{-\infty}^{\infty} \frac{1}{\sqrt{\pi}} e^{-x^2} f(u + 2x\sqrt{t})\,dx$$

$$= \frac{1}{\sqrt{\pi}} \int_{-\infty}^{\infty} e^{-x^2} (u + 2x\sqrt{t})^k\,dx,$$

where t is the scale factor and controls the amount of blurring. Solving the above integral yields

$$F(u) = \sum_{\substack{p=0 \\ p \text{ even}}} 1.3.5\ldots(p-1)\frac{(2t)^{p/2}k(k-1)\ldots(k-p+1)}{p!}u^{k-p}.$$

An expression for $\Gamma_{\sigma+\delta}$ in a neighborhood of the cusp point can be obtained by blurring each of its coordinate functions:

$$X(u) = u^m + c_1 t u^{m-2} + c_2 t^2 u^{m-4} + \cdots + c_{m-2/2} t^{m-2/2} u^2 c_{m/2} t^{m/2},$$

$$Y(u) = u^n + c_1' t u^{n-2} + c_2' t^2 u^{n-4} + \cdots + c_{n-1/2}' t^{n-1/2} u.$$

Note that all constants are positive, all powers of u in $X(u)$ are even, and all powers of u in $Y(u)$ are odd. It follows that all powers of u in

$$\dot{X}(u) = mu^{m-1} + (m-2)c_1 t u^{m-3} + \cdots + 2c_{m-2/2} t^{m-2/2} u$$

are odd, all powers of u in

$$\ddot{X}(u) = m(m-1)u^{m-2} + (m-2)(m-3)c_1 t u^{m-4}$$

$$+ \cdots + 2c_{m-2/2} t^{m-2/2}$$

are even, all powers of u in

$$\dot{Y}(u) = nu^{n-1} + (n-2)c_1' tu^{n-3} + \cdots + c_{n-1/2}' t^{n-1/2}$$

are even, and all powers of u in

$$\ddot{Y}(u) = n(n-1)u^{n-2} + (n-2)(n-3)c_1' tu^{n-4} + \cdots + c_{n-3/2}' t^{n-3/2}$$

are odd.

The curvature of $\Gamma_{\sigma+\delta}$ in a neighborhood of u_0 is given by:

$$\kappa(u) = \frac{\dot{X}(u)\ddot{Y}(u) - \dot{Y}(u)\ddot{X}(u)}{(\dot{X}(u)^2 + \dot{Y}(u)^2)^{3/2}}.$$

Since the denominator of $\kappa(u)$ never goes to zero in a neighborhood of u_0, the zero crossings of $\kappa(u)$ are the same as those of its numerator. Observe that the term with the highest power of u in $\dot{X}(u)\ddot{Y}(u)$ is $mn(n-1)u^{m+n-3}$, and the term with the highest power of u in $Y(u)\ddot{X}(u)$ is $m(m-1)nu^{m+n-3}$ and that in both $\dot{X}(u)\ddot{Y}(u)$ and $\dot{Y}(u)\ddot{X}(u)$, all powers of u are even, and all constants are positive. Furthermore, note that at $u = 0$, $\dot{X}(u)\ddot{Y}(u)$ is zero, and $\dot{Y}(u)\ddot{X}(u) > 0$. Therefore, at $u = 0$, $\kappa < 0$. As u grows larger in absolute value, the terms in $\dot{X}(u)\ddot{Y}(u)$ and $Y(u)\ddot{X}(u)$ with highest powers of u become dominant (all other terms have positive powers of $t = \delta$ in them). Since the dominant terms have equal powers of u, the one with the larger coefficient becomes the larger term. Since $n > m$, the largest term in $\dot{X}(u)\ddot{Y}(u)$ becomes larger than the largest term in $\dot{Y}(u)\ddot{X}(u)$. Therefore, as u grows in absolute value, κ becomes positive. It follows that there are two curvature zero crossings in the neighborhood of u_0 on $\Gamma_{\sigma+\delta}$. These zero crossings are new since it was shown that no zero crossings exist in the neighborhood of u_0 on Γ_σ. ⊠

Proof of Theorem 1.11. This theorem will be proven for regular evolution first followed by arc length evolution. To prove it for regular evolution, let

$$\Gamma_\sigma = (X(u, \sigma), Y(u, \sigma))$$

be an evolved version of Γ and let σ be large. Theorem 2.3 in [189] showed that $\ddot{X}(u, \sigma)$ and $\ddot{Y}(u, \sigma)$ each has only two zero-crossing points. It follows from theorem 3.2 in [189] that Γ_σ must be simple and convex. It follows from theorem 3.3 in [189] that Γ remains simple and convex.

To prove the theorem for arc length evolution, let

$$\Gamma_\sigma = (X(W, \sigma), Y(W, \sigma))$$

be an arc length evolved version of Γ and let σ be large. The proof of theorem 2.3 in [189] also applies to each of $X(W, \sigma)$ and $Y(W, \sigma)$ and therefore each of $\ddot{X}(W, \sigma)$ and $\ddot{Y}(W, \sigma)$ has only two zero-crossing points. To see this, recall that the process of arc length evolution of Γ is equivalent to convolving an arc length parametric representation of Γ with a Gaussian filter based on a small width σ_0, resampling the output curve by the arc length parameter, reconvolving the resampled curve with the same filter and repeating this procedure many times. It follows from theorem 2.3 in [189] that when Γ is convolved with a Gaussian, $X(W, t_0)$ can be represented as:

$$X(W, t_0) = \sum_{k=1}^{\infty} \frac{a_k}{e^{3k^2\pi^2 t_0}} \sin(2k\pi W) + \sum_{k=1}^{\infty} \frac{b_k}{e^{4k^2\pi^2 t_0}} \cos(2k\pi W),$$

where $t_0 = \sigma_0^2/2$. $Y(W, t_0)$ can be represented in a similar way. When the output curve is resampled by the arc length parameter, function $X(W, t_0)$ is mapped to a new function $X_r(W, t_0)$:

$$X_r(W, t_0) = \sum_{k=1}^{\infty} \frac{a_k c_k}{e^{4k^2\pi^2 t_0}} \sin(2k\pi W) + \sum_{k=1}^{\infty} \frac{b_k d_k}{e^{4k^2\pi^2 t_0}} \cos(2k\pi W),$$

where c_k and d_k are multiplied by the old constants to obtain the new ones. When the resampled curve is reconvolved with the same filter, we obtain:

$$X_r(W, t_1) = \sum_{k=1}^{\infty} \frac{a_k c_k}{e^{4k^2\pi^2 t_1}} \sin(2k\pi W) + \frac{b_k d_k}{e^{4k_2\pi^2 t_1}} \cos(2k\pi W), \qquad (A.14)$$

where $t_1 > t_0$. Observe that as this procedure is repeated many times, the t_1 in the denominator of (A.14) becomes larger and larger but the coefficients in the numerator remain bounded by upper and lower bounds determined by the original function (since resampling does not alter the shape of the curve). Therefore, as stated earlier, theorem 2.3 in [189] applies to each of $X(W, \sigma)$ and $Y(W, \sigma)$. From theorem 3.2 in [189], it follows that Γ_σ must be simple and convex, and from theorem 3.5 in [189], it follows that Γ remains simple and convex.   ⋈

Appendix B
Proofs of Theorems of Chapter 10

This appendix contains the proofs of the 12 theorems of chapter 10 regarding space curves.

Proof of theorem 10.1. We will show that evolution of a space curve is invariant under a general affine transform which includes transformations consisting of rotation, uniform scaling and translation.

Let $\Gamma = (x(u), y(u), z(u))$ be a space curve and let

$$\Gamma_\sigma = (X(u, \sigma), Y(u, \sigma), Z(u, \sigma))$$

be its evolved version. If Γ_σ undergoes an affine transformation, then its original coordinates and its new coordinates, $x_1(u, \sigma), y_1(u, \sigma)$ and $z_1(u, \sigma)$, will satisfy the following relationship:

$$
\begin{aligned}
x_1(u, \sigma) &= aX(u, \sigma) + bY(u, \sigma) + cZ(u, \sigma) + d \\
y_1(u, \sigma) &= eX(u, \sigma) + fY(u, \sigma) + gZ(u, \sigma) + h \\
z_1(u, \sigma) &= iX(u, \sigma) + jY(u, \sigma) + kZ(u, \sigma) + l.
\end{aligned}
$$

If Γ is first transformed according to an affine transform and then evolved, its new coordinates, $x_2(u, \sigma), y_2(u, \sigma)$ and $x_2(u, \sigma)$, will be:

$$
\begin{aligned}
x_2(u, \sigma) &= (ax(u) + by(u) + cz(u) + d) \otimes g(u, \sigma) \\
y_2(u, \sigma) &= (ex(u) + fy(u) + gz(u) + h) \otimes g(u, \sigma) \\
z_2(u, \sigma) &= (ix(u) + jy(u) + kz(u) + l) \otimes g(u, \sigma).
\end{aligned}
$$

It follows from the distributivity of the convolution operator [135] that

$$x_2(u, \sigma) = x_1(u, \sigma)$$

$$y_2(u, \sigma) = y_1(u, \sigma)$$

373

$$z_2(u, \sigma) = z_1(u, \sigma).$$

Note that this result also holds for convolution operators other than the Gaussian. The proof for arc length evolution is similar.   ⋈

Proof of theorem 10.2. A closed space curve parametrized by the normalized arc length parameter has:

$$(x(0), y(0), z(0)) = (x(1), y(1), z(1)).$$

It follows that

$$(X(0, \sigma), Y(0, \sigma), Z(0, \sigma)) = (X(1, \sigma), Y(1, \sigma), Z(1, \sigma)).$$

The proof for arc length evolution is similar.   ⋈

Proof of theorem 10.3. Let $\Gamma = (x(u), y(u), z(u))$ be a connected space curve and let

$$\Gamma_\sigma = (X(u, \sigma), Y(u, \sigma), Z(u, \sigma))$$

be its evolved version. We will show that Γ_σ is also a connected curve.

Since Γ is connected, $x(u), y(u)$ and $z(u)$ are continuous functions. Since these functions remain continuous after convolution with a Gaussian, $X(u, \sigma), Y(u, \sigma)$ and $Z(u, \sigma)$ are also continuous. Let u_0 be any value of parameter u and let x_0, y_0 and z_0 be the values of $X(u, \sigma), Y(u, \sigma)$ and $Z(u, \sigma)$ at u_0, respectively. If there is an infinitesimal change in u, then there will also be infinitesimal changes in $X(u, \sigma), Y(u, \sigma)$ and $Z(u, \sigma)$:

$$X(u, \sigma) \rightarrow x_0 + \delta$$
$$Y(u, \sigma) \rightarrow y_0 + \xi$$
$$Z(u, \sigma) \rightarrow z_0 + \zeta.$$

Therefore, point $P(x_0, y_0, z_0)$ on Γ_σ will go to point $P'(x_0 + \delta, y_0 + \xi, z_0 + \zeta)$. Let D be the distance between P and P' and let $|\delta|$ be the largest of $|\delta|, |\xi|$ and $|\zeta|$. Then

$$D = \sqrt{\delta^2 + \xi^2 + \zeta^2} \leq \sqrt{3\delta^2} = \delta\sqrt{3}.$$

Therefore, an infinitesimal change in parameter u results in an infinitesimal change in the position of point P. It follows that Γ_σ is a connected space curve. The proof for arc length evolution is similar.   ⋈

Proof of theorem 10.4. Let M be the center of mass of $\Gamma = (x(u), y(u), z(u))$ with coordinates (x_M, y_M, z_M). Then

$$x_M = \frac{\int_0^1 x(u)\, du}{\int_0^1 du} = \int_0^1 x(u)\, du.$$

The evolved curve $\Gamma_\sigma = (X(u, \sigma), Y(u, \sigma), Z(u, \sigma))$, where

$$X(u, \sigma) = \int_{-\infty}^{\infty} g(v, \sigma) x(u - v)\, dv$$

has M' with coordinates (X_M, Y_M, Z_M) as its center of mass. Now observe that

$$X_M = \int_0^1 X(u, \sigma) \, du = \int_0^1 \int_{-\infty}^{\infty} g(v, \sigma) x(u - v) \, dv \, du$$

$$= \int_{-\infty}^{\infty} g(v, \sigma) \left(\int_0^1 x(u - v) \, du \right) dv = x_M.$$

Similarly, $Y_M = y_M$ and $Z_M = z_M$. It follows that $M = M'$. The proof for arc length evolution is similar. ◻

Proof of theorem 10.5. Since $\mathcal{H}$ is simple and convex, every plane P tangent to $\mathcal{H}$ contains that object in the left (or right) half-space it creates. Since Γ is inside $\mathcal{H}$, Γ is also contained in the same half-space. Now rotate P and Γ so that P becomes parallel to the YZ-plane and therefore has equation $x = c$. Since P does not intersect Γ, it follows that $x(u_0) \geq c$ for every point up on Γ. Let Γ_σ be an evolved version of Γ. Every point of Γ_σ is a weighted average of all the points of Γ. Therefore, $X(u_0, \sigma) \geq c$ for every point u_0 on Γ_σ is also contained in the same half-space. This result holds for *every* plane tangent to $\mathcal{H}$, therefore Γ_σ is contained inside the intersection of all the left (or right) half-spaces created by the tangent planes of $\mathcal{H}$. So, Γ_σ is also inside $\mathcal{H}$. The proof for arc length evolution is similar. ◻

Proof of theorem 10.6. The proof has been divided into three sections. Section A shows that the derivatives at a point on a torsion zero-crossing contour provide homogeneous equations in the moments of the coordinate functions of the curve. Section B shows that the moments are related to the coefficients of expansion of the coordinate functions of the curve in functions related to the Hermite polynomials. Section C shows that the cubic equations obtained in section A can be converted into a homogeneous linear system of equations which can be solved uniquely for function $\tau(u)\kappa^2(u)$.

A. Constraints from the torsion zero-crossing Contours

Let $\Gamma = (x(u), y(u), z(u))$ be the arc-length parametrization of the curve with Fourier transform $\tilde{\Gamma} = (\tilde{x}(\omega), \tilde{y}(\omega), \tilde{z}(\omega))$. The Fourier transform of the Gaussian filter

$$G(u, t) = \frac{1}{\sqrt{2t}} e^{u^2/4t}$$

is $\tilde{G}(\omega) = e^{-\omega^2 t}$.

Let $\Gamma_{t_0} = (x(u, t_0), y(u, t_0), z(u, t_0))$ be a curve obtained by convolving $x(u)$, $y(u)$ and $x(u)$ with $G(u, t_0)$. Since Γ is in C_1, there exists a t_0 such that Γ_{t_0} is in C_∞. The torsion zero-crossings in a neighborhood of t_0 are given by solutions of $\beta(u, t) = 0$ where [95]

$$\beta(u, t) = \dot{x}(\ddot{y}\,\dddot{z} - \dddot{y}\,\ddot{z}) - \dot{y}(\ddot{x}\,\dddot{z} - \dddot{x}\,\ddot{z}) + \dot{z}(\ddot{x}\,\dddot{y} - \dddot{x}\,\ddot{y}) \qquad (\text{B.1})$$

where $\dot{}$ represents derivative with respect to u. Note that on Γ $(t = 0)$, $\beta(u, t)$ is given by

$$\beta(u,t) = \tau(u,t)\kappa^2(u,t). \tag{B.2}$$

Using the convolution theorem, $\dot{x}(u,t)$, $\dot{y}(u,t)$ and $\dot{z}(u,t)$ can be expressed as follows:

$$\dot{x} = \int e^{-\omega^2 t} e^{i\omega u} i\omega \tilde{x}(\omega)\, d\omega,$$

$$\dot{y} = \int e^{-\omega^2 t} e^{i\omega u} i\omega \tilde{y}(\omega)\, d\omega,$$

$$\dot{z} = \int e^{-\omega^2 t} e^{i\omega u} i\omega \tilde{z}(\omega)\, d\omega,$$

and therefore

$$\ddot{x} = \int e^{-\omega^2 t} e^{i\omega u} (-\omega^2) \tilde{x}(\omega)\, d\omega,$$

$$\ddot{y} = \int e^{-\omega^2 t} e^{i\omega u} (-\omega^2) \tilde{y}(\omega)\, d\omega,$$

$$\ddot{z} = \int e^{-\omega^2 t} e^{i\omega u} (-\omega^2) \tilde{z}(\omega)\, d\omega,$$

$$\dddot{x} = \int e^{-\omega^2 t} e^{i\omega u} (-\omega^3) \tilde{x}(\omega)\, d\omega,$$

$$\dddot{y} = \int e^{-\omega^2 t} e^{i\omega u} (-\omega^3) \tilde{y}(\omega)\, d\omega,$$

$$\dddot{z} = \int e^{-\omega^2 t} e^{i\omega u} (-\omega^3) \tilde{z}(\omega)\, d\omega.$$

Note that the moment of order k of the function $f(\omega) = e^{-\omega^2 t} e^{i\omega u} (i\omega) \tilde{x}(\omega)$ is defined by:

$$M_k = \int\limits_{-\infty}^{\infty} (i\omega)^k\, e^{-\omega^2 t}\, e^{i\omega u} (i\omega) \tilde{x}(\omega)\, d\omega$$

the moment of order k of the function $f'(\omega) = e^{-\omega^2 t} e^{i\omega u} (i\omega) \tilde{y}(\omega)$ is defined by:

$$M'_k = \int\limits_{-\infty}^{\infty} (i\omega)^k\, e^{-\omega^2 t}\, e^{i\omega u} (i\omega) \tilde{y}(\omega)\, d\omega$$

and the moment of order k of the function $f''(\omega) = e^{-\omega^2 t} e^{i\omega u} (i\omega) \tilde{z}(\omega)$ is defined by:

$$M''_k = \int\limits_{-\infty}^{\infty} (i\omega)^k\, e^{-\omega^2 t}\, e^{i\omega u} (i\omega) \tilde{z}(\omega)\, d\omega.$$

Therefore Equation (B.2) can be written as:

$$\begin{aligned}
\beta(u,t) \;=\; & M_0 M'_1 M''_2 - M_0 M''_1 M'_2 - M'_0 M_1 M''_2 \\
& + M'_0 M_2 M''_1 + M''_0 M_1 M'_2 - M''_0 M_2 M'_1.
\end{aligned}$$

The Implicit Function Theorem guarantees that the contours $u(t)$ are C_∞ in a neighborhood of t_0. Let ξ be a parameter of the torsion zero-crossing contour. Then

$$\frac{d}{d\xi} = \frac{du}{d\xi}\frac{\partial}{\partial u} + \frac{dt}{d\xi}\frac{\partial}{\partial t}.$$

On the torsion zero-crossing contour, $\beta = 0$ and $(d^k/d\xi^k)\beta = 0$ for all integers k. Furthermore, since the torsion zero-crossing contour is known, all the derivatives of u and t with respect to ξ are known as well. We now compute the derivatives of β with respect to ξ at (u_0, t_0). The first derivative is given by:

$$\frac{d}{d\xi}\beta(u_0, t_0) = \frac{du}{d\xi}\frac{\partial\beta(u_0, t_0)}{\partial u} + \frac{dt}{d\xi}\frac{\partial\beta(u_0, t_0)}{\partial t}, \tag{B.3}$$

where

$$\begin{aligned}
\frac{\partial\beta(u_0, t_0)}{\partial u} &= M_3'' M_0 M_1' - M_3' M_0 M_1'' - M_3'' M_0' M_1 \\
&\quad + M_3 M_0' M_1'' + M_3' M_0'' M_1 - M_3 M_0'' M_1'
\end{aligned}$$

and

$$\begin{aligned}
\frac{\partial\beta(u_0, t_0)}{\partial t} &= M_3' M_0 M_2'' + M_4'' M_0 M_1' - M_3'' M_0 M_2' \\
&\quad - M_4' M_0 M_1'' - M_3 M_0' M_2'' - M_4'' M_0' M_1 \\
&\quad + M_4 M_0' M_1'' + M'' + 3M_0' M_2 + M_3 M_0'' M_2' \\
&\quad + M_4' M_0'' M_1 - M_4 M_0'' M_1' - M_3' M_0'' M_2
\end{aligned}$$

and the second derivative is given by:

$$\begin{aligned}
\frac{\partial^2\beta}{\partial\xi^2} &= \frac{d^2u}{d\xi^2}\frac{\partial\beta}{\partial u} + \frac{d^2t}{d\xi^2}\frac{\partial\beta}{\partial t} + \left(\frac{du}{d\xi}\right)^2\frac{\partial^2\beta}{\partial u^2} \\
&\quad + 2\frac{du}{d\xi}\frac{dt}{d\xi}\frac{\partial^2\beta}{\partial u\partial t} + \left(\frac{dt}{d\xi}\right)^2\frac{\partial^2\beta}{\partial t^2}
\end{aligned} \tag{B.4}$$

where

$$\begin{aligned}
\frac{\partial^2\beta}{\partial u^2} &= M_4'' M_0 M_1' + M_2' M_3'' M_0 - M_4'' M_0 M_1'' \\
&\quad - M_2'' M_3' M_0 - M_4'' M_0' M_1 - M_2 M_3'' M_0' \\
&\quad + M_4 M_0' M_1'' + M_2'' M_3 M_0 + M_4' M_0'' M_1 \\
&\quad + M_2 M_3' M_0'' - M_4 M_0'' M_1' - M_2' M_3 M_0'',
\end{aligned}$$

$$\begin{aligned}
\frac{\partial^2\beta}{\partial u\partial t} &= M_5'' M_0 M_1' + M_2 M_3'' M_1' - M_5' M_0 M_1'' \\
&\quad - M_2 M_3' M_1'' - M_5'' M_0' M_1 - M_2' M_3'' M_1 \\
&\quad + M_5 M_0' M_1'' + M_2' M_3 M_1'' + M_5' M_0'' M_1 \\
&\quad + M_2'' M_3' M_1 - M_5 M_0'' M_1' - M_2'' M_3 M_1',
\end{aligned}$$

and

$$
\begin{aligned}
\frac{\partial^2 \beta}{\partial t^2} =\ & M_5' M_0 M_2'' + 2M_4'' M_3' M_0 + M_6'' M_0 M_1' \\
& + M_2 M_4'' M_1' - M_5'' M_0 M_2' - 2M_4' M_3'' M_0 \\
& - M_6' M_0 M_1'' - M_2 M_4' M_1'' - M_5 M_0' M_2'' \\
& - 2M_4'' M_3 M_0' - M_6'' M_0' M_1 - M_2' M_4'' M_1 \\
& + M_6 M_0' M_1'' + M_2' M_4 M_1'' + 2M_3'' M_4 M_0' \\
& + M_5'' M_0' M_2 + M_5 M_0'' M_2' + 2M_4' M_3 M_0'' \\
& + M_6' M_0'' M_1 + M_2'' M_4' M_1 - M_6 M_0'' M_1' \\
& - M_2'' M_4 M_1' - 2M_3' M_4 M_0'' - M_5' M_0'' M_2 .
\end{aligned}
$$

Since the parametric derivatives along the torsion zero-crossing contours are zero, Equations (B.3) and (B.4) are equal to zero. Note that equation (B.3) is in the first five moments of functions $f(\omega)$, $f'(\omega)$ and $f''(\omega)$ and equation (B.4) is in the first seven moments of those functions. In general; the $k+1$st equation, $(d^k/d\xi^k)\beta(u,t) = 0$ is a cubic equation in the first $2k+3$ moments of each of the functions $f(\omega)$, $f'(\omega)$ and $f''(\omega)$.

B. The moments and the coefficients of expansion of $\dot{x}(u)$, $\dot{y}(u)$ and $\dot{z}(u)$

This section shows that the moments and the moment-triples in equations

$$(d^k/d\xi^k)\beta(u,t)$$

are related respectively to the coefficients of the expression of the functions $\dot{x}(u)$, $\dot{y}(u)$ and $\dot{z}(u)$ and function $\beta(u)$ in functions related to the Hermite polynomials.

As shown in the proof of theorem 1.6 in Appendix A, $a_k(\sigma)$ and $b_k(\sigma)$, the coefficients of expansion of $\dot{x}(u)$ and $\dot{y}(u)$ in functions $\phi_k(u,\sigma)$, can be shown to be equal to M_k and M_k' modulo a factor $e^{i\omega u}$. Similarly, $c_k(\sigma)$, the coefficients of expansion of $\dot{z}(u)$ can be shown to be equal to M_k'' modulo a factor $e^{i\omega u}$. Furthermore, $a_k'(\sigma)$, $b_k'(\sigma)$ and $c_k'(\sigma)$, the coefficients of expansion of functions $\ddot{x}(u)$, $\ddot{u}(u)$ and $\ddot{z}(u)$ in $\phi_k(u,\sigma)$ respectively, can be seen to be related to $a_k(\sigma)$, $b_k(\sigma)$ and $c_k(\sigma)$ according to the following relationships:

$$
\begin{aligned}
a_{k-1}'(\sigma) &= a_k(\sigma), \\
b_{k-1}'(\sigma) &= b_k(\sigma), \\
c_{k-1}'(\sigma) &= c_k(\sigma)
\end{aligned}
\tag{B.5}
$$

and $a_k''(\sigma)$, $b_k''(\sigma)$ and $c_k''(\sigma)$, the coefficients of expansion of functions $\dddot{x}(u)$, $\dddot{y}(u)$ and $\dddot{z}(u)$ in $\phi_k(u,\sigma)$ respectively, can be seen to be related to $a_k(\sigma)$, $b_k(\sigma)$ and $c_k(\sigma)$ by the following relationships:

$$a_{k-2}''(\sigma) = a_k(\sigma),$$

$$b''_{k-2}(\sigma) = b_k(\sigma),$$

$$c''_{k-2}(\sigma) = c_k(\sigma). \tag{B.6}$$

Now observe that function $\tau(u)\kappa^2(u)$ can be expressed as:

$$
\begin{aligned}
\tau(u)\kappa^2(u) &= \dot{x}(u)\ddot{y}(u)\,\dddot{z}\,(u) - \dot{x}(u)\,\dddot{y}\,(u)\ddot{z}(u) - \dot{y}(u)\ddot{x}(u)\,\dddot{z}\,(u) \\
&\quad + \dot{y}(u)\,\dddot{x}\,(u)\ddot{z}(u) + \dot{z}(u)\ddot{x}(u)\,\dddot{y}\,(u) - \dot{z}(u)\,\dddot{x}\,(u)\ddot{y}(u) \\
&= \sum\sum\sum (a_i b'_j c''_k + b_i a''_j c'_k + c_i a'_j b''_k - a_i b''_j c'_k \\
&\quad - b_i a'_j c''_k - c_i a''_j b'_k)\phi_i\phi_j\phi_k.
\end{aligned}
$$

Using (B.5) and (B.6) we obtain

$$
\begin{aligned}
\tau(u)\kappa^2(u) &= \sum\sum\sum (a_i b_{j+1} c_{k+2} + b_i a_{j+2} c_{k+1} + c_i a_{j+1} b_{k+2} \\
&\quad - a_i b_{j+2} c_{k+1} - b_i a_{j+1} c_{k+2} - c_i a_{j+2} b_{k+1})\phi_i\phi_j\phi_k.
\end{aligned}
$$

It follows that if the triples $a_i(\sigma)b_j(\sigma)c_k(\sigma)$ in the equation above are all known, the function $\beta(u) = \tau(u)\kappa^2(u)$ can be reconstructed.

C. Reconstructing the Function $\tau(u)\kappa^2(u)$

Equations (B.3) and (B.4) can be converted into homogeneous linear equations by assuming that each moment-triple appearing in those equations is a new variable. Again we proceed to compute the first n derivatives at point (u_0, t_0) on one of the torsion zero-crossing contours. We now obtain $n+1$ homogeneous linear equations in some of the moment-triples $M_i M'_j M''_k$. Since this system is in terms of the first $2n+3$ moments of functions $f(\omega)$, $f'(\omega)$ and $f''(\omega)$; it will contain $O(n^3)$ moment-triples. Therefore additional equations are required to constrain the system. To obtain those equations, we proceed as follows:

Assume that moments of order higher than $2n + 2$ are zero. Compute derivatives of order higher than n at (u_0, t_0) but set moments of order higher than $2n+2$ to zero in the resulting equations. If a sufficient number of derivatives are computed at (u_0, t_0), the number of equations obtained will be equal to the number of moment-triples and our linear system will be constrained. Again note that this procedure enables us to use only one point to obtain all the needed equations.

It follows from an assumption of generality that the system will have a unique zero eigenvector and therefore a unique solution modulo scaling. Once the moment-triples in the system are known, all other moment-triples can be computed from the known ones using the following relationships:

$$
\begin{aligned}
M_i M'_j M''_k &= \frac{M_i M'_{j-1} M''_k \cdot M_{i+1} M'_j M''_k}{M_{i+1} M'_{j-1} M''_k} = \frac{M_i M'_{j-1} M''_k \cdot M_{i-1} M'_j M''_k}{M_{i-1} M'_{j-1} M''_k} \\
&= \frac{M_{i-1} M'_j M''_k \cdot M_i M'_{j+1} M''_k}{M_{i-1} M'_{j+1} M''_k} = \frac{M_{i+1} M'_j M''_k \cdot M_i M'_{j+1} M''_k}{M_{i+1} M'_{j+1} M''_k}
\end{aligned}
$$

$$= \frac{M_i M'_{j-1} M''_k \cdot M_i M'_j M''_{k-1}}{M_i M'_{j-1} M''_{k-1}} = \frac{M_i M'_{j+1} M''_k \cdot M_i M'_j M''_{k-1}}{M_i M'_{j+1} M''_{k-1}}$$

$$= \frac{M_i M'_{j-1} M''_k \cdot M_i M'_j M''_{k+1}}{M_i M'_{j-1} M''_{k+1}} = \frac{M_i M'_j M''_{k+1} \cdot M_i M'_{j+1} M''_k}{M_i M'_{j+1} M''_{k+1}}$$

$$= \frac{M_i M'_j M''_{k-1} \cdot M_{i-1} M'_j M''_k}{M_{i-1} M'_j M''_{k-1}} = \frac{M_{i-1} M'_j M''_k \cdot M_i M'_j M''_{k+1}}{M_{i-1} M'_j M''_{k+1}}$$

$$= \frac{M_i M'_j M''_{k+1} \cdot M_{i+1} M'_j M''_k}{M_{i+1} M'_j M''_{k+1}} = \frac{M_{i+1} M'_j M''_k \cdot M_i M'_j M''_{k-1}}{M_{i+1} M'_j M''_{k-1}}.$$

Since all the moment-triples $M_i M'_j M''_k$ together determine function of $\beta(u)$, it follows that function $\beta(u)$ can be determined modulo constant scaling.

It has been shown in [333] that a 1-D signal can be reconstructed using two points from its scale space image. Note that our result implies that only one point is sufficient for the reconstruction of that signal. To prove the same result about the resampled TSS image, recall that derivatives at one point (at any scale) on any torsion zero-crossing contour in the TSS image of Γ were computed and it was shown that the resulting equations can be solved for the coefficients of expansion of $\tau^*(u)$ in functions related to the Hermite polynomials. As before, choose a point on a zero-crossing contour at any scale of the resampled TSS image of Γ and compute the necessary derivatives. The value of σ in the resulting equations is then set to zero. Consequently, the arc length evolved curve Γ_σ, where σ corresponds to the scale at which the derivatives were computed, is reconstructed modulus uniform scaling, rotation and translation. The next step is to recover the original curve Γ modulo function $\tau^*(u)$. This is done by applying *reverse* arc length evolution to Γ_σ. Let arc length evolved curve Γ_σ be defined by:

$$\Gamma_\sigma = \{(\mathcal{X}(\mathcal{W}, \sigma), \mathcal{Y}(\mathcal{W}, \sigma), \mathcal{Z}(\mathcal{W}, \sigma) \mid \mathcal{W} \in [0, 1]\}$$

A reverse arc length evolved curve Γ is defined by:

$$\Gamma = \{(x(w), y(w), z(w)) \mid w \in [0, 1]\}$$

where

$$x(w) = \mathcal{X}(w, \sigma) \otimes D_N(w, \sigma)$$
$$y(w) = \mathcal{Y}(w, \sigma) \otimes D_N(w, \sigma)$$
$$z(w) = \mathcal{Z}(w, \sigma) \otimes D_N(w, \sigma)$$

where D_N is a deblurring operator defined in [120], and

$$w(\mathcal{W}, t) = \int_0^t \int_0^w \kappa^2(w, t) dw dt$$

where $t = \sigma^2/2$. As a result, Γ is recovered modulo function $\tau^*(u)$.

To prove the same result about the renormalized TSS image, evolved curve Γ_σ is again reconstructed, then each of its coordinate functions is deblurred by convolution with D_N. Once again Γ is recovered modulo function $\tau^*(u)$.   ⋈

Proof of theorem 10.7. Since by assumption all evolved and arc-length evolved curves Γ_σ are in C_3, the conditions of the implicit function theorem are satisfied on contours $\tau(u, \sigma)$ in the torsion scale space image of Γ. Since the proofs for evolution and arc-length evolution are similar, the theorem will be proven here for the regular torsion scale space image.

On any contour in the TSS image of Γ, $\tau(u, \sigma) = 0$. Since all Γ_σ are in C_3, the conditions of the implicit function theorem are satisfied on contours $\beta(u, t) = 0$, where $\beta(u, t)$ is the numerator of the torsion function given in chapter 10:

$$t = t(u),$$

$$t_u(u) = \frac{dt}{du} = \frac{-\beta_u}{\beta_t}.$$

The functions $X(u, t)$, $Y(u, t)$ and $Z(u, t)$ satisfy the heat equation:

$$X_{uu}(u, t) = X_t(u, t),$$

$$Y_{uu}(u, t) = Y_t(u, t),$$

$$Z_{uu}(u, t) = Z_t(u, t).$$

The theorem will be proven if it is shown that if $t_u(u) = 0$ at any point on a torsion zero-crossing contour, then $t_{uu}(u) < 0$ at that point. Observe that $t_u(u) = 0$ if and only if $\beta_u(u, t) = 0$. It follows that at a point where $t_u(u) = 0$:

$$t_{uu}(u) = \frac{d}{du}\left(\frac{-\beta_u}{\beta_t}\right) = \frac{\partial}{\partial u}\left(\frac{-\beta_u}{\beta_t}\right) + \frac{\partial}{\partial t}\left(\frac{-\beta_u}{\beta_t}\right)\frac{dt}{du} = \frac{-\beta_{uu}}{\beta_u}.$$

Therefore it must be shown that if $\beta(u, t) = \beta_u(u, t) = 0$ then $\beta_{uu}/\beta_t > 0$.

We will now derive explicit expressions for β_{uu} and β_t. Differentiating $\beta(u, t)$ with respect to u and simplifying yields:

$$\beta_u(u, t) = Z_{tt}X_uY_t - Z_{tt}Y_uX_t + Y_{tt}Z_uX_t$$
$$- Y_{tt}X_uZ_t + X_{tt}Y_uZ_t - X_{tt}Z_uY_t.$$

Differentiating β_u with respect to u and simplifying yields:

$$\beta_{uu} = \Psi_1 + \Psi_2,$$

where

$$\Psi_1 = Z_{ttu}X_uY_t - Z_{ttu}Y_uX_t + Y_{ttu}Z_uX_t$$
$$- Y_{ttu}X_uZ_t + X_{ttu}Y_uZ_t - X_{ttu}Z_uY_t$$

and

$$\Psi_2 = Y_{tu}Z_{tt}X_u - X_{tu}Z_{tt}Y_u + X_{tu}Y_{tt}Z_u$$
$$- Z_{tu}Y_{tt}X_u + Z_{tu}X_{tt}Y_u - Y_{tu}X_{tt}Z_{tt}.$$

Differentiating the expression for $\beta(u, t)$ with inspect to t and simplifying yields:

$$\beta_t = \Psi_1 - \Psi_2.$$

Let P be a point on an evolved curve Γ_σ where $\beta(u,t) = \beta_u(u,t) = 0$. The coordinate functions of Γ_σ can be locally approximated at P using polynomial functions. Furthermore, assume that $u = 0$ at point P. It follows that (u^m, u^n, u^p) is a local approximation to Γ_σ around P where m, n and p are the lowest non-zero powers of the polynomials approximating functions $X(u,t)$, $Y(u,t)$ and $Z(u,t)$ respectively. Also assume without loss of generality that $p > n > m$. Observe that

$$X_u = mu^{m-1},$$

$$X_t = m(m-1)u^{m-2},$$

$$X_{tu} = m(m-1)(m-2)u^{m-3},$$

$$X_{tt} = m(m-1)(m-2)(m-3)u^{m-4},$$

$$X_{ttu} = m(m-1)(m-2)(m-3)(m-4)u^{m-5}$$

and that

$$Y_u = nu^{m-1},$$

$$Y_t = n(n-1)u^{n-2},$$

$$Y_{tu} = n(n-1)(n-2)u^{n-3},$$

$$Y_{tt} = n(n-1)(n-2)(n-3)u^{n-4},$$

$$Y_{ttu} = n(n-1)(n-2)(n-3)(n-4)u^{n-5}$$

and that

$$Z_u = pu^{p-1},$$

$$Z_t = p(p-1)u^{p-2},$$

$$Z_{tu} = p(p-1)(p-2)u^{p-3},$$

$$Z_{tt} = p(p-1)(p-2)(p-3)u^{p-4},$$

$$Z_{ttu} = p(p-1)(p-2)(p-3)(p-4)u^{p-5}.$$

It now follows that at point P:

$$\frac{\beta_{uu}}{\beta_t} = \frac{\Xi_1 u^{m+n+p-8} + \Xi_2 u^{m+n+p-8}}{\Xi_1 u^{m+n+p-8} - \Xi_2 u^{m+n+p-8}} = \frac{\Xi_1 + \Xi_2}{\Xi_1 - \Xi_2},$$

where

$$\Xi_1 = (p-1)(p-2)(p-3)(p--4)(n-m)$$
$$+ (n-1)(n2)(n-3)(n-4)(m-p)$$
$$+ (m-1)(m-2)(m-3)(m-4)(p-n)$$

and

$$\Xi_2 = (p-1)(p-2)(p-3)[(n-1)(n-2)-(m-1)(m-2)]$$
$$+ (n-1)(n-2)(n-3)[(m-1)(m-2)-(p-1)(p-2)]$$
$$+ (m-1)(m-2)(m-3)[(p-1)(p-2)-(n-1)(n-2)].$$

It can be shown that:

$$\Xi_1 = (p-n)(p-m)(n-m)(p^2 + (n+m-10)p + n^2 + m^2 + mn$$
$$- 10n - 10m + 35)$$

and that:

$$\Xi_2 = (p-n)(p-m)(n-m)(p(n+m) - 3(n+m) - 3p + mn + 7).$$

It can now be concluded that to prove the theorem, it must be shown that:

$$|\Xi_1| \geq |\Xi_2|$$

or

$$|\Delta_1| \geq |\Delta_2|,$$

where

$$\Delta_1 = p^2 + n^2 + m^2 + np + mp + mn - 10p - 10n - 10m + 35$$

and

$$\Delta_2 = np + mp + mn - 3p - 3n - 3m + 7.$$

We shall now use a case analysis to prove that the inequality above holds for all valid triples of values of m, n and p. The analysis below shows that only triples of values which satisfy the inequality:

$$p > m + n$$

are valid:

Recall that (u^m, u^n, u^p) was used to approximate the curve around point P. It follows that in a neighborhood of P, torsion is given by:

$$\tau(u) = \frac{\lambda u^{p+n+m-6}}{\lambda_1 u^{2(p+n-3)} + \lambda_2 u^{2(p+m-3)} + \lambda_3 u^{2(m+n-3)}},$$

where λ, λ_1, λ_2 and λ_3 are constants. The expression above is ambiguous at $u = 0$. To resolve the ambiguity, l'Hopital's rule can be applied repeatedly. Since both the numerator and the denominator are polynomials, to have $\tau(u) = 0$ at $u = 0$, repeated application of l'Hopital's rule should result in:

$$\lim_{u \to 0} \tau(u) = \frac{\psi u^i}{\xi + f(u)},$$

where ψ and ξ are constants, $i > 0$ and $f(u) = 0$ at $u = 0$. This can only happen when one of the following three conditions are met:

a. $p + n + m - 6 > 2(p + n - 3)$,
b. $p + n + m - 6 > 2(p + m - 3)$,
c. $p + n + m - 6 > 2(m + n - 3)$.

Conditions **a** and **b** are not possible since they violate the assumption that $p > n > m$. However, condition **c** is possible. It follows from this condition that:

$$p > m + n.$$

We can now proceed with the case analysis. All triples of values for m, n and p in which m is even correspond to cusp points [187, 186] which are excluded by the assumption that all evolved curves Γ_σ are in C_3. Therefore we will consider only odd values of m.

Case 1. Suppose $m \geq 7$. Recall that $p > n > m$. It is easily seen that both Δ_1 and Δ_2 are positive. So the absolute value signs can be dropped and the inequality:

$$\Delta_1 \geq \Delta_2$$

can be simplified. As a result, we must now show that the following inequality holds:

$$p^2 + n^2 + m^2 \geq 7p + 7n + 7m - 28.$$

Note that $m^2 \geq 7m$, $n^2 > 7n$ and $p^2 > 7p$. It follows that the inequality does hold.

Case 2. Suppose $m = 5$. Again, it can be seen that both Δ_1 and Δ_2 are positive. We must again show that:

$$p^2 + n^2 + m^2 \geq 7p + 7n + 7m - 28.$$

Substitute $m = 5$ in the above inequality. We now have:

$$p^2 + n^2 \geq 7p + 7n - 18.$$

Since $n \geq 6$, $n^2 \geq 7n - 18$ and since $p > 11$, $p^2 > 7p$. Hence the inequality again holds.

Case 3. Suppose $m = 3$. Substitute this value for m in Δ_1. As a result, Δ_1 simplifies to:

$$p^2 + n^2 + np - 7p - 7n + 14.$$

Note that $n \geq 4$ and $p \geq 8$. So $p^2 > 7p$. Hence to show that Δ_1 is positive, it is sufficient to show that:

$$n^2 + np - 7n + 14 > 0.$$

Since $p \geq 8$, $np \geq 8n$. Therefore:

$$n^2 + np - 7n + 14 \geq n^2 + 8n - 7n + 14 = n^2 + n + 14 > 0.$$

Now substitute $m = 3$ in Δ_2. As a result Δ_2 simplifies to:

$$np + 3p + 3n - 3p - 3n - 9 + 7 = np - 2$$

which is always positive. Therefore we must again show that:

$$p^2 + n^2 + m^2 \geq 7p + 7n + 7m - 28.$$

Substituting $m = 3$ in the above inequality yields:

$$p^2 + n^2 \geq 7p + 7n - 16.$$

Since $p \geq 8$, $p^2 > 7p$ and it is sufficient to show that:

$$n_2 \geq 7_n - 16.$$

It is easily seen that this inequality is satisfied for $n \geq 4$.

Case 4. Suppose $m = 1$. Substituting this value in Δ_1 simplifies it to:

$$p^2 + n^2 + np - 9p - 9n + 26.$$

Since $p \geq 4$, $p^2 - 9p \geq -20$. Hence to show that Δ_1 is non-negative, it is sufficient to show that:

$$n^2 + np - 9n + 6 \geq 0.$$

Again since $p \geq 4$:

$$n^2 + np - 9n + 6 \geq n^2 + 4n - 9n + 6 = n^2 - 5n + 6$$

which is non-negative for $n \geq 2$. Now substitute for $m = 1$ in Δ_2 to obtain:

$$pn - 2p - 2n + 4 = p(n - 2) - 2n + 4.$$

Since $p \geq 4$

$$p(n - 2) - 2n + 4 \geq 4(n - 2)2n + 4 = 4n - 8 - 2n + 4 = 2n - 4$$

which is non-negative since $n \geq 2$. So Δ_2 is also non-negative. Therefore we must again show that:

$$p^2 + n^2 + m^2 \geq 7p + 7n + 7m - 28.$$

Substitute for $m = 1$ in the above expression to obtain:

$$p^2 + n^2 \geq 7p + 7n - 22$$

which is equivalent to:

$$(p^2 - 7p) + (n^2 - 7n) - 22 \geq 0.$$

If $n = 2$, then $n^2 - 7n = -10$ and $p \geq 4$. It follows from $p \geq 4$ that $p^2 - 7p \geq -12$. As a result, the inequality above is satisfied. If $n > 2$, then $n^2 - 7n \geq -12$ and $p \geq 5$. It follows from $p \geq 5$ that $p^2 - 7p \geq -10$. Therefore, the inequality above is again satisfied.

This completes the case analysis. We have shown that the inequality:

$$|\Delta_1| \geq |\Delta_2|$$

and therefore the inequality:

$$|\Xi_1| \geq |\Xi_2|$$

is satisfied for all valid triples of values of m, n and p. Therefore β_{uu}/β_t is always positive. Hence all extrema of contours in all torsion scale space images of Γ are maxima [227].   ⋈

Proof of theorem 10.8. It has been shown in [120] that the class of polynomial functions is closed under convolution with a Gaussian. Therefore, $X(u, \sigma), Y(u, \sigma)$ and $Z(u, \sigma)$ are also polynomial functions:

$$
\begin{aligned}
X(u, \sigma) &= a_0 + a_1 u + a_2 u^2 + a_3 u^3 + \cdots, \\
Y(u, \sigma) &= b_0 + b_1 u + b_2 u^2 + b_3 u^3 + \cdots, \\
Z(u, \sigma) &= c_0 + c_1 u + c_2 u^2 + c_3 u^3 + \cdots.
\end{aligned}
$$

Let Γ_σ go through the origin of the coordinate system at $u_0 = 0$. It follows that:

$$a_0 = b_0 = c_0 = 0.$$

Every cusp point is also a singular point of the curve. Therefore Γ_σ has a singularity at u_0. Now observe that

$$
\begin{aligned}
X_u(u,\sigma) &= a_1 + 2a_2u + 3a_3u^2 + 4a_4u^3 + \cdots, \\
Y_u(u,\sigma) &= b_1 + 2b_2u + 3b_3u^2 + 4b_4u^3 + \cdots, \\
Z_u(u,\sigma) &= c_1 + 2c_2u + 3c_3u^2 + 4c_4u^3 + \cdots.
\end{aligned}
$$

$X_u(u,\sigma)$, $Y(u,\sigma)$ and $Z_u(u,\sigma)$ are zero at a singular point. It follows that:

$$
a_1 = b_1 = c_1 = 0.
$$

As a result, all powers of u in $X(u,\sigma), Y(u,\sigma)$ and $Z(u,\sigma)$ are larger than one.

The following case analysis identifies the cases in which the singular point at u_0 is also a cusp point. Since Γ_σ is examined in a small neighborhood of point $u_0 = 0$, it can be approximated using the lowest degree terms in $X(u,\sigma), Y(u,\sigma)$ and $Z(u,\sigma)$:

$$
\Gamma_\sigma = (u^m, u^n, u^p).
$$

Assume without loss of generality that $p > n > m$. Observe that

$$
r_u(u,\sigma) = (X_u(u,\sigma), Y_u(u,\sigma), Z_u(u,\sigma)) = (mu^{m-1}, nu^{n-1}, pu^{p-1}).
$$

Therefore

$$
\begin{aligned}
r_u(\epsilon,\sigma) &= (m\epsilon^{m-1}, n\epsilon^{n-1}, p\epsilon^{p-1}) = \epsilon^{m-1}(m, n\epsilon^{n-m}, p\epsilon^{p-m}) \\
r_u(-\epsilon,\sigma) &= (m(-\epsilon)^{m-1}, n(-\epsilon)^{n-1}, p(-\epsilon)^{p-1}).
\end{aligned}
$$

Since m, n and p can be odd or even, the singular point at u_0 must be analyzed in each of eight possible cases:

1. m, n and p are even: $m-1, n-1$ and $p-1$ are odd. So,

$$
r_u(-\epsilon,\sigma) = (-m\epsilon^{m-1}, -n\epsilon^{n-1}, -p\epsilon^{p-1}) = -\epsilon^{m-1}(m, n\epsilon^{n-m}, p\epsilon^{p-m})
$$

2. m and n are even, p is odd: $m-1$ and $n-1$ are odd and $p-1$ is even. Hence,

$$
r_u(-\epsilon,\sigma) = (-m\epsilon^{m-1}, -n\epsilon^{n-1}, p\epsilon^{p-1}) = \epsilon^{m-1}(-m, -n\epsilon^{n-m}, p\epsilon^{p-m})
$$

3. m is even, n is odd and p is even: $m-1$ is odd, $n-1$ is even and $p-1$ is odd. Therefore,

$$
r_u(-\epsilon,\sigma) = (-m\epsilon^{m-1}, n\epsilon^{n-1}, -p\epsilon^{p-1}) = \epsilon^{m-1}(-m, n\epsilon^{n-m}, -p\epsilon^{p-m})
$$

4. m is even, n and p are odd: $m-1$ is odd, $n-1$ and $p-1$ are even. Hence,

$$
r_u(-\epsilon,\sigma) = (-m\epsilon^{m-1}, n\epsilon^{n-1}, p\epsilon^{p-1}) = \epsilon^{m-1}(-m, n\epsilon^{n-m}, p\epsilon^{p-m})
$$

In each of the cases 1–4, a comparison of $r_u(\epsilon,\sigma)$ and $r_u(-\epsilon,\sigma)$ shows that an infinitesimal change in parameter u in a neighborhood of the singular point results in a large change in the direction of the tangent vector. Therefore, the singular points in cases 1–4 are also cusp points.

5. m is odd, n and p are even: $m - 1$ is even, $n - 1$ and $p - 1$ are odd. Therefore,

$$r_u(-\epsilon, \sigma) = (m\epsilon^{m-1}, -n\epsilon^{n-1}, -p\epsilon^{p-1}) = \epsilon^{m-1}(m, -n\epsilon^{n-m}, -p\epsilon^{p-m})$$

6. m is odd, n is even, p is odd: $m - 1$ is even, $n - 1$ is odd and $p - 1$ is even. So,

$$r_u(-\epsilon, \sigma) = (m\epsilon^{m-1}, -n\epsilon^{n-1}, p\epsilon^{p-1}) = \epsilon^{m-1}(m, -n\epsilon^{n-m}, p\epsilon^{p-m})$$

7. m and n are odd, p is even: $m - 1$ and $n - 1$ are even and $p - 1$ is odd. Hence,

$$r_u(-\epsilon, \sigma) = (m\epsilon^{m-1}, n\epsilon^{n-1}, -p\epsilon^{p-1}) = \epsilon^{m-1}(m, n\epsilon^{n-m}, -p\epsilon^{p-m})$$

8. m, n and p are odd: $m - 1, n - 1$ and $p - 1$ are even. So,

$$r_u(-\epsilon, \sigma) = (m\epsilon^{m-1}, n\epsilon^{n-1}, p\epsilon^{p-1}) = \epsilon^{m-1}(m, n\epsilon^{n-m}, p\epsilon^{p-m})$$

In each of the cases 5–8, comparing $r_u(\epsilon, \sigma)$ to $r_u(-\epsilon, \sigma)$ shows that an infinitesimal change in u in the neighborhood of the singular point also results in an infinitesimal change in the tangent direction. Therefore the singular points in cases 5–8 are *not* cusp points. It follows from the case analysis above that only the singular points in cases 1–4 are cusp points.

We now derive analytical expressions for the curve $\Gamma_{\sigma-\delta}$ so that it can be analyzed in a small neighborhood of the cusp point. To deblur function $f(u) = u^k$, a rescaled version of that function is convolved with the function $(2/\sqrt{\pi})\,e^{-u^2}(1 - u^2)$. This function is an approximation to the deblurring operator derived in [120] and is good for small amounts of deblurring. The convolution is expressed as:

$$(D_t f)(u) = \frac{2}{\sqrt{\pi}} \int_{-\infty}^{\infty} e^{-v^2}(1 - v^2)(u + 2v\sqrt{t})^k \, dv,$$

where t is the scale factor and controls the amount of deblurring. Solving the integral above yields

$$
(D_t f)(u) \;=\; \sum_{\substack{p=0 \\ (p \text{ even})}}^{k} 1.3\ldots(p-1)
$$

$$
\times \;\frac{(2t)^{p/2} k(k-1)\ldots(k-p+1)}{p!}(1-p)u^{k-p}. \tag{B.7}
$$

$\Gamma_{\sigma-\delta}$ can now be analyzed in each of the cases 1–4:

Case 1: Γ_σ is approximated by (u^m, u^n, u^p) where m, n and p are even.

It can be shown [187] that this kind of cusp point must also exist on Γ itself. This is a contradiction of the assumption that Γ is in C_1. Therefore, Γ_σ can not have a cusp point of this kind at u_0.

Case 2: Γ_σ is approximated by (u^m, u^n, u^p) where m and n are even and p is odd. $\Gamma_{\sigma-\delta}$ is obtained by deblurring each of the coordinate functions of Γ_σ:

$$(D_t x)(u) = u^m - c_1 t u^{m-2} - \cdots - c_{(m-2)/2} t^{(m-2)/2} u^2 - c_{m/2} t^{m/2} \tag{B.8}$$

$$(D_t y)(u) = u^n - c_1' t u^{n-2} - \cdots - c_{(n-2)/2}' t^{(n-2)/2} u^2 - c_{n/2}' t^{n/2} \qquad \text{(B.9)}$$

$$(D_t z)(u) = u^p - c_1'' \, t u^{p-2} - \cdots - c_{(p-1)/2}'' t^{(p-1)/2} u. \qquad \text{(B.10)}$$

Note that $(D_t x)$ and $(D_t y)$ contain even powers of u only, $(D_t z)$ contains odd powers of u only and all constants are positive. The deblurred curve intersects itself if there are two values of u, u_1 and u_2, such that

$$x(u_1) = x(u_2), \quad y(u_1) = y(u_2), \quad z(u_1) = z(u_2).$$

It follows from the first two constraints above that $u_= - u_2$. Substituting for u_2 in the third constraint, letting $t = \delta$ and simplifying yields:

$$u_1^p - c_1'' \delta u_1^{p-2} - \cdots - c_{(p-1)/2}'' \delta^{(p-1)/2} u_1 = 0. \qquad \text{(B.11)}$$

The LHS of (B.11) is negative for very small values of u_1 since the first term will be smaller than all other terms, which are negative. As u_1 grows, the first term becomes larger than the sum of all other terms and the LHS of (B.11) becomes positive. Therefore, there is a positive value of u_1 at which (B.11) is satisfied. It can be seen that the tangents of $\Gamma_{\sigma-\delta}$ at u_1 and u_2 are not equal. Hence, $\Gamma_{\sigma-\delta}$ intersects itself in a neighborhood of u_0.

Case 3: Γ_σ is approximated by (u^m, u^n, u^p) where m is even, n is odd and p is even. The proof is analogous to that of case 2, and the same result follows.

Case 4: Γ_σ is approximated by (u^m, u^n, u^p) where m is even and n and p are odd. Expressions for $(D_t x)(u)$ and $(D_t z)(u)$ are again given by formulae (B.8) and (B.10), respectively and $(D_t y)(u)$ is given by:

$$(D_t y)(u) = u^n - c_1' t u^{n-2} - \cdots - c_{(n-1)/2}' t^{(n-1)/2} u.$$

All powers of u in $(D_t x)$ are even and all powers of u in $(D_t y)$ and $(D_t z)$ are odd. As before, $\Gamma_{\sigma-\delta}$ intersects itself if the three constraints:

$$x(u_1) = x(u_2)$$

$$y(u_1) = y(u_2)$$

$$z(u_1) = z(u_2)$$

are satisfied simultaneously. It follows from the first constraint that $u_1 = -u_2$. Now substitute for u_2 in the second and third constraints, let $t = \delta$ and simplify:

$$u_1^n - c_1' \delta u_1^{n-2} - \cdots - c_{(n-1)/2}' \delta^{(n-1)/2} u_1 = 0 \qquad \text{(B.12)}$$

$$u_1^p - c_1'' \delta u_1^{p-2} - \cdots - c_{(p-1)/2}'' \delta^{(p-1)/2} u_1 = 0. \qquad \text{(B.13)}$$

Each of the formulae (B.12) and (B.13) is satisfied at a positive value of u_1 but, in general, the same value of u_1 will *not* satisfy both. It follows that, in this case, $\Gamma_{\sigma-\delta}$ does *not* intersect itself. However, an argument similar to the one in case 3 shows that the planar curves, defined by $(D_t x)$ and $(D_t y)$ and by $(D_t x)$ and $(D_t z)$, that is, the projections of $\Gamma_{\sigma-\delta}$ on the XY and XZ planes, respectively, intersect themselves in a neighborhood of u_0.

This completes the proof of this theorem. ▪

Proof of theorem 10.9. Using a case analysis similar to the one in the proof of the previous theorem to characterize all the possible singularities of Γ_σ at u_0, we again conclude that only the singular points in cases 1–4 are cusp points.

We now derive analytical expressions for $\Gamma_{\sigma+\delta}$ so that it can be analyzed in a neighborhood of u_0. To blur function $f(u) = u^k$, we convolve a resealed version of that function with the function $1/\sqrt{\pi}\,e^{-u^2}$, the blurring operator, as follows:

$$F(u) = \frac{1}{\sqrt{\pi}} \int_{-\infty}^{\infty} e^{-v^2} f(u + 2v\sqrt{t})\, dv = \frac{1}{\sqrt{\pi}} \int_{-\infty}^{\infty} e^{-v^2}(u + 2v\sqrt{t})^k\, dv,$$

where t is the scale factor and controls the amount of blurring. Solving the integral above yields

$$F(u) = \sum_{\substack{p=0 \\ (p \text{ even})}}^{k} 1.3\ldots(p-1)\frac{(2t)^{p/2}k(k-1)\ldots(k-p+1)}{p!}u^{k-p}. \tag{B.14}$$

An expression for $\Gamma_{\sigma+\delta}$ in a neighborhood of the cusp point can be obtained by blurring each of its coordinate functions. Furthermore, expressions for $\Gamma_{\sigma-\delta}$ in a neighborhood of the cusp point can be obtained by deblurring each of its coordinate functions according to (B.7).

Each of the cases 1–4 can now be analyzed in turn:

Case 1: Γ_σ is approximated by (u^m, u^n, u^p) where m, n and p are even.
It can be shown [187] that this type of cusp point must also exist on Γ. This is a contradiction of the assumption that Γ is in C_1. It follows that this kind of cusp point can not arise during evolution:

Case 2: Γ_σ is approximated by (u^m, u^n, u^p) where m and n are even and p is odd.
Observe that

$$\dot{x} = mu^{m-1}, \quad \ddot{x} = m(m-1)u^{m-2}, \quad \dddot{x} = m(m-1)(m-2)u^{m-3}$$

$$\dot{y} = nu^{n-1}, \quad \ddot{y} = n(n-1)u^{n-2}, \quad \dddot{y} = n(n-1)(n-2)u^{n-3}$$

$$\dot{z} = pu^{p-1}, \quad \ddot{z} = p(p-1)u^{p-2}, \quad \dddot{z} = p(p-1)(p-2)u^{p-3}.$$

Torsion on Γ_σ is given by [95]:

$$\tau(u) = \frac{\dddot{z}\,\dot{x}\ddot{y} - \dddot{z}\,\dot{y}\ddot{x} + \dddot{y}\,\dot{z}\ddot{x} - \dddot{y}\,\dot{x}\ddot{z} + \dddot{x}\,\dot{y}\ddot{z} - \dddot{x}\,\dot{z}\ddot{y}}{(\dot{y}\ddot{z} - \dot{z}\ddot{y})^2 + (\dot{z}\ddot{x} - \dot{x}\ddot{z})^2 + (\dot{x}\ddot{y} - \dot{y}\ddot{x})^2}$$

or

$$\tau(u) = \frac{mnpK u^{p+n+m-6}}{A + B + C} \tag{B.15}$$

where

$$A = ((np(p-1) - pn(n-1))u^{p+n-3})^2$$

$$B = ((pm(m-1) - mp(p-1))u^{p+m-3})^2$$

$$C = ((mn(n-1) - nm(m-1))u^{m+n-3})^2.$$

At $u = 0$ (cusp point), τ is undefined. When u positive or negative, the sign of $\tau(u)$ depends on the sign of K. Observe that

$$
\begin{aligned}
K &= (p-1)(p-2)(n-m) + (n-1)(n-2)(m-p) \\
&\quad + (m-1)(m-2)(p-n) = (n-m)p^2 + (m^2 - n^2)p \\
&\quad + mn^2 - nm^2 \\
&= (n-m)(p^2 - (m+n)p + mn) = (n-m)(p-m)(p-n).
\end{aligned}
$$

K is positive because of the assumption that $p > n > m$. Since $p + n + m - 6$, the power of a in the numerator, is odd, it follows that $\tau(u)$ is positive for positive u and negative for negative u.

We now investigate $\tau(u)$ on $\Gamma_{\sigma+\delta}$. It follows from (B.14), that $\Gamma_{\sigma+\delta}$ is given by:

$$
X(u) = u^m + d_1 tu^{m-2} + \cdots + d_{(m-2)/2} t^{(m-2)/2} u^2 + d_{m/2} t^{m/2} \tag{B.16}
$$

$$
\begin{aligned}
Y(u) &= u^n + d_1' tu^{n-2} + \cdots + d_{(n-2)/2}' t^{(n-2)/2} u^2 + d_{n/2}' t^{n/2} \\
Z(u) &= u^p + d_1'' tu^{p-2} + \cdots + d_{(p-1)/2}'' t^{(p-1)/2} u
\end{aligned}
\tag{B.17}
$$

where all constants are positive, all powers of u in $X(u)$ and $Y(u)$ are even, all powers of u in $Z(u)$ are odd and t equals δ, a small constant. Note also that the last terms in $X(u)$ and $Y(u)$ do not contain any positive powers of u but all terms in $Z(u)$ contain positive powers of u. It follows that the last terms in $\ddot{X}(u), \dddot{Y}(u), \dot{Z}(u)$ and $\dddot{Z}(u)$ do not contain positive powers of u whereas all terms in $\dot{X}(u), \dddot{X}(u), \dot{Y}(u), \dddot{Y}(u)$ and $\ddot{Z}(u)$ contain positive powers of u. Therefore, at $u = 0$,

$$
\dot{X}(u) = \dddot{X}(u) = \dot{Y}(u) = \dddot{Y}(u) = \ddot{Z}(u) = 0
$$

and $\tau = 0$. As u, grows, the terms in $\dot{X}(u), \ddot{X}(u), \dddot{X}(u), \dot{Y}(u), \ddot{Y}(u), \dddot{Y}(u), \dot{Z}(u), \ddot{Z}(u)$ and $\dddot{Z}(u)$ with the largest power of u (which are also the only terms without δ) become dominant and torsion is again given by (B.15). It follows that $\tau(u)$ is positive for positive u and negative for negative u on $\Gamma_{\sigma+\delta}$. Since τ is zero at $u = 0$, $\Gamma_{\sigma+\delta}$ has a torsion zero-crossing point at $u = 0$.

We next investigate $\tau(u)$ on $\Gamma_{\sigma-\delta}$. From (B.7) it follows that $(D_t x)(u), (D_t y)(u)$, and $(D_t z)(u)$ are given by equations (B.8–B.10) respectively. It again follows that $\tau = 0$ at $u = 0$, τ is positive for positive u and negative for negative u. Therefore there is also a torsion zero-crossing point at $u = 0$ on $\Gamma_{\sigma-\delta}$. It follows that there is a torsion zero-crossing point at u_0 on $\Gamma_{\sigma-\delta}$ before the formation of the cusp point and on $\Gamma_{\sigma+\delta}$ after the formation of the cusp point.

Case 3: Γ_σ is approximated by (u^m, u^n, u^p), where m is even, n is odd and p is even. The proof is analogous to that of case 2, and the same result follows.

Case 4: Γ_σ is approximated by (u^m, u^n, u^p), where m is even, and n and p are odd. At $u = 0$, the cusp point, τ is undefined. At all other points, $\tau(u)$ is given by (B.15). Since the coefficient of the numerator of (B.15) is positive (as shown in the proof of case 2) and $p + n + m - 6$, the power of u in the numerator, is even, $\tau(u)$

is positive for positive and negative values of u in the neighborhood of u_0 on Γ_σ. Therefore, there are *no* torsion zero-crossing points in the neighborhood of u_0 on Γ_σ.

We now investigate $\tau(u)$ on $\Gamma_{\sigma+\delta}$. Expressions for $X(u)$ and $Z(u)$ are again given by equations (B.16) and (B.17), respectively and $Y(u)$ is given by:

$$Y(u) = u^n + d_1' t u^{n-2} + \cdots + d_{(n-1)/2}' t^{(n-1)/2} u$$

where all constants are positive, all powers of u in $X(u)$ are even, all powers of u in $Y(u)$ and $Z(u)$ are odd and t equals δ, a small constant. Furthermore, note that the last term in $X(u)$ does not contain a positive power of u but all terms in $Y(u)$ and $Z(u)$, contain positive powers of u. Therefore, the last terms in $\ddot{X}(u), \dot{Y}(u), \ddot{Y}(u), \dot{Z}(u)$ and $\dddot{Z}(u)$ do not contain positive powers of u, whereas all terms in $\dot{X}(u), \ddot{X}(u), \dddot{Y}(u)$ and $\ddot{Z}(u)$ contain positive powers of u. Hence, at $u = 0$,

$$\dot{X}(u) = \ddot{X}(u) = \ddot{Y}(u) = \ddot{Z}(u) = 0$$

and

$$\tau(u) = \frac{\ddot{X}(u)(\dddot{Y}(u)\dot{Z}(u) - \dddot{Z}(u)\dot{Y}(u))}{(\dot{Z}(u)\ddot{X}(u))^2 + (\dot{Y}(u)\ddot{X}(u))^2}.$$

Since the denominator is positive and $\ddot{X}(u)$ is positive, to determine the sign, of $\tau(u)$, we must determine the sign of the expression $\dddot{Y}(u)\dot{Z}(u) - \dddot{Z}(u)\dot{Y}(u)$. At $u = 0$, using (B.14) we conclude that the non-zero term of $\dot{Y}(u)$ is:

$$c_{(n-1)/2}' t^{(n-1)/2} = 1.3\ldots(n-2)\frac{(2t)^{(n-1)/2} n!}{(n-1)!} = 1.3\ldots n(2t)^{(n-1)/2}.$$

Similarly, at $u = 0$, the non-zero term of $\dot{Z}(u)$ is:

$$c_{(p-1)/2}'' t^{(p-1)/2} = 1.3\ldots p(2t)^{(p-1)/2}.$$

Using (B.14), it follows that at $u = 0$, the non-zero term of $\dddot{Y}(u)$ is:

$$6c_{(n-3)/2}' t^{(n-3)/2} = 6(1.3\ldots(n-4))\frac{(2t)^{(n-3)/2} n!}{6(n-3)!} + (1.3\ldots n)(n-1)(2t)^{n-3}.$$

Similarly, at $u = 0$, the non-zero term of $\dddot{Z}(u)$ is:

$$6c_{(p-3)/2}'' t^{(p-3)/2} = (1.3\ldots p)(p-1)(2t)^{(p-3)/2}.$$

Therefore,

$$\dddot{Y}(u)\dot{Z}(u) - \dddot{Z}(u)\dot{Y}(u) = (2t)^{(p+n-4)/2}(1.3\ldots n)(1.3\ldots p)(n-p) < 0$$

since $n < p$. Therefore, $\tau(u)$ is negative at $u = 0$ on $\Gamma_{\sigma+\delta}$. As u grows the terms in $\dot{X}(u), \ddot{X}(u), \dddot{X}(u), \dot{Y}(u), \ddot{Y}(u), \dddot{Y}(u), \dot{Z}(u), \ddot{Z}(u)$ and $\dddot{Z}(u)$ with the largest power of u (which are also the only terms without δ) become dominant and $\tau(u)$ is again given by (B.15). Since $p + n + m - 6$, the power of u in the numerator, in now eves, $\tau(u)$ becomes positive as u grows in absolute value. Therefore, there exist two new torsion zero-crossings in a neighborhood of u_0 on $\Gamma_{\sigma+\delta}$.

This completes the proof of this theorem. ⋈

Proof of theorem 10.10. It will be shown that this theorem holds for an arbitrary parametrization of Γ_σ. Therefore it must also be true of arc length parametrization or close approximations.

Let $\Gamma = (x(u), y(u), z(u))$ be a space curve and let $x(u)$, $y(u)$ and $z(u)$ be polynomial functions of u. Let $\Gamma_\sigma = (X(u, \sigma), Y(u, \sigma), Z(u, \sigma))$ be an evolved version of Γ with a point of zero curvature at u_0. Assume without loss of generality that $u_0 = 0$ and that at u_0, Γ_σ goes through the origin of the coordinate system. It follows that Γ_σ can be approximated in a neighborhood of u_0 by (u^m, u^n, u^p) where u^m, u^n and u^p are the lowest degree terms in $X(u, \sigma)$, $Y(u, \sigma)$ and $Z(u, \sigma)$ respectively. Assume without loss of generality that $p > n > m$. Observe that

$$\dot{x}(u) = mu^{m-1},$$

$$\ddot{x}(u) = m(m-1)u^{m-2},$$

$$\dddot{x}(u) = m(m-1)(m-2)u^{m-3},$$

$$\dot{y}(u) = nu^{n-1},$$

$$\ddot{y}(u) = n(n-1)u^{n-2},$$

$$\dddot{y}(u) = n(n-1)(n-2)u^{n-3},$$

$$\dot{z}(u) = pu^{p-1},$$

$$\ddot{z}(u) = p(p-1)u^{p-2},$$

$$\dddot{z}(u) = p(p-1)(p-2)u^{p-3},$$

Torsion on Γ_σ is given by [187]:

$$\tau(u) = \frac{\dddot{z}\,\dot{x}\ddot{y} - \dddot{z}\dot{y}\ddot{x} + \dddot{y}\dot{x}\ddot{z} + \ddot{x}\dot{y}\dddot{z} - \ddot{x}\dot{z}\dddot{y}}{(\dot{y}\ddot{z} - \dot{z}\ddot{y})^2 + (\dot{z}\ddot{x} - \dot{x}\ddot{z})^2 + (\dot{x}\ddot{y} - \dot{y}\ddot{x})^2}$$

or

$$\tau(u) = \frac{mnpKu^{p+n+m-6}}{A + B + C}, \qquad (B.18)$$

where

$$A = ((np(p-1) - pn(n-1))u^{p+n-3})^2,$$

$$B = ((pm(m-1) - mp(p-1))u^{p+m-3})^2,$$

$$C = ((mn(n-1) - nm(m-1))u^{m+n-3})^2,$$

and

$$K = (p-1)(p-2)(n-m) + (n-1)(n-2)(m-p)$$
$$+ (m-1)(m-2)(p-n).$$

We now derive analytical expressions for $\Gamma_{\sigma+\delta}$ so that it can be analyzed in a neighborhood of u_0. To blur $f(u) = u^k$, we convolve a resealed version of it with $e^{-u^2}/\sqrt{\pi}$, the blurring operator, as follows:

$$F(u) = \int_{-\infty}^{\infty} \frac{1}{\sqrt{\pi}} e^{-v^2} f(u + 2v\sqrt{t})\, dv$$

or

$$F(u) = \frac{1}{\sqrt{\pi}} \int\limits_{-\infty}^{\infty} e^{-v^2}(u + 2v\sqrt{t})^k \, dv,$$

where t is the scale factor and controls the amount of blurring. Solving the integral above yields:

$$F(u) = \sum_{\substack{p=0 \\ (p \text{ even})}}^{k} D(2t)^{p/2}k(k-1)\cdots(k-p+l)u^{k-p}, \qquad (B.19)$$

where

$$D = \frac{1.3.5\ldots(p-1)}{p!}.$$

An expression for $\Gamma_{\sigma+\delta}$ in a neighborhood of u_0 can be obtained by blurring each of its coordinate functions.

Since m, n and p can be odd or even, point u_0 must be analyzed in each of eight possible cases. The analysis in the proof of Theorem 2.5 showed that when m is even, a cusp point exists on Γ_σ at u_0. We will therefore look at the remaining four cases in which m is odd:

Case 1. m is odd and n and p are even.

Torsion on Γ_σ is given by equation (B.18). Since $p + n + m - 6$ is odd, torsion is positive for positive u and negative for negative u in a neighborhood of u_0. We now investigate torsion on $\Gamma_{\sigma+\delta}$, where δ is a small, positive number. Expressions for $X(u,\sigma)$, $Y(u,\sigma)$ and $Z(u,\sigma)$ can be obtained using equation (B.19). Note that all powers of u in $X(u,\sigma)$ are odd and all powers of u in $Y(u,\sigma)$ and $Z(u,\sigma)$ are even. It follows that all powers of u in $\dot{X}(u)$, $\ddot{X}(u)$, $\ddot{Y}(u)$ and $\ddot{Z}(u)$ are even and all powers of u in $\ddot{X}(u)$, $\dot{Y}(u)$, $\dddot{Y}(u)$, $\dot{Z}(u)$ and $\dddot{Z}(u)$ are odd. Note also that those terms in which all powers of u are odd, are equal to zero at u_0. Therefore torsion is zero at u_0 on $\Gamma_{\sigma+\delta}$. As u grows, u^m, u^n and u^p, that is the terms in $X(u,\sigma)$, $Y(u,\sigma)$ and $Z(u,\sigma)$ with the largest powers of u, become dominant and torsion is again given by equation (B.18). It follows that torsion is positive for positive u and negative for negative u on $\Gamma_{\sigma+\delta}$ in a neighborhood of u_0. Hence no new torsion zero-crossings have been created.

Case 2. m is odd, n is even and p is odd.

Torsion on Γ_σ is again given by (B.18). Since $p+n+m-6$ is even, torsion is positive for positive and negative u on Γ_σ. We now investigate torsion on $\Gamma_{\sigma+\delta}$. Note that all powers of u in $X(u,\sigma)$ are odd, all powers of u in $Y(u,\sigma)$ are even and all powers of u in $Z(u,\sigma)$ are odd. It follows that all powers of u in $\dot{X}(u)$, $\ddot{X}(u)$, $\ddot{Y}(u)$, $\dot{Z}(u)$ and $\dddot{Z}(u)$ are even and all powers of u in $\ddot{X}(u)$, $\dot{Y}(u)$, $\dddot{Y}(u)$ and $\ddot{Z}(u)$ are odd. Note also that those terms in which all powers of u are odd, are equal to zero at u_0. It follows that torsion on $\Gamma_{\sigma+\delta}$ at u_0 is given by:

$$\tau(u) = \frac{\dddot{Z}\,\dot{X}\ddot{Y} - \ddot{X}\,\dot{Z}\ddot{Y}}{(\dot{Z}\ddot{Y})^2 + (\dot{X}\ddot{y})^2} = \frac{\ddot{Y}(\dddot{Z}\,\dot{X} - \dddot{X}\,\dot{Z})}{(\dot{Z}\ddot{Y})^2 + (\dot{X}\ddot{Y})^2}.$$

Since the denominator of the expression above is positive and $\ddot{Y}$ is positive, the sign of $\tau(u)$ is the same as the sign of the expression: $\dddot{Z}\,\dot{X} - \dddot{X}\,\dot{Z}$. At u_0, using (B.19) it can be shown that:

$$\dddot{Z}\,\dot{X} - \dddot{X}\,\dot{Z} = M(2t)^{(p+m-4)/2}$$

where

$$M = (1.3.5\cdots p)(1.3.5\cdots m)(p - m).$$

Note that $\dddot{Z}\,\dot{X} - \dddot{X}\,\dot{Z}$ is positive at u_0. As u grows larger, torsion is again given by (B.18) in a neighborhood of u_0 and is therefore positive in that neighborhood. Again no new torsion zero-crossings have been created.

Case 3. m and n are odd and p is even.

Torsion is again given by (B.18) on Γ_σ. Since $p + n + m - 6$ is even, torsion is positive for positive and negative u on Γ_σ. We now investigate torsion on $\Gamma_{\sigma+\delta}$. Note that all powers of u in $X(u,\sigma)$ and $Y(u,\sigma)$ are odd and all powers of u in $Z(u,\sigma)$ are even. Hence all powers of u in $\dot{X}(u)$, $\dddot{X}(u)$, $\dot{Y}(u)$, $\dddot{Y}(u)$ and $\ddot{Z}(u)$ are even and all powers of u in $\ddot{X}(u)$, $\ddot{Y}(u)$, $\dot{Z}(u)$ and $\dddot{Z}(u)$ are odd. Note also that those terms in which all powers of u are odd, are equal to zero at u_0. Therefore torsion on $\Gamma_{\sigma+\delta}$ at u_0 is given by:

$$\tau(u) = \frac{\ddot{X}\,\dot{Y}\ddot{Z} - \ddot{Y}\,\dot{X}\ddot{Z}}{(\dot{Y}\ddot{Z})^2 + (\dot{X}\ddot{Z})^2} = \frac{\ddot{Z}(\ddot{X}\,\dot{Y} - \ddot{Y}\,\dot{X})}{(\dot{Y}\ddot{Z})^2 + (\dot{X}\ddot{Z})^2}.$$

Since the denominator of the expression above is positive and $\ddot{Z}$ is positive, the sign of $\tau(u)$ is the same as the sign of the expression: $\ddot{X}\,\dot{Y} - \ddot{Y}\,\dot{X}$. At u_0, using (B.19) it can be shown that:

$$\ddot{X}\,\dot{Y} - \ddot{Y}\,\dot{X} = N(2t)^{(n+m-4)/2},$$

where

$$N = (1.3.5\cdots n)(1.3.5\cdots m)(m - n).$$

Note that $\ddot{X}\,\dot{Y} - \ddot{Y}\,\dot{X}$ is negative since $n > m$. Therefore torsion is negative at u_0 on $\Gamma_{\sigma+\delta}$. As u grows larger, torsion is again given by (B.18) in a neighborhood of u_0 and is therefore positive for positive and negative u. It follows that there are two *new* torsion zero-crossings in a neighborhood of u_0 on $\Gamma_{\sigma+\delta}$.

Case 4. m, n and p are odd.

Torsion on Γ_σ is again given by equation (B.18). Since $p + n + m - 6$ is odd, torsion is positive for positive u and negative for negative u on Γ_σ. We now investigate torsion on $\Gamma_{\sigma+\delta}$. Note that all powers of u in $X(u,\sigma)$, $Y(u,\sigma)$ and $Z(u,\sigma)$ are odd. Hence all powers of u in $\dot{X}(u)$, $\dot{Y}(u)$, $\dot{Z}(u)$, $\dddot{X}(u)$, $\dddot{Y}(u)$ and $\dddot{Z}(u)$ are even and all powers of u in $\ddot{X}(u)$, $\ddot{Y}(u)$ and $\ddot{Z}(u)$ are odd. Note also that those terms in which all powers of u are odd, are equal to zero at u_0. It follows that torsion is unbounded at u_0 on $\Gamma_{\sigma+\delta}$. As u grows larger, torsion is again given by (B.18) in a neighborhood of u_0 and is therefore positive for positive u and negative for negative u. Hence there are no new torsion zero-crossings in a neighborhood of u_0 on $\Gamma_{\sigma+\delta}$ [191]. ◻

Proof of theorem 10.11. Since Γ is closed, its coordinate functions, $x(u)$, $y(u)$ and $z(u)$ are periodic functions. Let

$$X(u,t) = x(u) \otimes g(u,t),$$

$$Y(u,t) = y(u) \otimes g(u,t)$$

and

$$Z(u, t) = z(u) \otimes g(u, t).$$

where $t = \sigma^2/2$. As shown in the proof of theorem 2.3 in [227], when t is large, $X'(u, t)$ and $Y'(u, t)$ can be expressed as:

$$X'(u, t) = \frac{2\pi a_1}{e^{4\pi^2 t}} \cos(2\pi u) - \frac{2\pi b_1}{e^{4\pi^2 t}} \sin(2\pi u)$$

and

$$Y'(u, t) = \frac{2\pi c_1}{e^{4\pi^2 t}} \cos(2\pi u) - \frac{2\pi d_1}{e^{4\pi^2 t}} \sin(2\pi u).$$

Similarly

$$Z'(u, t) = \frac{2\pi e_1}{e^{4\pi^2 t}} \cos(2\pi u) - \frac{2\pi f_1}{e^{4\pi^2 t}} \sin(2\pi u).$$

Furthermore

$$X''(u, t) = -\frac{4\pi^2 a_1}{e^{4\pi^2 t}} \sin(2\pi u) - \frac{4\pi^2 b_1}{e^{4\pi^2 t}} \cos(2\pi u),$$

$$Y''(u, t) = -\frac{4\pi^2 c_1}{e^{4\pi^2 t}} \sin(2\pi u) - \frac{4\pi^2 d_1}{e^{4\pi^2 t}} \cos(2\pi u),$$

$$Z''(u, t) = -\frac{4\pi^2 e_1}{e^{4\pi^2 t}} \sin(2\pi u) - \frac{4\pi^2 f_1}{e^{4\pi^2 t}} \cos(2\pi u),$$

$$X'''(u, t) = -\frac{8\pi^3 a_1}{e^{4\pi^2 t}} \cos(2\pi u) + \frac{8\pi^3 b_1}{e^{4\pi^2 t}} \sin(2\pi u),$$

$$Y'''(u, t) = -\frac{8\pi^3 c_1}{e^{4\pi^2 t}} \cos(2\pi u) + \frac{8\pi^3 d_1}{e^{4\pi^2 t}} \sin(2\pi u),$$

and

$$Z'''(u, t) = -\frac{8\pi^3 e_1}{e^{4\pi^2 t}} \cos(2\pi u) + \frac{8\pi^3 f_1}{e^{4\pi^2 t}} \sin(2\pi u).$$

torsion $\tau(u, t)$ on Γ_t is given by [95]:

$$\tau = \frac{X''(Z'Y''' - Y'Z''') + Y''(X'Z''' - Z'X''') + Z''(Y'X''' - X'Y''')}{(Y'Z'' - Z'Y'')^2 + (Z'X'' - X'Z'')^2 + (X'Y'' - Y'X'')^2}.$$

Since the denominator of the expression above is always positive, we will investigate the numerator only. Note that

$$\begin{aligned}Z'Y''' = Y'Z''' &= \xi[-c_1 e_1 \cos^2(2\pi u) + c_1 f_1 \sin(2\pi u) \cos(2\pi u) \\ &\quad + d_1 e_1 \sin(2\pi u) \cos(2\pi u) - d_1 f_1 \sin^2(2\pi u)]\end{aligned}$$

and

$$\begin{aligned}X'Z''' = Z'X''' &= \xi[-a_1 e_1 \cos^2(2\pi u) + b_1 e_1 \sin(2\pi u) \cos(2\pi u) \\ &\quad + a_1 f_1 \sin(2\pi u) \cos(2\pi u) - b_1 f_1 \sin^2(2\pi u)]\end{aligned}$$

and

$$\begin{aligned}Y'X''' = X'Y''' &= \xi[-a_1 c_1 \cos^2(2\pi u) + a_1 d_1 \sin(2\pi u) \cos(2\pi u) \\ &\quad + b_1 c_1 \sin(2\pi u) \cos(2\pi u) - b_1 d_1 \sin^2(2\pi u)]\end{aligned}$$

where

$$\xi = \frac{16\pi^4}{e^{8\pi^2 t}}.$$

Therefore torsion goes to zero at every point of Γ_t as t grows large. It follows that as t becomes large, Γ_t tends to a planar curve. Note that the proof for arc-length evolution is similar.   ⋈

Proof of theorem 10.12. Let $\Gamma = (x(u), y(u), z(u))$ be a closed space curve and let $\Gamma_\sigma = (X(u, \sigma), Y(u, \sigma), Z(u, \sigma))$ be an evolved version of Γ. By theorem 2.1 in [191], when σ is large, functions $\ddot{X}(u, \sigma)$, $\ddot{Y}(u, \sigma)$ and $\ddot{Z}(u, \sigma)$ (which are periodic) will each have only two zero-crossing points. Therefore by theorem 2.2 in [191], the planar curves defined by the coordinate function-pairs $(X(u, \sigma), Y(u, \sigma))$, $(X(u, \sigma), Z(u, \sigma))$ and $(Y(u, \sigma), Z(u, \sigma))$ are simple and convex. By theorem 3.1 in [191], new torsion zero-crossing points can appear on a space curve during evolution in a neighborhood of a point of zero curvature. On Γ_σ, the magnitude of curvature is given by:

$$|\kappa| = \frac{\sqrt{A^2 + B^2 + C^2}}{((\dot{x})^2 + (\dot{y})^2 + (\dot{z})^2)^{3/2}}$$

where

$$A = \dot{y}\ddot{z} - \dot{z}\ddot{y}$$

$$B = \dot{z}\ddot{x} - \dot{x}\ddot{z}$$

$$C = \dot{x}\ddot{y} - \dot{y}\ddot{x}.$$

Note that A, B and C determine the signs of the curvatures of the planar curves defined by the coordinate functions of Γ_σ. It follows that if a point of zero curvature exists on Γ_σ, there are also curvature zero-crossing points on the planar curves defined by the coordinate functions of Γ_σ. Since those curves are simple and convex, Γ_σ cannot have any zeros of curvature either. Theorem 2 in [186] states that the only other time torsion zero-crossings can appear on a space curve during evolution is right after the formation of a cusp point. Theorem 1 in [186] states that a space curve or two of the planar curves defined by the coordinate functions of that space curve must intersect themselves just before the formation of a cusp point during evolution. If the space curve intersects itself, then all three planar curves defined by the coordinate functions of that curve intersect themselves. However, all the planar curves defined by the coordinate functions of Γ_σ are simple and convex. It follows that cusp points cannot exist on Γ_σ either. Hence none of the conditions necessary for the creation of new torsion zero-crossings are realized on Γ_σ. Furthermore, for larger values of σ, $\ddot{X}(u, \sigma)$, $\ddot{Y}(u, \sigma)$ and $\ddot{Z}(u, \sigma)$ will continue to have only two zero-crossings. Hence the arguments above will continue to apply and Γ_σ will remain in a state in which new torsion zero-crossings will not be created.   ⋈

Appendix C
Proofs of Theorems of Chapter 11

This appendix contains the proofs of the 7 theorems of chapter 11 regarding free-form 3-D surfaces.

Proof of theorem 11.1. Assume by contradiction that the procedure outlined in the theorem does not construct a geodesic. Let g_1 be the segment of the geodesic on T_1 and let g_2 be the segment of the geodesic on T_2. Rotate T_2 about e so that it becomes coplanar with T_1. By assumption, g_1 and g_2 will not be co-linear. Hence, for a point P_1 on g_1 and a point P_2 on g_2, there will be a shorter path from P_1 to P_2. This is the straight line joining P_1 to P_2. Now rotate T_2 back to its original position. The length of the path just constructed remains the same, so it will still be shorter than the geodesic from P_1 to P_2. A contradiction has been reached. Therefore the procedure described correctly constructs a geodesic (see figure C.1). ◻

Proof of theorem 11.2. The curvature vector k of the path obtained by the procedure outlined in the theorem lies in Q. k is also perpendicular to tangent plane T (which is defined as perpendicular to $\mathbf{n}$ at V). The vector of geodesic curvature of the path is obtained by projecting k on the tangent plane. It follows that geodesic curvature of the path is zero. Hence the path is a geodesic line (see figure C.2). ◻

Proof of theorem 11.3. Suppose surface S is evolved into S_σ. Every point of S_σ is a weighted average of a subset of points of S. Therefore evolution at each point Q of S can be expressed as the convolution of a neighbourhood of Q with a 2-D function (not Gaussian) with unknown values.

$$P(X, Y, Z) = (x(u, v) \otimes f(u, v),\ y(u, v) \otimes f(u, v),\ z(u, v) \otimes f(u, v))$$

$\otimes$ denotes convolution. Now apply an affine transform to point P to obtain $P_1(X_1, Y_1, Y_1)$ where:

$$X_1 = a_1\ X + b_1\ Y + c_1\ Z + d_1$$

$$Y_1 = a_2\ X + b_2\ Y + c_2\ Z + d_1$$

$$Z_1 = a_3\ X + b_3\ Y + c_3\ Z + d_1$$

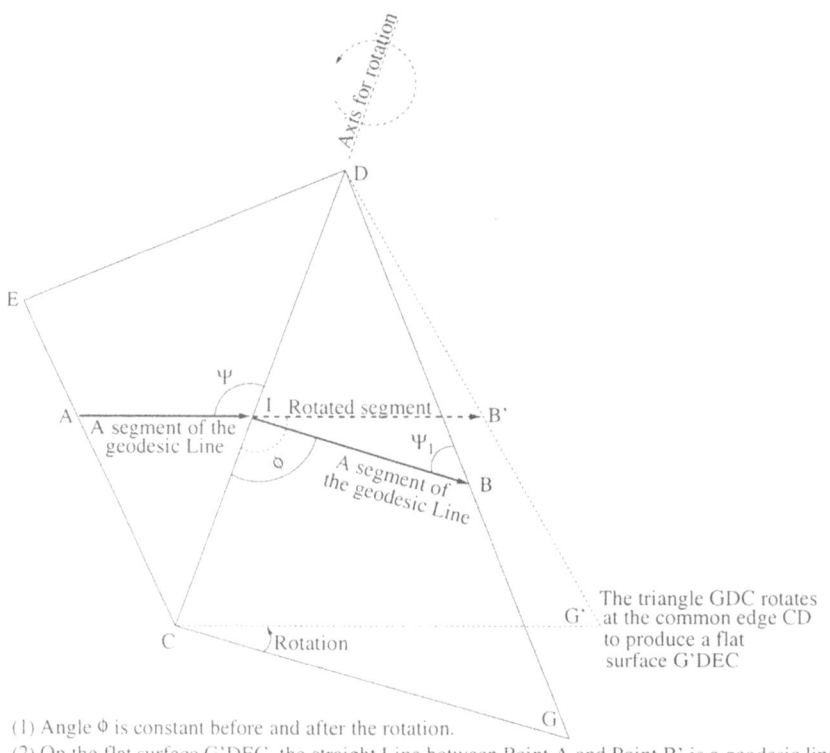

Axis for rotation
D
E
Ψ
A
I
Rotated segment
B'
A segment of the geodesic Line
Ψ1
φ
A segment of the geodesic Line
B
The triangle GDC rotates at the common edge CD to produce a flat surface G'DEC
G'
C
Rotation
G
(1) Angle φ is constant before and after the rotation.
(2) On the flat surface G'DEC, the straight Line between Point A and Point B' is a geodesic line.

Figure C.1. Geodesic line at a triangle edge

W
N(B)
n i
F
Second neighbour
α i
S 0
B
P0
L
S 1
P1
r
C
I
D
Anti-clockwise

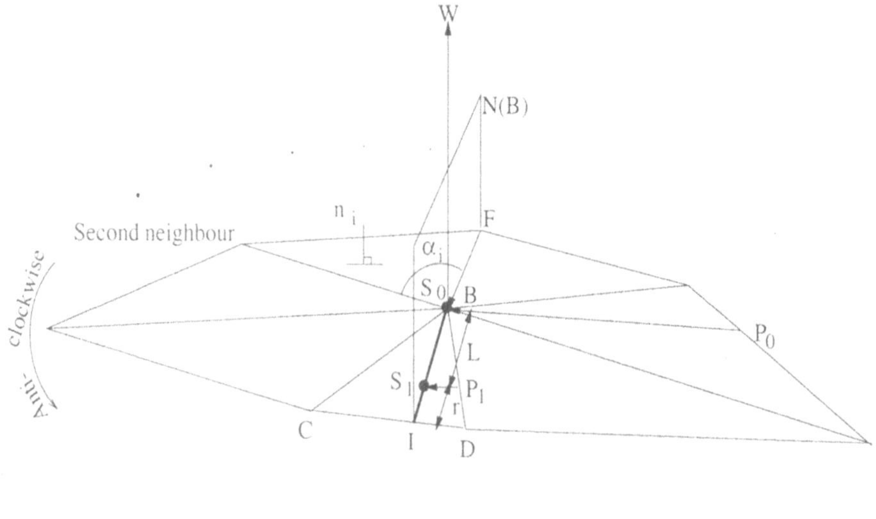

Figure C.2. Geodesic line at a vertex

Alternatively, apply an affine transform to point Q first; and then evolve:

$$X_2 = (a_1\, x(u,v) + b_1\, y(u,v) + c_1\, z(u,v) + d_1) \otimes f(u,v)$$

$$Y_2 = (a_2\, x(u,v) + b_2\, y(u,v) + c_2\, z(u,v) + d_2) \otimes f(u,v)$$

$$Z_2 = (a_3\, x(u,v) + b_3\, y(u,v) + c_3\, z(u,v) + d_3) \otimes f(u,v)$$

so: $X_2 = X_1$, $Y_2 = Y_1$ and $Z_2 = Z_1$. Note that affine transforms also include shape preserving transforms.   ⋈

Proof of theorem 11.4. Since $\mathcal{H}$ is a convex surface, every plane P tangent to $\mathcal{H}$ contains that surface in the left (or right) half-space it creates. Since S is inside $\mathcal{H}$, S is also contained in the same half-space. Now rotate P and S so that P becomes parallel to the xy plane. P is now described by the equation $z = c$. Since P does not intersect S, it follows that $Q_z \geq c$ for every point Q on S. Let S_σ be an evolved version of S. Every point of S_σ is a weighted average of a subset of points of S. Therefore, $R_z \geq c$ for every point R on S_σ, and S_σ is also contained in the same half-space. This result holds for every plane tangent to $\mathcal{H}$; therefore S_σ is contained inside the intersection of all the left (or right) half-spaces created by the tangent planes of $\mathcal{H}$. It follows that S_σ is also inside $\mathcal{H}$.   ⋈

Proof of theorem 11.5. Let ϵ be the maximum error in the location of any point of surface S when the heat diffusion of S is approximated through Gaussian filtering with standard deviation $\Delta\sigma$. Observe that at a point P of S

$$\epsilon = |(\mathbf{r} + Hn) - (\mathbf{r} + \Delta\mathbf{r}_g)| = |Hn - \Delta\mathbf{r}_g|$$

where H is mean curvature, n is the normal vector at P, $\mathbf{r}$ is the position vector of P and $\Delta\mathbf{r}_g$ is the amount of change in the position vector of P after Gaussian filtering. According to heat diffusion equation

$$\frac{\partial \mathbf{r}}{\partial t} = Hn$$

Let

$$\Delta\mathbf{r}_g = H_g n_g$$

where n_g is a unit vector with the same direction as that of $\Delta\mathbf{r}_g$, and H_g is equal to length of $\Delta\mathbf{r}_g$. Let k_1 and k_2 be the principal curvatures at P. Assume that k_1 and k_2 are constant in a small neighborhood of P. The following cases can be distinguished:

- k_1 and k_2 are both zero: the surface is locally planar.
- One of k_1 and k_2 is zero: the surface is locally cylindrical.
- k_1 and k_2 are both positive or both negative: the surface is locally ellipsoidal.
- One of k_1 and k_2 is positive and the other is negative: the surface is locally saddle-shaped.

In each case, it can be confirmed that Gaussian filtering is equivalent to diffusion smoothing of the surface. It follows that for a small $\Delta\sigma$, $H_g \to H$ and $n_g \to n$, and therefore $\epsilon \to 0$. After j iterations of smoothing, total error is given by $j\epsilon$ which is small.   ⋈

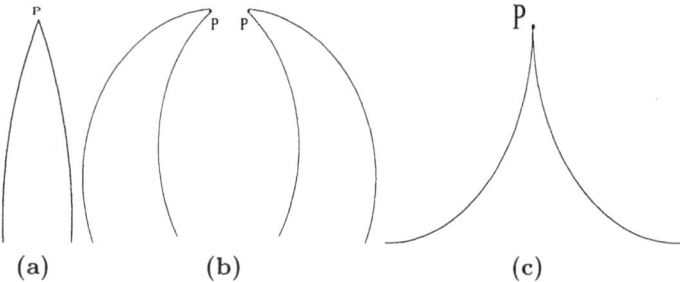

Figure C.3.   Cross-section of the surface

Proof of theorem 11.6. It follows from the equation for heat diffusion

$$\frac{\partial r}{\partial t} = Hn$$

that for two points with infinitesimal distance on $\mathcal{S}$, application of infinitesimal diffusion will result in two new points also with infinitesimal distance. This is because at two nearby points P_1 and P_2, $H_1 \approx H_2$ and $n_1 \approx n_2$ so

$$\frac{\partial r_1}{\partial t} \approx \frac{\partial r_2}{\partial t}$$

It follows further that the tangent planes T_1 and T_2 will also be at infinitesimal distance. Now suppose there is a cusp point on S_σ at point P. That cusp point did not exist on the original surface. Consider two points P_1 and P_2 on $S_{\sigma-\delta}$ in a small neighborhood of P. The cusp point at P on S_σ can not be of the forms shown in Figures C.3(a) and (b), since the difference between the tangent planes at P_1 and P_2 on S_σ would be large (which is not possible). It follows that only cusp points of the form shown in Figure C.3(c) are possible since only in this case, the difference between tangent planes at P_1 and P_2 near P is small. Applying reverse diffusion to this object in a neighborhood of P results in a surface that intersects itself near P. It follows that $S_{\sigma-\delta}$ is self-intersecting in a neighborhood of the cusp point.   ⊠

Proof of theorem 11.7. Assume by contradiction that $\mathcal{S}$ is a simple surface that intersects itself during evolution. The location vector of each point of $\mathcal{S}$ is a continuous function of σ during evolution; therefore $\mathcal{S}$ must touch itself at point P before self-intersection. Let S_{σ_0} be such a surface. Consider two neighborhoods S_1 and S_2 of S_{σ_0} that have point P only in common. Hence S_1 and S_2 are non-overlapping. Note that S_1 and S_2 have the same tangent plane at P. Denote this tangent plane by T. The tangent plane exists since it follows from theorem 11.6 that P can not be a cusp point on either S_1 or S_2 since S_σ does not self-intersect for $\sigma \leq \sigma_0$. Recall that the infinitesimal movement during arc length evolution of each point of S_1 and S_2 is determined by the equation

$$\frac{\partial \mathbf{r}}{\partial t} = Hn$$

where H is the mean curvature, n is normal and t is time. Therefore during arc length evolution, every point will move in the direction of the normal vector by an amount equal to the curvature at that point. Similarly, during reverse arc length evolution, every point will move in the opposite direction of the normal vector by an amount equal to the curvature at that point. It follows that if S_1 and S_2 are on opposite sides of T, after an infinitesimal amount of reverse arc length evolution they will intersect. This is a contradiction of the assumption that S was simple before touching itself. Assume then that S_1 and S_2 are on the same side of T. Note that S_1 and S_2 can not be overlapping since they would still be overlapping after an infinitesimal amount of reverse arc length evolution, which is also a contradiction of the assumption that S was simple before touching itself. Let S_1 be the segment *inside* S_2, i.e., the tangent to S_2 always has S_1 to the same side. It can be seen that S_1 has a larger curvature at P than S_2. Therefore, after an infinitesimal amount of reverse arc length evolution, point P on S_1 and point P on S_2 will move in the same direction, but point P on S_1 will move by a larger amount. It follows that after an infinitesimal amount of reverse arc length evolution, S_1 and S_2 will intersect, which is again a contradiction. It follows that S remains simple during arc length evolution. ⋈

References

[1] S. Abbasi and F. Mokhtarian. Curvature scale space with affine length parametrisation. In *Proc International Conference on Scale-Space Theory in Computer Vision*, Corfu, Greece, 1999.

[2] S. Abbasi and F. Mokhtarian. Robustness of shape similarity retrieval under affine transformation. In *Proc Challenge of Image Retrieval*, Newcastle upon Tyne, UK, 1999.

[3] S. Abbasi and F. Mokhtarian. Shape similarity retrieval under affine transform: Application to multi-view object representation and recognition. In *Proc International Conference on Computer Vision*, pages 450–455, Corfu, Greece, 1999.

[4] S. Abbasi and F. Mokhtarian. Automatic view selection in multi-view object recognition. In *Proc International Conference on Pattern Recognition*, pages vol I: 13–16, Barcelona, Spain, 2000.

[5] S. Abbasi and F. Mokhtarian. Matching shapes with self-intersections. In *Proc International Conference on Visual Information Systems*, pages 233–243, Lyon, France, 2000.

[6] S. Abbasi and F. Mokhtarian. Multi-view object representation and recognition through curvature scale space. In *Proc Computer Society of Iran Computer Conference*, pages 1–6, Tehran, Iran, 2000.

[7] S. Abbasi and F. Mokhtarian. Affine-similar shape retrieval: Application to multiview 3-D object recognition. *IEEE Transactions on Image Processing*, 10(1):131–139, 2001.

[8] S. Abbasi, F. Mokhtarian, and J. Kittler. The modified curvature scale space image for shape similarity retrieval. In *Proc International Annual Computer Society of Iran Computer Conference*, pages 164–170, Tehran, Iran, 1997.

[9] S. Abbasi, F. Mokhtarian, and J. Kittler. Reliable classification of chrysanthemum leaves through curvature scale space. In *Proc International Conference on Scale-Space Theory in Computer Vision*, pages 284–295, Utrecht, The Netherlands, 1997.

[10] S. Abbasi, F. Mokhtarian, and J. Kittler. Shape similarity retrieval using a height adjusted curvature scale space image. In *Proc International Conference on Visual Information Systems*, pages 173–180, San Diego, CA, 1997.

[11] S. Abbasi, F. Mokhtarian, and J. Kittler. Curvature scale space image in shape similarity retrieval. *Springer Journal of MultiMedia Systems*, 7(6):467–476, 1999.

[12] S. Abbasi, F. Mokhtarian, and J. Kittler. Enhancing curvature scale space based shape retrieval for objects with shallow concavities. *Image and Vision Computing*, 18:199–211, 2000.

[13] A. Ahmed, F. Mokhtarian, and A. Hilton. Parametric motion blending through wavelet analysis. In *Proc Eurographics Conference*, pages 347–353, Manchester, UK, 2001.

[14] S. G. Akl and G. T. Toussaint. A fast convex hull algorithm. *Information Processing Letters*, 7:219–222, 1978.

[15] T. D. Alter and W. E. L. Grimson. Fast and robust 3-d recognition by alignment. In *Proc International Conference on Computer Vision*, pages 113–120, Berlin, 1993.

[16] L. Alvarez, F. Guichard, P. L. Lions, and J. M. Morel. Axioms and fundomental equations in image processing. *Arch. Rational Mech.*, 123:199–257, 1993.

[17] L. Alvarez, P. L. Lions, and J. M. Morel. Image selective smoothing and edge detection by nonlinear diffusion. *SIAM Journal of Numerical Analysis*, 29:845–866, 1992.

[18] A. Amini, S. Tehrani, and T. E. Weymouth. Using dynamic programming for minimizing the energy of active contours in the presence of hard constraints. In *Proc International Conference on Computer Vision*, pages 95–99, Tampa, Florida, 1988.

[19] A. Amini, T. E. Weymouth, and R. C. Jain. Using dynamic programming for solving variational problems in vision. *IEEE Transactions on Pattern Analysis and Machine Intelligence*, 12(9):855–867, 1990.

[20] K. Arbter, W. E. Snyder, H. Burkhardt, and G. Hirzinger. Applications of affine-invariant fourier descriptors to recognition of 3-d objects. *IEEE Trans. Pattern Analysis and Machine Intelligence*, 12(7):640–646, 1990.

[21] F. Arrebola, A. Bandera, P. Camacho, and F. Sandoval. Corner detection by local histograms of contour chain code. *Electronics Letters*, 33(21):1769–1771, 1997.

[22] K. S. Arun, T. S. Huang, and S. D. Blostein. Least-squares fitting of two 3-d point sets. *IEEE Trans Pattern Analysis and Machine Intelligence*, 9(5):698–700, 1987.

[23] H. Asada and M. Brady. The curvature primal sketch. *IEEE Trans Pattern Analysis and Machine Intelligence*, 8(1):2–14, 1986.

[24] N. Ayache and O. D. Faugeras. Hyper: A new approach for the recognition and positioning of 2-d objects. *IEEE Trans Pattern Analysis and Machine Intelligence*, 8(1):44–54, 1986.

[25] D. H. Ballard. Generalizing the hough transform to detect arbitrary shapes. *Pattern Recognition*, 13:111–122, 1981.

[26] D. H. Ballard. Strip trees: A hierarchical representation for curves. *Communications of the ACM*, 24(5):310–321, 1981.

[27] D. H. Ballard and C. M. Brown. *Computer Vision*. Prentice-Hall, Englewood Cliffs, NJ, 1982.

[28] P. R. Beaudet. Rotationally invariant image operators. In *Int. Joint Conf. Pattern recognition*, pages 579–583, 1978.

[29] E. J. Bellegarda, J. R. Bellegarda, D. Nahamoo, and K. S. Nathan. A probabilistic framework for the recognition of on-line handwriting. In *Proc International Workshop on Frontiers in Handwriting Recognition*, 1993.

[30] S. Belongie, J. Malik, and J. Puzicha. Shape matching and object recognition using shape context. *IEEE Transactions on Pattern Analysis and Machine Intelligence*, 24:509–522, 2002.

[31] J. L. Bentley, K. L. Clarkson, and D. B. Levine. Fast linear expected-time algorithms for computing maxima and convex hulls. *Algorithmica*, 9:168–183, 1993.

[32] J. L. Bentley, M. G. Faust, and F. P. Preparata. Approximation algorithms for convex hulls. *Communications of the ACM*, 25(1):64–68, 1982.

[33] P. J. Besl. The free-form surface matching problem. *In Machine Vision for Three-dimensional Scenes (H. Freeman, ed.)*, pages 25–71, 1990.

[34] P. J. Besl and R. C. Jain. Three dimentional object recognition. *ACM Computing Surveys*, 17:75–145, 1985.

[35] P. J. Besl and R. C. Jain. Invariant surface characteristics for 3d object recognition in range images. *Computer Vision, Graphics, and Image Processing*, 33:33–80, 1986.

[36] P. J. Besl and N. D. McKay. A method for registration of 3-d shapes. *IEEE Trans Pattern Analysis and Machine Intelligence*, 14(2):239–256, 1992.

[37] A. Blake and M. Isard. *Active Contours*. Springer, London, 1998.

[38] H. Blum. Biological shape and visual science (part i). *Journal of Theoretical Biology*, 38:205–287, 1973.

[39] M. Bober. Mpeg-7 visual shape descriptors. *IEEE Transactions on Circuits and Systems for Video Technology*, 11:716–719, 2001.

[40] M. Bober, K. Asai, and A. Divakaran. An mpeg-47-based internet video and still image browsing system. *Proceedings of SPIE - Multimedia Systems and Applications, Part III*, 4209:33–38, 2000.

[41] G. Bradski and S. Grossberg. Fast-learning viewnet architectures for recognizing 3-D objects from multiple 2-D views. *Neural Networks*, 8:1053–1080, 1995.

[42] M. Brady, J. Ponce, A. Yuille, and H. Asada. Describing surfaces. *Computer Vision, Graphics, and Image Processing*, 32:1–28, 1985.

[43] C. Brechbuhler, G. Gerig, and O. Kubler. Parametrization of closed surfaces for 3-d shape description. *Computer Vision and Image Understanding*, 61(2):154–170, 1995.

[44] R. A. Brooks. Symbolic reasoning among 3-D models and 2-D images. *Artificial Intelligence*, 17:285–348, 1981.

[45] J. F. Canny. A computational approach to edge detection. *IEEE Trans on Pattern Analysis and Machine Intelligence*, 8(6):679–698, 1986.

[46] V. Caselles, R. Kimmel, and G. Sapiro. Geodesic active contours. In *Proc International Conference on Computer Vision*, 1995.

[47] T. A. Cass. A robust parallel implementation of 2d model-based recognition. In *Proc IEEE Conference on Computer Vision and Pattern Recognition*, pages 879–884, 1988.

[48] T. A. Cass. Polynomial-time object recognition in the presence of clutter, occlusion, and uncertainty. In *Proc European Conference on Computer Vision*, pages 834–842, Santa Margherita Ligure, Italy, 1992.

[49] F. Chabat, G. Yang, and D. Hansell. A corner orientation detector. *Image and Vision Computing*, 17:761–769, 1999.

[50] S. A. Chang and A. Hsu. Image information systems, where do we go from here? *IEEE Trans. on Knowledge and Data Engineering*, 4(5):431–442, 1992.

[51] B. B. Chaudhuri and A. Rosenfeld. On the computation of the digital convex hull and circular hull of a digital region. *Pattern Recognition*, 31(12):2007–2015, 1998.

[52] W. C. Chen and P. Rockett. Bayesian labelling of corners using a grey-level corner image model. In *Proc IEEE International Conf Image Processing*, pages vol I: 687–690, 1997.

[53] W. T. Chen and T. R. Chou. Hierarchical deformation model for online cursive script recognition. *Pattern Recognition*, 27(2):205–219, 1994.

[54] S. S. Chern, P. Hartman, and A. Wintner. On isometric coordinates. In *Commentaries Mathematical Helvetici*, volume 28, 1954.

[55] R. T. Chin and C. R. Dyer. Model-based recognition in robot vision. *ACM Computing Surveys*, 18(1):67–108, 1986.

[56] K. Choo, I. D. Yun, and S. U. Lee. Edge-based approach to mesh simplification. In *International conference on 3-D digital imaging and modeling*, 1999.

[57] C. S. Chua and R. Jarvis. Point signatures: A new representation for 3d object recognition. *International Journal of Computer Vision*, 25(1):63–85, 1997.

[58] S. Clergeau-Tournemire and R. Plamondon. Integration of lexical and syntactical knowledge in a handwriting recognition system. *Machine Vision and Applications*, 8:249–259, 1995.

[59] J. Cohen. Simplification envelopes. In *Proc SIGGRAPH*, 1996.

[60] T. Cohignac, C. Lopez, and J. M. Morel. Integral and local affine invariant parameter and application to shape recognition. In *Proc. International Conference on Pattern Recognition*, volume 1, pages 164–168, 1994.

[61] D. Cyganski, T. A. Cott, J. A. Orr, and R. J. Dodson. Development, implementation, testing, and application of an affine transform invariant curvature function. In *Proc International Conference on Computer Vision*, pages 496–500, London, England, 1987.

[62] D. Cyganski and R. F. Vaz. A linear signal decomposition approach to affine invariant contour identification. In *Proc SPIE - Intelligent Robots and Computer Vision X: Algorithms and Techniques*, volume 1607, pages 98–109, 1991.

[63] L. da Fontoura Costa and R. M. C. Jr. *Shape Analysis and Classification, Theory and Practice*. CRC press, 2001.

[64] P. E. Danielsson. A new shape factor. *Computer Graphics and Image Processing*, 7:292–299, 1978.

[65] E. R. Davies. Application of the generalized hough transform to corner detection. *IEE Proceedings*, 135:49–54, 1988.

[66] L. S. Davis. Understanding shape: Angles and sides. *IEEE Trans Computers*, C-26:236–242, 1977.

[67] A. Del-Bimbo, P. Pala, and S. Santini. Image retrieval by elastic matching of shapes and image patterns. In *Proceedings of the 1996 International Conference on Multimedia Computing and Systems*, pages 215–218, Hiroshima, Japan, 1996. IEEE, Los Alamitos, CA.

[68] S. J. Dickinson, A. P. Pentland, and A. Rosenfeld. 3-D shape recovery using distributed aspect matching. *IEEE Trans. on Pattern Analysis and Machine Intelligence*, 14:174–198, 1992.

[69] L. Dreschler and H. H. Nagel. Volumetric model and 3d trajectory of a moving car derived from monocular tv frame sequences of a street scene. In *IJCAI-81*, pages 692–697, 1981.

[70] R. O. Duda and P. E. Hart. Use of the hough transformation to detect lines and curves in pictures. *Communications of the ACM*, 15:11–15, 1972.

[71] J. Dunker, G. Hartmann, and M. Stohr. Single view recognition and pose estimation of 3-D objects using sets of prototypical views and spatially tolerant contour representations. In *Proc International Conference on Pattern Recognition*, volume 4, pages 4–18, 1996.

[72] S. Dusani, K. Breeding, and R. B. Mcghee. Aircraft identification by moment invariants. *IEEE Transactions on Computers*, C-26:39–45, 1977.

[73] W. F. Eddy. A new convex hull algorithm for planar sets. *ACM Trans on Circuits and Systems*, 3:398–403, 1977.

[74] P. Eggleston. Constraint-based feature indexing and retrieval for image databases. *SPIE*, 1819:27–39, 1992.

[75] L. F. Estrozi, A. G. Campos, L. G. Rios, R. M. C. Jr., and L. da F Costa. Comparing curvature estimation techniques. In *Proc. 4th Simpósio Brasileiro de Automação Inteligente*, pages 58–63, São Paulo, Brazil, 1999.

[76] L. F. Estrozi, L. G. Rios, A. G. Campos, R. M. C. Jr., and L. da F Costa. 1D and 2D Fourier-based approaches to numeric curvature estimation and their comparative performance assessment. *submitted to Graphical Models*, 2001.

[77] A. C. Evans, N. A. Thacker, and J. E. W. Mayhew. A practical view-based 3-D object recognition system. In *Proc International Conference on Artificial Networks*, pages 6–15, 1993.

[78] T. J. Fan, G. Medioni, and R. Nevatia. Recognising 3-d objects using surface descriptions. *IEEE Trans Pattern Analysis and Machine Intelligence*, 11:1140–1157, 1989.

[79] J. Q. Fang and T. S. Huang. A corner finding algorithm for image analysis and registration. In *Proc. AAAI Conf*, pages 46–49, 1982.

[80] O. D. Faugeras and M. Hebert. The representation, recognition, and locating of 3-d objects. *International Journal of Robotics Research*, 5(3):27–52, 1986.

[81] J. T. Favata. General word recognition using approximate segment-string matching. In *Proc International Conference on Document Analysis and Recognition*, pages 92–95, Ulm, Germany, 1997.

[82] J. Flusser and T. Suk. Pattern recognition by affine moment invariants. *Pattern Recognition*, 26(1):167–174, 1993.

[83] P. J. Flynn and A. K. Jain. On reliable curvature estimation. In *Proc IEEE Conference on Computer Vision and Pattern Recognition*, pages 110–116, 1989.

[84] P. J. Flynn and A. K. Jain. Bonsai: 3-d object recognition using constrained search. *IEEE Trans Pattern Analysis and Machine Intelligence*, 10:1066–1075, 1991.

[85] P. J. Flynn and A. K. Jain. Three dimensional object recognition. *Handbook of Pattern Recognition and Image Processing (T.Y.Young, ed.)*, 2:497–541, 1994.

[86] J. D. Foley and A. VanDam. *Fundamentals of Interactive Computer Graphics*. Addison-Wesley, 1984.

[87] J. D. Foley, A. VanDam, S. K. Feiner, J. F. Hughes, and R. L. Phillips. *Introduction to Computer Graphics*. Addison-Wesley, 1997.

[88] J. Franke, J. M. Gloger, A. Kaltenmeier, and E. Mandler. Comparison of gaussian distribution and polynomial classifiers in a hidden markov model based system for the recognition of cursive script. In *Proc International Conference on Document Analysis and Recognition*, pages 515–518, 1997.

[89] H. Freeman. Computer processing of line-drawing images. *Computing Surveys*, 6, 1974.

[90] B. Furht, S. W. Smoliar, and H. Zhang. *Video and Image Processing in MultiMedia Systems*. Kluwer Academic, 1996.

[91] M. Gage and R. S. Hamilton. The heat equation shrinking convex plane curves. *Journal of Differential Geometry*, 23:69–96, 1986.

[92] M. Gilloux, M. Leroux, and J. M. Bertille. Strategies for cursive script recognition using hidden markov models. *Machine Vision and Applications*, 8:197–205, 1995.

[93] G. A. Giraldi, E. Strauss, and A. Oliveira. Boundary extraction method based on dual-t-snakes and dynamic programming. *Proc IEEE Conference on Computer Vision and Pattern Recognition*, 1:44–49, 2000.

[94] C. Goad. Special purpose automatic programming for 3-D model-based vision. In *Proc Image Understanding Workshop*, pages 94–104, 1983.

[95] A. Goetz. *Introduction to Differential Geometry*. Addison-Wesley, Reading, MA, 1970.

[96] M. Golin and R. Sedgewick. Analysis of a simple yet efficient convex hull algorithm. In *Proc Annual Symposium on Computational Geometry*, pages 153–163, 1988.

[97] R. L. Graham. An efficient algorithm for determining the convex hull of a finite planar set. *Information Processing Letters*, 1:132–133, 1972.

[98] W. Grimson and T. Lozano-Perez. Localizing overlapping parts by searching the interpretation tree. *IEEE Trans Pattern Analysis and Machine Intelligence*, 9:469–482, 1987.

[99] W. E. L. Grimson and T. Lozano-Perez. Model-based recognition and localization from sparse range or tactile data. *International Journal of Robotics Research*, 3(3):3–35, 1984.

[100] R. P. Grzeszuk and D. N. Levin. Brownian strings; images segmentation with stochastically deformable contours. *IEEE Transaction on Pattern Analaysis and Machine Intelligence*, 1995.

[101] H. W. Guggenheimer. *Differential Geometry*. McGraw-Hill, New York, 1963.

[102] D. Guillevic and C. Y. Suen. Cursive script recognition: A fast reader scheme. In *Proc International Conference on Document Analysis and Recognition*, pages 311–314, 1993.

[103] I. Guskov, W. Sweldens, and P. Schroder. Multiresolution signal processing for meshes. In *Proc SIGGRAPH*, 1999.

[104] C. G. Harris. Determination of ego-motion from matched points. In *Proc. Alvey Vision Conf.*, Cambridge, UK, 1987.

[105] M. Hebert and T. Kanade. The 3-d profile method for object recognition. In *Proc IEEE Conference on Computer Vision and Pattern Recognition*, pages 458–463, San Francisco, 1985.

[106] S. Heitzz. *Coordinates in Geodesy*. Addison-Wesley, Berlin, 1985.

[107] A. Hilton, A. J. Stoddart, J. Illingworth, and T. Windeatt. Marching triangles: Range image fusion for complex object modelling. In *Proc IEEE International Conference on Image Processing*, pages 381–384, Lausanne, Switzerland, 1996.

[108] A. Hilton, A. J. Stoddart, J. Illingworth, and T. Windeatt. Reliable surface reconstruction from multiple range images. In *Proc European Conference on Computer Vision*, pages 117–126, Cambridge, UK, 1996.

[109] J. Hoffman, J. Skrzypek, and J. J. Vidal. Cluster network for recognition of handwritten, cursive script characters. *Neural Networks*, 6(1):69–78, 1993.

[110] J. Hong and X. Tan. Recognize the similarity between shapes under affine transformation. In *Proc International Conference on Computer Vision*, pages 489–493, Tarpon Springs, Florida, 1988.

[111] K. S. Hong and K. Ikeuchi. Generating a strategy for configuration determination: a module of vision algorithm compiler for object localization programs. In *Proc International Symposium of Robotics Research*, pages 164–171, 1989.

[112] H. Hoppe. Mesh optimization. In *Proc SIGGRAPH*, 1993.

[113] H. Hoppe. Progressive meshes. In *Proc SIGGRAPH*, pages 99–106, 1996.

[114] B. K. P. Horn. Closed-form solution of absolute orientation using unit quaternions. *Journal of Optical Society of America*, A4:629–642, 1987.

[115] B. K. P. Horn, H. M. Hilden, and S. Negahdaripour. Closed-form solution of absolute orientation using orthonormal matrices. *Journal of Optical Society of America*, A5:1128–1135, 1988.

[116] B. K. P. Horn and E. J. Weldon. Filtering closed curves. *IEEE Trans Pattern Analysis and Machine Intelligence*, 8:665–668, 1986.

[117] P. V. C. Hough. Method and means for recognizing complex patterns. US Patent 3,069,654, 1962.

[118] M. K. Hu. Visual pattern recognition by moments invariants. *IRE Trans. Information Theory,*, IT-8:179–187, 1962.

[119] Z. Huang and F. S. Cohen. Affine-invariant B-spline moment for curve matching. In *Proc IEEE Computer Society Conference on Computer Vision and Pattern Recognition*, pages 490–495, 1994.

[120] R. A. Hummel, B. Kimia, and S. W. Zucker. Deblurring gaussian blur. *Computer Vision, Graphics and Image Processing*, 38:66–80, 1987.

[121] D. P. Huttenlocher and S. Ullman. Object recognition using alignment. In *Proc International Conference on Computer Vision*, pages 102–111, London, England, 1987.

[122] D. P. Huttenlocher and S. Ullman. Recognizing solid objects by alignment. In *Proc Image Understanding Workshop*, pages 1114–1124, Cambridge, MA, 1988.

[123] K. Ikeuchi. Precompiling a geometrical model into an interpretation for object recognition in bin-picking tasks. In *Proc Image Understanding Workshop*, pages 321–339, Los Angeles, CA, 1987.

[124] H. Ip and D. Shen. Affine-invariant active contour model for model-based segmentation. *Image and Vision Computing*, 16(2):135–146, 1998.

[125] ISO/IEC 15938-8. Extraction and use of mpeg-7 descriptions. Technical Report ISO/IEC JTC 1/SC 29/WG 11 N4579, INTERNATIONAL ORGANIZATION FOR STANDARDIZATION, ISO/IEC JTC 1/SC 29/WG 11, CODING OF MOVING PICTURES AND AUDIO, 2002.

[126] ISO/IEC FDIS 15938-3. Text of ISO/IEC 15938-3/FCD information technology multimedia content description interface part 3 visual. Technical report, 2001.

[127] R. A. Jarvis. On the identification of the convex hull of a finite set of points in the plane. *Information Processing Letters*, 2:18–21, 1973.

[128] Q. Ji and R. M. Haralick. Corner detection with covariance propagation. In *Proc IEEE Conf Computer Vision and Pattern Recognition*, pages 362–367, 1997.

[129] J. S. Jin and R. Kurniawati. A scheme for intelligent image retrieval in multimedia databases. *Journal of Visual Communication and Image Representation*, 7(4):369–377, 1996.

[130] A. E. Johnson and M. Hebert. Using spin images for efficient object recognition in cluttered 3D scenes. *IEEE Trans Pattern Analysis and Machine Intelligence*, 21(5):433–449, 1999.

[131] M. Kadirkamanathan and P. J. W. Rayner. A unified approach to on-line cursive script segmentation and feature extraction. In *IEEE International Conference on Acoustics, Speech, and Signal Processing*, pages 1659–1662, 1989.

[132] D. J. Kang. Stable snake algorithm for convex tracking of MRI sequences. *Electronics Letters*, 35(13):1070–1071, 1999.

[133] S. B. Kang and K. Ikeuchi. The complex egi, a new representation for 3-d pose determination. *IEEE Trans Pattern Analysis and Machine Intelligence*, 15(7):707–721, 1993.

[134] M. Kass, A. Witkin, and D. Terzopoulos. Snakes: active contour models. In *Proc International Conference on Computer Vision*, pages 259–268, 1987.

[135] W. Kecs. *The Convolution Product and Some Applications*. D Reidel, Boston, MA, 1982.

[136] N. Khalili, F. Mokhtarian, and P. Yuen. Free-form surface description in multiple scales: Extension to incomplete surfaces. In *Proc International Conference on Computer Analysis of Images and Patterns*, pages 293–300, Ljubljana, Slovenia, 1999.

[137] N. Khalili, F. Mokhtarian, and P. Yuen. Recovery of curvature and torsion features from free-form 3-D meshes at multiple scales. In *Proc Asian Conference on Computer Vision*, pages 1070–1075, Taipei, Taiwan, 2000.

[138] A. Khotanzad and Y. H. Hong. Invariant image recognition by zernike moments. *IEEE Trans Pattern Analysis and Machine Intelligence*, 12(5):489–497, 1990.

[139] J. D. Kim and H. K. Kim. Shape descriptor based on multi-layer eigenvector. Technical Report ISO/IEC P517, 1999.

[140] W. Kim, S. G. Hong, and J. J. Lee. Active contour model using image flow for tracking a moving object. *IEEE International Conference on Intelligent Robots and Systems*, 1:216–221, 1999.

[141] B. Kimia, A. Tannenbaum, and S. W. Zucker. Toward a computational theory of shape: an overview. In *Proc European Conference on Computer Vision*, pages 402–407, Antibes, France, 1990. Springer.

[142] B. B. Kimia and K. Siddiqi. Geometric heat equation and nonlinear diffusion of shapes and images. *Computer Vision and Image Understanding*, 64(3):305–332, 1996.

[143] L. Kitchen and A. Rosenfeld. Gray level corner detection. *Pattern Recognition Letters*, pages 95–102, 1982.

[144] J. J. Koenderink. *Solid Shape*. MIT Press, Cambridge, MA, 1990.

[145] J. J. Koenderink and A. J. vanDoorn. Internal representation of solid shape with respect to vision. *Biological Cybernetics*, 32(4):211–216, 1979.

[146] J. J. Koenderink and A. J. vanDoorn. Dynamic shape. *Biological Cybernetics*, 53:383–396, 1986.

[147] K. Kohlmann. Corner detection in natural images based on the 2-d hilbert transform. *Signal Processing*, 48(3):225–234, 1996.

[148] A. Kornai, K. M. Mohiuddin, and S. D. Connell. Recognition of cursive writing on personal checks. In *Proc International Workshop on Frontiers in Handwriting Recognition*, pages 373–378, 1996.

[149] E. Kreyszig. *Differential Geometry*. Oxford University Press, 1959.

[150] Y. Kuno, Y. Okamoto, and S. Okada. Object recognition using a feature search strategy generated from a 3-D model. In *Proc IEEE International Conference on Computer Vision*, pages 626–635, 1990.

[151] K. K. Lai and P. S. Y. Wu. Effective edge-corner detection method for defected images. In *Proc International Conference on Signal Processing*, pages II: 1151–1154, 1996.

[152] Y. Lamdan, J. T. Schwartz, and H. J. Wolfson. Object recognition by affine invariant matching. In *Proc of CVPR*, pages 335–344, Ann Arbor, Michigan, 1988.

[153] Y. Lamdan, J. T. Schwartz, and H. J. Wolfson. On recognition of 3-d objects from 2-d images. In *Proc IEEE International Conference on Robotics and Automation*, pages 1407–1413, Philadelphia, Pa., 1988.

[154] Y. Lamdan and H. J. Wolfson. Geometric hashing: A general and efficient model-based recognition scheme. In *Proc International Conference on Computer Vision*, pages 238–249, Tampa, Florida, 1988.

[155] B. Lamiroy and P. Gros. Rapid object indexing and recognition using enhanced geometric hashing. In *Proc European Conference on Computer Vision*, pages 59–70, Cambridge, UK, 1996.

[156] L. J. Latecki and R. Lakamper. Contour-based shape similarity. In *Proc International Conference on Visual Information Systems*, pages 617–624, Amsterdam, 1999.

[157] L. J. Latecki and R. Lakamper. Shape similarity measure based on correspondence of visual parts. *IEEE Trans. Pattern Analysis and Machine Intelligence*, 22(10):1185–1190, 2000.

[158] E. Lecolinet and J. V. Moreau. Off-line recognition of handwritten cursive script for the automatic reading of city names on real mail. In *Proc International Conference on Pattern Recognition*, pages 674–676, 1990.

[159] K. J. Lee and Z. Bien. Grey-level corner detector using fuzzy logic. *Pattern Recognition Letters*, 17(9):939–950, 1996.

[160] S. H. Lee and J. H. Kim. Augmenting the discrimination power of hmm by nn for on-line cursive script recognition. *Applied Intelligence*, 7(4):305–314, 1997.

[161] Y. Leung, J. S. Zhang, and Z. B. Xu. Neural networks for convex hull computation. *IEEE Transactions on Neural Networks*, 8(3):601–611, 1997.

[162] P. Liang and C. H. Taubes. Orientation-based differential geometric representations for computer vision applications. *IEEE Trans Pattern Analysis and Machine Intelligence*, 16(3):249–258, 1994.

[163] P. Liang and J. S. Todhunter. Representation and recognition of surface shapes in range images: A differential geometry approach. *Computer Vision, Graphics, and Image Processing*, 52(1):78–109, 1990.

[164] S. Linnainmaa, D. Harwood, and L. Davis. Pose determination of a three-dimensional object using triangle pairs. *IEEE Trans Pattern Analysis and Machine Intelligence*, 10(5):649–647, 1988.

[165] S. Loncaric. Survey of shape analysis techniques. *Pattern Recognition*, 31(8):983–1001, 1998.

[166] D. G. Lowe. Organization of smooth image curves at multiple scales. In *Proc International Conference on Computer Vision*, pages 558–567, Tampa, FL, 1988.

[167] D. G. Lowe. Fitting parametrized 3-d models to images. *IEEE Trans Pattern Analysis and Machine Intelligence*, 13(5):441–450, 1991.

[168] H. Luo, Q. Lu, R. Acharya, and R. Gaborski. Robust snake model. *Proc IEEE Conference on Computer Vision and Pattern Recognition*, 1:452–457, 2000.

[169] A. K. Mackworth and F. Mokhtarian. Scale-based description of planar curves. In *Proc Canadian Society for Computational Studies of Intelligence*, pages 114–119, London, Ontario, 1984.

[170] A. K. Mackworth and F. Mokhtarian. The renormalized curvature scale space and the evolution properties of planar curves. In *Proc IEEE Conference on Computer Vision and Pattern Recognition*, pages 318–326, Ann Arbor, Michigan, 1988.

[171] R. Malladi and J. A. Sethian. Level set and fast marching methods in image processing and computer vision. In *Proc IEEE International Conference on Computer Vision*, pages 489–492, 1996.

[172] R. Malladi and J. A. Sethian. Unified approach to noise removal, image enhancement, and shape recovery. *IEEE Trans on Image Processing*, 5(11):1554–1568, 1996.

[173] G. M. T. Man and J. C. H. Poon. Cursive script segmentation and recognition by dynamic programming. *SPIE Character Recognition Technologies*, 1906:184–194, 1993.

[174] B. Manjunath, P. Salembier, and T. Sikora. *Introduction to MPEG-7, Multimedia Content Description Interface*. Wiley, England, 2002.

[175] D. Marr. Representing visual information. memo 415, MIT AI Lab, Cambridge, 1977.

[176] D. Marr and E. C. Hildreth. Theory of edge detection. *Proc Royal Society London B*, 207:187–217, 1980.

[177] D. Marr and H. K. Nishihara. Representation and recognition of the spatial organization of 3-d structures. *Proc Royal Society London B*, 200:269–294, 1978.

[178] J. W. McKee and J. K. Aggarwal. Computer recognition of partial views of curved objects. *IEEE Trans Computers*, C-26:790–800, 1977.

[179] R. Mehrotra and J. E. Gary. Feature-based retrieval of similar shapes. In *Proceedings of 9th International Conference on Data Engineering*, pages 108–115, Vienna, Austria, 1993. IEEE Computer Society Press, Los Alamitos, CA.

[180] R. Mehrotra, S. Nichani, and N. Ranganathan. Corner detection. *Pattern Recognition*, 23(11):1223–1233, 1990.

[181] F. Mohanna and F. Mokhtarian. Improved curvature estimation for accurate localization of active contours. In *Proc International Conference on Image Processing*, pages vol II: 781–784, Thessaloniki, Greece, 2001.

[182] F. Mohanna and F. Mokhtarian. Performance evaluation of corner detection algorithms under similarity and affine transforms. In *Proc British Machine Vision Conference*, pages 353–362, Manchester, UK, 2001.

[183] F. Mohanna and F. Mokhtarian. An efficient active contour model through curvature scale space filtering. *MultiMedia Tools and Applications*, 2002.

[184] F. Mohanna and F. Mokhtarian. Fusion of corners from multiple scales for robust tracking. In *Proc International Conference on Computer Vision, Pattern Recognition, and Image Processing*, Durham, North Carolina, 2002.

[185] F. Mohanna and F. Mokhtarian. Robust corner tracking for multimedia applications. In *Proc International Conference on Image Processing*, Rochester, New York, 2002.

[186] F. Mokhtarian. Evolution properties of space curves. In *Proc International Conference on Computer Vision*, pages 100–105, Tarpon Springs, Florida, 1988.

[187] F. Mokhtarian. Multi-scale description of space curves and 3-d objects. In *Proc IEEE Conference on Computer Vision and Pattern Recognition*, pages 298–303, Ann Arbor, Michigan, 1988.

[188] F. Mokhtarian. Multi-scale, torsion-based shape representations for space curves. In *Proc IEEE Conference on Computer Vision and Pattern Recognition*, pages 660–661, New York City, 1993.

[189] F. Mokhtarian. Convergence properties of curvature scale space representations. In *Proc British Machine Vision Conference*, pages 357–366, Birmingham, UK, 1995.

[190] F. Mokhtarian. Silhouette-based isolated object recognition through curvature scale space. *IEEE Trans Pattern Analysis and Machine Intelligence*, 17(5), 1995.

[191] F. Mokhtarian. Zero-crossings of curvature and torsion in the limit. In *Proc Asian Conference on Computer Vision*, pages III:457–461, Singapore, 1995.

[192] F. Mokhtarian. Silhouette-based object recognition with occlusion through curvature scale space. In *Proc European Conference on Computer Vision*, pages I:566–578, Cambridge, England, 1996.

[193] F. Mokhtarian. Curvature scale space representation for robust, silhouette-based object recognition with occlusion. *Scientia Iranica*, 3(4):121–135, 1997.

[194] F. Mokhtarian. Multi-scale contour segmentation. In *Proc International Conference on Scale-Space Theory in Computer Vision*, pages 296–307, Utrecht, The Netherlands, 1997.

[195] F. Mokhtarian. Multi-scale object segmentation for robust recognition. In *Proc Vision Interface*, pages 8–15, Kelowna, British Columbia, Canada, 1997.

[196] F. Mokhtarian. Silhouette-based occluded object recognition through curvature scale space. *Machine Vision and Applications*, 10:87–97, 1997.

[197] F. Mokhtarian. A theory of multi-scale, torsion-based shape representation for space curves. *Computer Vision and Image Understanding*, 68(1):1–17, 1997.

[198] F. Mokhtarian. Torsion scale space images: Robust representations for space curves. In *Proc Scandinavian Conference on Image Analysis*, pages 515–522, Lappeenranta, Finland, 1997.

[199] F. Mokhtarian. Segmenting objects at multiple scales: A robust approach. In *Proc Asian Conference on Computer Vision*, pages II: 73–80, Hong Kong, 1998.

[200] F. Mokhtarian. Content-based similarity retrieval from large image databases. In M. Varga and M. Petrou, editors, *Practical Applications of Image Processing and Interpretation Techniques*. Wiley, 2002.

[201] F. Mokhtarian. Shape analysis through the curvature scale space technique. In M. Varga and M. Petrou, editors, *Practical Applications of Image Processing and Interpretation Techniques*. Wiley, 2002.

[202] F. Mokhtarian and S. Abbasi. Curvature scale space for shape similarity retrieval under affine transforms. In *Proc. International Conf on Computer Analysis of Images and Patterns*, pages 65–72, Ljubljana, Slovenia, 1999.

[203] F. Mokhtarian and S. Abbasi. Retrieval of similar shapes under affine transformation. In *Proc International Conference on Visual Information Systems*, Amsterdam, The Netherlands, 1999.

[204] F. Mokhtarian and S. Abbasi. Shape-based indexing using curvature scale space with affine curvature. In *Proc European Workshop on Content-Based Multi-Media Indexing*, pages 255–262, Toulouse, France, 1999.

[205] F. Mokhtarian and S. Abbasi. Affine curvature scale space with affine length parametrisation. *Pattern Analysis and Applications*, 4(1):1–8, 2001.

[206] F. Mokhtarian and S. Abbasi. Multi-view free-form 3-d object retrieval with incomplete data. In *Proc IEEE International Workshop on MultiMedia Signal Processing*, pages 287–292, Cannes, France, 2001.

[207] F. Mokhtarian and S. Abbasi. Shape similarity retrieval under affine transforms. *Pattern Recognition*, 35(1), 2002.

[208] F. Mokhtarian, S. Abbasi, and J. Kittler. Efficient and robust retrieval by shape content through curvature scale space. In *Proc International Workshop on Image DataBases and MultiMedia Search*, pages 35–42, Amsterdam, The Netherlands, 1996.

[209] F. Mokhtarian, S. Abbasi, and J. Kittler. Indexing an image database by shape content using curvature scale space. In *Digest of IEE Colloquium on Intelligent Image DataBases*, pages 4/1–4/6, London, UK, 1996.

[210] F. Mokhtarian, S. Abbasi, and J. Kittler. Robust and efficient shape indexing through curvature scale space. In *Proc British Machine Vision Conference*, pages 53–62, Edinburgh, UK, 1996.

[211] F. Mokhtarian, S. Abbasi, and J. Kittler. A new approach to computation of curvature scale space image for shape similarity retrieval. In *Proc International Conference on Image Analysis and Processing*, pages 140–147, Florence, Italy, 1997.

[212] F. Mokhtarian, S. Abbasi, and J. Kittler. Efficient curvature-based shape representation for similarity retrieval. In *European Signal Processing Conference*, pages 597–600, Island of Rhodes, Greece, 1998.

[213] F. Mokhtarian, N. Khalili, and P. Yuen. Multi-scale 3-d free-form surface smoothing. In *Proc British Machine Vision Conference*, pages 730–739, Southampton, UK, 1998.

[214] F. Mokhtarian, N. Khalili, and P. Yuen. Multi-scale free-form surface description. In *Proc Indian Conference on Computer Vision, Graphics and Image Processing*, pages 70–75, New Delhi, India, 1998.

[215] F. Mokhtarian, N. Khalili, and P. Yuen. Free-form 3-d object recognition at multiple scales. In *Proc British Machine Vision Conference*, pages 446–455, Bristol, 2000.

[216] F. Mokhtarian, N. Khalili, and P. Yuen. Curvature computation on free-form 3-d meshes at multiple scales. *Computer Vision and Image Understanding*, 83(2):118–139, 2001.

[217] F. Mokhtarian, N. Khalili, and P. Yuen. Free-form 3-d surface description in multiple scales. *Scientia Iranica*, 8(4):1–12, 2001.

[218] F. Mokhtarian, N. Khalili, and P. Yuen. Multi-scale free-form 3-D object recognition using 3-D models. *Image and Vision Computing*, 19:271–281, 2001.

[219] F. Mokhtarian, N. Khalili, and P. Yuen. Estimation of error in curvature computation on multi-scale free-form surfaces. *International Journal of Computer Vision*, 48(2):131–149, 2002.

[220] F. Mokhtarian and A. K. Mackworth. Scale-based description and recognition of planar curves and two-dimensional shapes. *IEEE Trans Pattern Analysis and Machine Intelligence*, 8(1):34–43, 1986.

[221] F. Mokhtarian and A. K. Mackworth. Scale-based description and recognition of planar curves and two-dimensional shapes. In R. Kasturi and R. C. Jain, editors, *Computer Vision: Advances & Applications*, pages 154–163. IEEE Computer Society Press, Los Alamitos, CA, 1991.

[222] F. Mokhtarian and A. K. Mackworth. A theory of multi-scale, curvature-based shape representation for planar curves. *IEEE Trans Pattern Analysis and Machine Intelligence*, 14(8):789–805, 1992.

[223] F. Mokhtarian and F. Mohanna. Enhancing the curvature scale space corner detector. In *Proc Scandinavian Conference on Image Analysis*, pages 145–152, Bergen, Norway, 2001.

[224] F. Mokhtarian and F. Mohanna. Fast active contour convergence through curvature scale space filtering. In *Proc Image and Vision Computing New Zealand*, Dunedin, New Zealand, 2001.

[225] F. Mokhtarian and F. Mohanna. Fast active contour convergence through adaptive curvature scale space smoothing. In *Proc International Conference on Computer Vision, Pattern Recognition, and Image Processing*, Durham, North Carolina, 2002.

[226] F. Mokhtarian and H. Murase. Silhouette-based object recognition through curvature scale space. In *Proc International Conference on Computer Vision*, pages 269–274, Berlin, 1993.

[227] F. Mokhtarian and S. Naito. Scale properties of curvature and torsion zero-crossings. In *Proc Asian Conference on Computer Vision*, pages 303–308, Osaka, Japan, 1993.

[228] F. Mokhtarian and R. Suomela. Curvature scale space based image corner detection. In *European Signal Processing Conference*, pages 2549–2552, Island of Rhodes, Greece, 1998.

[229] F. Mokhtarian and R. Suomela. Curvature scale space for robust image corner detection. In *Proc International Conference on Pattern Recognition*, pages 1819–1821, Brisbane, Australia, 1998.

[230] F. Mokhtarian and R. Suomela. Robust image corner detection through curvature scale space. *IEEE Trans on Pattern Analysis and Machine Intelligence*, 20(12):1376–1381, 1998.

[231] F. Mokhtarian and R. Suomela. Curvature scale space for image point feature detection. In *Proc International Conference on Image Processing and its Applications*, pages 206–210, Manchester, UK, 1999.

[232] F. Mokhtarian, R. Suomela, and K. C. Chan. Image point feature detection through curvature scale space. In *Proc Indian Conference on Computer Vision, Graphics and Image Processing*, pages 261–266, New Delhi, India, 1998.

[233] F. Mokhtarian and Y. Ung. Automatic fitting of digitised contours at multiple scales through curvature scale space. In *Proc EuroGraphics Conference*, pages 99–106, Interlaken, Switzerland, 2000.

[234] P. G. Morasso, M. Limoncelli, and M. Morchio. Incremental learning experiments with scriptor: an engine for on-line recognition of cursive handwriting. *Machine Vision and Applications*, 8:206–214, 1995.

[235] H. P. Moravec. Towards automatic visual obstacle avoidance. In *Proc Int. Joint Conf. Artificial Intelligence*, page 584, 1977.

[236] K. Muller and J. R. Ohm. Description of core experiment for motion/shape. Technical Report ISO/IEC N4740, 1999.

[237] D. Mumford. The problem of robust shape descriptions. In *International Conference on Computer Vision*, pages 602–606, London, England, 1987.

[238] H. Murase and S. K. Nayar. Visual learning and recognition of 3-D objects from appearance. *International Journal of Computer Vision*, 14(1):5–24, 1995.

[239] A. A. Y. Mustafa, L. G. Shapiro, and M. A. Ganter. 3-D object recognition from color intensity images. In *Proc International Conference on Pattern Recognition*, volume 1, pages 627–631, 1996.

[240] S. Nassif, D. Capson, and A. Vaz. Robust real-time corner location measurement. In *Proc IEEE Conf Instrumentation and Measurement Technology*, pages 106–111, 1997.

[241] S. Nayar, S. Nene, and H. Murase. Real-time 100 object recognition system. In *Proc ARPA Image Understanding Workshop*, 1996.

[242] K. Ngoi and J. Jia. Active contour model for colour region extraction in natural scenes. *Image and Vision Computing*, 17(13):955–966, 1999.

[243] W. Niblack, R. Barber, W. Equitz, M. Flickner, E. Glasman, D. Petkovic, P. Yanker, C. Faloutsos, and G. Taubin. The qbic project: Querying images by content using color, texture and shape. *SPIE*, 1908:173–187, 1993.

[244] H. K. Nishihara. Intensity, visible-surface, and volumetric representations. *Artificial Intelligence*, 17:265–284, 1981.

[245] A. Noble. Finding corners. *Image and Vision Computing*, 6:121–128, 1988.

[246] C. M. Orange and F. C. A. Groen. Model based corner detection. In *Proc IEEE Conf on Computer Vision and Pattern Recognition*, 1993.

[247] K. Paler, J. Foglein, J. Illingworth, and J. Kittler. Local ordered grey levels as an aid to corner detection. *Pattern Recognition*, 17(5):535–543, 1984.

[248] M. Parizeau and R. Plamondon. A fuzzy-syntactic approach to allograph modeling for cursive script recognition. *IEEE Transactions on Pattern Analysis and Machine Intelligence*, 17(7):702–712, 1995.

[249] T. Pavlidis. Polygonal approximations by newton's method. *IEEE Trans Computers*, C-26:800–807, 1977.

[250] A. Pentland and S. Sclaroff. Closed-form solutions for physically-based shape modelling and recognition. *IEEE Trans Pattern Analysis and Machine Intelligence*, 13:715–729, 1991.

[251] A. P. Pentland. Perceptual organisation and the representation of natural form. *Artificial Intelligence*, 28:293–331, 1986.

[252] E. Persoon. Shape discrimination using fourier descriptors. *IEEE Trans Systems, Man and Cybernetics*, 7:170–179, 1977.

[253] N. Peterfreund. Velocity snake: Deformable contour for tracking in spatio-velocity space. *Computer Vision and Image Understanding*, 73(3):346–356, 1999.

[254] R. W. Picard and T. Kabir. Finding similar patterns in large image databases. In *Proc IEEE Conference on Acoustics, Speech and Signal Processing*, Minneapolis, MN, 1993.

[255] M. Pilu and R. Fisher. Recognition of geons by parametric deformable contour models. In *Proc European Conference on Computer Vision*, pages 71–82, Cambridge, UK, 1996.

[256] J. Ponce, D. J. Kriegman, S. Petitjaan, S. Sullivan, G. Taubin, and B. Vijayakumar. Representations and algorithms for 3-D curved object recognition. In *In Three-dimensional Object recognition Systems*, pages 17–56, Amsterdam, Netherlands, 1993.

[257] F. P. Preparata. An optimal real-time algorithm for planar convex hull. *Communications of ACM*, 22:402–405, 1979.

[258] F. P. Preparata and S. J. Hong. Convex hulls of finite sets of points in two and three dimensions. *Communications of ACM*, 20:87–93, 1977.

[259] F. P. Preparata and M. I. Shamos. *Computational Geometry: An Introduction.* Springer-Verlag, 1985.

[260] S. Proctor, J. Illingworth, and F. Mokhtarian. Cursive handwriting recognition using hidden markov models and a lexicon-driven level building algorithm. *IEE Proceedings on Vision, Image, and Signal Processing*, 147(4):332–339, 2000.

[261] A. Quddus and M. Fahmy. Fast wavelet-based corner detection technique. *Electronics Letters*, 35(4):287–288, 1999.

[262] K. Rangarajan, M. Shah, and D. V. Brackle. Optimal corner detector. *Computer Vision, Graphics and Image Processing*, 48:230–245, 1989.

[263] R. P. N. Rao and D. H. Ballard. Object indexing using an iconic sparse distributed memory. In *Proc International Conference on Computer Vision*, Cambridge, MA, 1995.

[264] W. Richards, B. Dawson, and D. Whittingham. Encoding contour shape by curvature extrema. In W. Richards, editor, *Natural Computations*, pages 83–98. MIT Press, Cambridge, MA, 1988.

[265] K. Rohr. Recognizing corners by fitting parametric models. *Computer Vision*, 9(3):213–230, 1992.

[266] T. G. Rose and L. J. Evett. The use of context in cursive script recognition. *Machine Vision and Applications*, 8:241–248, 1995.

[267] P. L. Rosin. Representing curves at their natural scales. *Pattern Recognition*, 25:1315–1325, 1992.

[268] P. L. Rosin. Non-parametric multi-scale curve smoothing. *International Journal of Pattern Recognition and Artificial Intelligence*, 1994.

[269] Y. Rui, A. C. She, and T. S. Huang. Modified fourier descriptors for shape representation - a practical approach. In *International Workshop on Image*

DataBases and Multi-Media Search, pages 115–122, Amsterdam, The Netherlands, 1996.

[270] E. Saber and A. M. Tekalp. Image query-by-example using region-based shape matching. In *Proceedings of SPIE - The International Society for Optical Engineering*, volume 2666, pages 200–211, 1996.

[271] H. Samet. *The design and analysis of spatial data structures*. Addison-Wesley, 1990.

[272] G. Sapiro and A. Tannenbaum. Affine invariant scale-space. *International Journal of Computer Vision*, 11(1):25–44, 1993.

[273] E. Saund. Adding scale to the primal sketch. In *Proc IEEE Conference on Computer Vision and pattern Recognition*, pages 70–78, San Diego, CA, 1989.

[274] M. Schenkel, I. Guyon, and D. Henderson. On-line cursive script recognition using time-delay neural networks and hidden markov models. *Machine Vision and Applications*, 8:215–223, 1995.

[275] C. Schmid and R. Mohr. Matching by local invariants. Technical report, The French National Institute for research in computer science and control, 1995.

[276] P. J. Schneider. An algorithm for automatically fitting digitized curves. In A. S. Glassner, editor, *Graphics Gems*, pages 612–626. Academic Press Professional, 1990.

[277] L. Schomaker. Using stroke- or character-based self-organizing maps in recognition of on-line, connected cursive script. *Pattern Recognition*, 26(3):443–450, 1993.

[278] W. Schroeder, K. Martin, and B. Lorensen. *The Visualization Toolkit: An Object Oriented Approach to 3-D Graphics*. Prentice Hall, 1996.

[279] S. Sclaroff and A. P. Pentland. Modal matching for corresponding and recognition. *IEEE Transactions on Pattern Analysis and Machine Intelligence*, 17(6):545–561, 1995.

[280] M. Seibert and A. M. Waxman. Adaptive 3-d object recognition from multiple views. *IEEE Trans Pattern Analysis and Machine Intelligence*, 14:107–124, 1992.

[281] G. Seni, R. K. Srihari, and N. Nasrabadi. Large vocabulary recognition of on-line handwritten cursive words. *IEEE Trans Pattern Analysis and Machine Intelligence*, 18:757–762, 1996.

[282] J. A. Sethian. *Level Set Methods*. Cambridge University Press, 1996.

[283] B. Shahraray and D. J. Anderson. Optimal estimation of contour properties by cross-validated regularization. *IEEE Trans Pattern Analysis and Machine Intelligence*, 11:600–610, 1989.

[284] E. Shilat, M. Werman, and Y. Gdalyahu. Ridge's corner detection and correspondence. In *Proc IEEE Conf Computer Vision and Pattern Recognition*, pages 976–981, 1997.

[285] I. Shimshoni and J. Ponce. Probabilistic 3-d object recognition. In *Proc International Conference on Computer Vision*, pages 488–493, Cambridge, MA, 1995.

[286] K. Siddiqi, B. B. Kimia, A. Tannenbaum, and S. W. Zucker. Shapes, shocks and wiggles. *Image and Vision Computing*, 17(5):365–373, 1999.

[287] B. K. Sin and J. H. Kim. Ligature modelling for online cursive script recognition. *IEEE Trans Pattern Analysis and Machine Intelligence*, 19(6):623–633, 1997.

[288] S. S. Sinha and R. Jain. Range image analysis. In *Handbook of Pattern Recognition and Image Processing: Computer Vision (T Y Young, ed.)*, volume 2, pages 185–237, 1994.

[289] S. M. Smith and J. M. Brady. A new approach to low-level image processing. *International Journal of Computer Vision*, 23(1):45–78, 1997.

[290] K. Sohn, J. H. Kim, and W. E. Alexander. Mean field annealing approach to robust corner detection. *IEEE Transactions on Systems, Man, and Cybernetics, Part B: Cybernetics*, 28(1):82–90, 1998.

[291] F. Solina and R. Bajcsy. Recovery of parametric models from range images: The case for superquadrics with global deformations. *IEEE Trans. on Pattern Analysis and Machine intelligence*, 12:131–147, 1990.

[292] M. Soucy and D. Laurendeau. A general surface approach to the integration of a set of range views. *IEEE Trans Pattern Analysis and Machine Intelligence*, 17:344–358, 1995.

[293] M. Soucy and D. Laurendeau. Multiresolution surface modeling based on hierarchical triangulation. *Computer Vision and Image Understanding*, 1996.

[294] J. Sporring, O. Olsen, M. Nielsen, and J. weickert. Smoothing images creates corners. *Image and Vision Computing*, 18:261–266, 2000.

[295] J. Stansfield. Conclusions from the commodity expert project. memo 601, MIT AI Lab, Cambridge, 1980.

[296] F. Stein and G. Medioni. Structural indexing: Efficient 3-d object recognition. *IEEE Trans Pattern Analysis and Machine Intelligence*, 14:125–145, 1992.

[297] G. Stockman. Object recognition and localization via pose clustering. *Journal of Computer Vision, Graphics and Image Processing*, 40:361–387, 1987.

[298] A. J. Stoddart and M. Baker. Reconstruction of smooth surfaces with arbitrary topology adaptive splines. In *Proc ECCV*, 1998.

[299] A. J. Stoddart, J. Illingworth, and T. Windeatt. Optimal parameter selection for derivative estimation from range images. *Image and Vision Computing*, 13(8):629–635, 1995.

[300] D. L. Swets. The self-organising hierarchical optimal subspace learning and inference framework for object recognition. In *Ph.D. thesis, Michigan State University, Dept of Computer Science*, East Lansing, Michigan, 1996.

[301] C. C. Tappert, C. Y. Suen, and T. Wakahara. The state of the art in on-line handwriting recognition. *IEEE Trans Pattern Analysis and Machine Intelligence*, 12(8):787–808, 1990.

[302] G. Taubin. Curve and surface smoothing without shrinkage. In *Proc ICCV*, pages 852–857, 1995.

[303] G. Taubin, R. M. Bolle, and D. B. Cooper. Representing and comparing shapes using shape polynomials. In *Proc IEEE Conference on Computer Vision and Pattern Recognition*, pages 510–516, 1989.

[304] G. Taubin, T. Zhang, and G. Golub. Optimal surface smoothing as filter design. In *Proc ECCV*, pages 283–292, 1996.

[305] C. H. Teh and R. T. Chin. On image analysis by the methods of moments. *IEEE Trans Pattern Analysis and Machine Intelligence*, 10(4):496–513, 1980.

[306] B. M. ter Haar Romeny. *Geometry Driven Diffusion in Computer Vision*. Kluwer Academic, 1994.

[307] C. Tomasi and T. Kanade. Shape and motion from image streams under orthography - a factorization method. *International Journal of Computer Vision*, 9(2):137–154, 1992.

[308] M. Trajkovic and M. Hedley. Fast corner detection. *Image and Vision Computing*, 16(2):75–87, 1998.

[309] D. M. Tsai. Boundary based corner detection using neural networks. *Pattern Reognition*, 30(1):85–97, 1997.

[310] S. Umeyama. Parametrized point pattern matching and its application to recognition of object families. *IEEE Trans Pattern Analysis and Machine Intelligence*, 15(2):136–144, 1993.

[311] Y. Ung and F. Mokhtarian. Multi-scale spline-based contour data compression and reconstruction through curvature scale space. In *Proc International Conference on Acoustics, Speech, and Signal Processing*, pages 2123–2126, Istanbul, Turkey, 2000.

[312] B. C. Vemuri, A. Mitiche, and J. K. Aggarwal. Curvature-based representation of objects from range data. *Image and Vision Computing*, 4(2):107–114, 1986.

[313] T. Wakahara. On-line cursive script recognition using local affine transform. In *Proc International Conference on Pattern Recognition*, pages 1133–1137, 1988.

[314] M. W. Walker, L. Shao, and R. A. Volz. Estimating 3-d location parameters using dual number quaternions. *CVGIP: Image Understanding*, 54(3):358–367, 1991.

[315] T. P. Wallace and P. Wintz. An efficient three-dimensional aircraft recognition algorithm using normalised Fourier descriptors. *Computer Graphics and Image Processing*, 13:99–126, 1980.

[316] H. Wang and M. Brady. A practical solution to corner detection. In *Proc International Conference on Image Processing*, volume 1, 1994.

[317] J. Y. Wang and F. S. Cohen. 3-D object recognition and shape estimation from image contours using B-splines, unwarping techniques and neural networks. In *Proc IEEE International Joint Conference on Neural Networks*, volume 3, pages 2318–2324, 1991.

[318] M. Wang, J. Evans, L. Hassebrook, and C. Knapp. Multi-stage, optimal active contour model. *IEEE Trans on Image Processing*, 5(11):1586–1591, 1996.

[319] E. Wennmyr. A convex hull algorithm for neural networks. *IEEE Transactions on Circuits and Systems*, 36:1478–1484, 1989.

[320] D. J. Williams and M. Shah. A fast algorithm for active contours and curvature estimation. *CVGIP: Image understanding*, 55(1):14–26, 1991.

[321] P. H. Winston. *Artificial Intelligence*. Addison-Wesley, Reading, MA, 1979.

[322] A. P. Witkin. Scale space filtering. In *Proc International Joint Conference on Artificial Intelligence*, pages 1019–1023, Karlsruhe, Germany, 1983.

[323] A. K. C. Wong, S. W. Lu, and M. Rioux. Recognition and shape synthesisof 3-d objects based on attributed hypergraphs. *IEEE Trans Pattern Analysis and Machine Intelligence*, 11:279–290, 1989.

[324] Z. O. Wu and A. Rosenfeld. Filtered projections as an aid to corner detection. *Pattern Recognition*, 16(31), 1983.

[325] X. Xie, R. Sudhakar, and H. Zhuang. Corner detection by a cost minimization approach. *Pattern Recognition*, 26(8):1235–1243, 1993.

[326] C. Xu and J. Prince. Snakes, shapes, and gradient vector flow. *IEEE Trans on Image Processing*, 7(3):359–369, 1998.

[327] B. A. Yanikoglu and P. Sandon. Recognizing off-line cursive handwriting. In *Proc International Conference on Computer Vision and Pattern Recognition*, Seattle, Washington, 1994.

[328] H. Yoshitaka and B. Keserci. Bayesian wavelet snake for computer-aided diagnosis of lung nodules. *Integrated Computer-Aided Engineering*, 7(3):253–269, 2000.

[329] P. Yuen, G. C. Feng, and P. J. Zhou. Contour detection method: initialization and contour model. *Pattern Recognition Letters*, 20(2):141–148, 1999.

[330] P. Yuen, N. Khalili, and F. Mokhtarian. Curvature estimation on smoothed 3-d meshes. In *Proc British Machine Vision Conference*, pages 133–142, Nottingham, 1999.

[331] P. Yuen, F. Mokhtarian, and N. Khalili. Multi-scale 3-d surface description: Open and closed surfaces. In *Proc Scandinavian Conference on Image Analysis*, pages 303–310, Kangerlussuaq, Greenland, 1999.

[332] P. Yuen, F. Mokhtarian, N. Khalili, and J. Illingworth. Curvature and torsion feature extraction from free-form 3-D meshes at multiple scales. *IEE Proceedings on Vision, Image, and Signal Processing*, 147(5):454–462, 2000.

[333] A. L. Yuille and T. A. Poggio. Fingerprint theorems for zero-crossings. memo 730, MIT AI Lab, Cambridge, 1983.

[334] C. T. Zahn and R. Z. Roskies. Fourier descriptors for plane closed curves. *IEEE Transactions on Computers*, C-21:269–281, 1972.

[335] X. Zhang and D. Zhao. Parallel algorithm for detecting dominant points on multiple digital curves. *Pattern Recognition*, 30(2):239–244, 1997.

[336] A. Zhao and J. Chen. Affine curve moment invariants for shape recognition. *Pattern Recognition*, 30(6):895–901, 1997.

[337] Z. Zheng, H. Wang, and E. K. Teoh. Analysis of gray level corner detection. *pattern recognition Letters*, 20:149–162, 1999.

[338] Y. Zimmer, R. Tepper, and S. Akselrod. An improved method to compute the convex hull of a shape in a binary image. *Pattern Recognition*, 30(3):397–402, 1997.

[339] O. A. Zuniga and R. M. Haralick. Corner detection using the facet model. In *Proc Conference on Pattern Recognition and Image Processing*, pages 30–37, 1983.

Index

Computational Imaging and Vision

1. B.M. ter Haar Romeny (ed.): *Geometry-Driven Diffusion in Computer Vision.* 1994
    ISBN 0-7923-3087-0
2. J. Serra and P. Soille (eds.): *Mathematical Morphology and Its Applications to Image Processing.* 1994 ISBN 0-7923-3093-5
3. Y. Bizais, C. Barillot, and R. Di Paola (eds.): *Information Processing in Medical Imaging.* 1995 ISBN 0-7923-3593-7
4. P. Grangeat and J.-L. Amans (eds.): *Three-Dimensional Image Reconstruction in Radiology and Nuclear Medicine.* 1996 ISBN 0-7923-4129-5
5. P. Maragos, R.W. Schafer and M.A. Butt (eds.): *Mathematical Morphology and Its Applications to Image and Signal Processing.* 1996 ISBN 0-7923-9733-9
6. G. Xu and Z. Zhang: *Epipolar Geometry in Stereo, Motion and Object Recognition. A Unified Approach.* 1996 ISBN 0-7923-4199-6
7. D. Eberly: *Ridges in Image and Data Analysis.* 1996 ISBN 0-7923-4268-2
8. J. Sporring, M. Nielsen, L. Florack and P. Johansen (eds.): *Gaussian Scale-Space Theory.* 1997 ISBN 0-7923-4561-4
9. M. Shah and R. Jain (eds.): *Motion-Based Recognition.* 1997 ISBN 0-7923-4618-1
10. L. Florack: *Image Structure.* 1997 ISBN 0-7923-4808-7
11. L.J. Latecki: *Discrete Representation of Spatial Objects in Computer Vision.* 1998
    ISBN 0-7923-4912-1
12. H.J.A.M. Heijmans and J.B.T.M. Roerdink (eds.): *Mathematical Morphology and its Applications to Image and Signal Processing.* 1998 ISBN 0-7923-5133-9
13. N. Karssemeijer, M. Thijssen, J. Hendriks and L. van Erning (eds.): *Digital Mammography.* 1998 ISBN 0-7923-5274-2
14. R. Highnam and M. Brady: *Mammographic Image Analysis.* 1999
    ISBN 0-7923-5620-9
15. I. Amidror: *The Theory of the Moiré Phenomenon.* 2000 ISBN 0-7923-5949-6;
    Pb: ISBN 0-7923-5950-x
16. G.L. Gimel'farb: *Image Textures and Gibbs Random Fields.* 1999 ISBN 0-7923-5961
17. R. Klette, H.S. Stiehl, M.A. Viergever and K.L. Vincken (eds.): *Performance Characterization in Computer Vision.* 2000 ISBN 0-7923-6374-4
18. J. Goutsias, L. Vincent and D.S. Bloomberg (eds.): *Mathematical Morphology and Its Applications to Image and Signal Processing.* 2000 ISBN 0-7923-7862-8
19. A.A. Petrosian and F.G. Meyer (eds.): *Wavelets in Signal and Image Analysis. From Theory to Practice.* 2001 ISBN 1-4020-0053-7
20. A. Jaklič, A. Leonardis and F. Solina: *Segmentation and Recovery of Superquadrics.* 2000 ISBN 0-7923-6601-8
21. K. Rohr: *Landmark-Based Image Analysis.* Using Geometric and Intensity Models. 2001 ISBN 0-7923-6751-0
22. R.C. Veltkamp, H. Burkhardt and H.-P. Kriegel (eds.): *State-of-the-Art in Content-Based Image and Video Retrieval.* 2001 ISBN 1-4020-0109-6
23. A.A. Amini and J.L. Prince (eds.): *Measurement of Cardiac Deformations from MRI: Physical and Mathematical Models.* 2001 ISBN 1-4020-0222-X

24. M.I. Schlesinger and V. Hlaváč: *Ten Lectures on Statistical and Structural Pattern Recognition*. 2002   ISBN 1-4020-0642-X
25. F. Mokhtarian and M. Bober: *Curvature Scale Space Representation: Theory, Applications, and MPEG-7 Standardization*. 2003   ISBN 1-4020-1233-0

Kluwer Academic Publishers – Dordrecht / Boston / London